# DEVELOPMENT AND PLASTICITY OF THE BRAIN
## AN INTRODUCTION

# DEVELOPMENT AND
# PLASTICITY OF THE BRAIN
## AN INTRODUCTION

**R. D. LUND**
Departments of Biological Structure
and Neurological Surgery
University of Washington

New York • OXFORD UNIVERSITY PRESS • 1978

Copyright © 1978 by Oxford University Press, Inc.

Printed in the United States of America

Library of Congress Cataloging in Publication Data

Lund, Raymond D.
  Development and plasticity of the brain.

  Bibliography: p.
  Includes index.
  1. Brain—Growth.  2. Developmental neurology.
3. Adaptation (Physiology)  4. Mammals—Physiology.
I. Title.
QP376.L79      596 ′.01 ′88      77-26936
ISBN 0-19-502307-2
ISBN 0-19-502308-2 pbk.

# PREFACE

This book is largely concerned with the mechanisms involved in the development and maintenance of predictable organization in the central nervous system of vertebrates, particularly of mammals. Some experiments undertaken on the peripheral nervous system are also described, since certain principles derive from them which may also be relevant to central neural problems. Emphasis is given throughout to the dynamic properties of nerve cells, their interactions, their specificities of behavior, and the aberrations which result from lesions, genetic variation, and environmental factors.

Up to the late 1960s most of the research on development and plasticity of the nervous system had been done on regeneration in amphibians. There were relatively few experimental studies of development and, with some notable exceptions, the mammalian central nervous system was largely ignored. Despite this and despite technical limitations, a number of general principles were developed which have, for the most part, held good.

In recent years, interest in neural development and plasticity has heightened, and a wide variety of strategies and techniques has been applied to major problems in the field. More emphasis has been given to development per se. Basic principles of cell biology have been applied to and extracted from studies of the nervous system, and a whole new literature has emerged on development and plasticity in mammalian brains, some of which is relevant to clinical conditions. The general principles established in earlier work have been redefined, qualified, and extended as a result of all this effort.

One of the frustrations of teaching developmental neurobiology has been the absence of a basic text that reviews these new developments and their interrelations and implications. It is hoped that this book will serve that purpose. The text is structured so that, although reasonably comprehensive, it should give a novice to neurobiology an overview of the field, perhaps through selective reading.

While parts of the book go into considerable detail, Chapters 1, 2, and 18, the introductions to sections and chapters, and the summaries and discussions at the ends of chapters present essentials without the encumbrance of experimental details.

The book is planned in several sections, beginning with an introduction to the topics to be discussed and to the basic organization of the brain. After a survey of the cell biology of neurons, subsequent sections describe the sequence of events which leads to the development of a normal brain and shows how patterns of development can go astray under particular conditions. The problems associated with maintaining the order of the brain once it is mature are then discussed. Finally, the basic mechanisms which may underlie the whole process are reviewed in the last chapter.

One problem which emerges in the literature, especially in many of the reviews written in recent years, is that in the desire to present a coherent story, the need for critical assessment of data is often overlooked. Furthermore, all experiments are limited by the methods available, and while it is permissible to hypothesize beyond the data, it is important to know the boundary between fact and conjecture. I have included some criticism of experiments and interpretations where it seemed especially important, in part because I feel that with the current trend toward making science easy to digest, students are never challenged to question dogma and think for themselves. The purpose of education (*educare*— to lead out) is to encourage enquiry, not to stifle it. I hope my criticisms will not be misconstrued as attacks on the scientific integrity of the workers concerned. Frequently it is the case that results may be interpreted differently in the light of subsequent studies or that further experiments are required for their verification.

I regret not including more of the early literature, but this has been done very well by Gaze (1970) and Jacobson (1970). I also regret that because of space limitations most of the invertebrate developmental literature has been excluded, even though it may be relevant to the matters under discussion.

The bulk of the text was written during two consecutive summers in Seattle and Canberra and reflects the literature available during that period. With the exception of some papers to which I had access prior to publication, only published works available in Canberra up to the end of March 1977 have been included.

*Canberra*
*April 1977*

# ACKNOWLEDGMENTS

This book could not have been written without the help and encouragement of a large number of people. I am particularly grateful to Dr. Jenny Lund, who discussed many sections, read a lot of the preliminary drafts, advised on pictures, provided some of her own, and prevented the manuscript from being thrown into Lake Burley Griffin, prior to completion. I am also indebted to Ms. Renée Wise, who besides running my laboratory for the years during which my own research in this area evolved, has helped in many ways in the completion of the text. I am also pleased to acknowledge the expert help offered by Drs. G. Henry, I. Hendry, and D. Mitchell who read and criticized particular chapters. The help given me by Mr. Ken Collins in doing a large number of the illustrations has contributed immensely to this book, as have the diagrams which people have kindly provided for me either unpublished or from impending and existing publications. To D. Bray, M. Bunge, V. Caviness, E. Elekessy, S. LeVay, P. Rakic, P. Snow, and I. Hendry, I am especially grateful. I thank the various publishers for permission to reproduce published illustrations. Specific acknowledgement is given with the particular diagrams. The final book is the product of an army of secretaries in England, Seattle, and Canberra and to all of them I offer thanks. It is doubtful whether this book would have been finished without the generous facilities afforded me by Professor P. O. Bishop of the John Curtin School of Medical Research, Australian National University, Canberra. The excellent support services and libraries of this institution make a project such as this much easier. Many of my approaches to development and plasticity of the nervous system have been developed while doing research supported by the National Institutes of Health and for that I am especially thankful. Finally, I should like to thank Mr. Jeffrey House of O.U.P. who started this project moving, has attempted to maintain its momentum, and has steered it to its conclusion.

Needless to say, despite all this help, the errors and misinterpretations which must inevitably have found their way into the book are entirely my own responsibility. I would hasten to add however, that authors who feel that their efforts have not been given appropriate attention should realize that it is becoming close to impossible to provide a comprehensive synthesis in this fast-moving and broad-based subject.

# CONTENTS

# INTRODUCTION

It is important when studying individual experiments on neural development and plasticity to appreciate the broader concepts which may have prompted the experiment or which bear on its interpretation. In addition it is essential to know some details of the normal adult organization of the region under investigation, since most research in the field is directed at identifying deviations from normal and these can only be recognized if the normal patterns are fully understood.

In Chapter 1, some of the broader issues facing developmental neurobiologists are introduced and these will be enlarged upon in the succeeding chapters. Chapter 2 discusses normal gross development and normal organization of the brain, paying particular attention to the regions discussed in more detail throughout the rest of the book.

# 1

# SOME PROBLEMS OF NEURAL
# DEVELOPMENT

The tissues of the body are derived from a single cell—the fertilized ovum. Throughout the series of cell divisions necessary to make a multicellular organism, the daughter cells develop an identity which distinguishes them not only from cells of other tissues but also from other types of cell in the same tissue. In most tissues this identity is expressed largely in terms of how the cell interacts with its neighbors and what special molecules it synthesizes. The position of a particular cell in any tissue of the body is significant to the extent that the cell is located in the appropriate organ and may occupy a certain niche among a complex of cells making up an organ. Beyond that the exact location of a cell is relatively unimportant for its essential function in all tissues other than the nervous system: fibroblasts do the same things wherever they are located and muscle cells taken from the same type of striated muscle of the leg or arm are indistinguishable.

Like other tissues, the nervous system is composed of relatively few basic cell types—neurons and three distinct classes of supporting (neuroglial) cells—and these, like other cells, express their identity by their interactions with one another and by their special secretions. What is different in the nervous system is that each of the more than $10^{10}$ neurons making up a mammalian brain has a unique function determined not only by what class of neuron it is but also by its particular spatial distribution, which enables the individual neuron to receive information from a restricted and unique set of sources, integrate that information, and then transmit it to another select group of cells.

In order to consider this individuality further we must first examine a typical neuron such as that shown in Figure 1-1. There is a central region termed the cell body, or soma, in which the nucleus is situated. From the cell body a number of processes radiate. Of these, the dendrites tend to be of large diameter at their origin from the cell body, to show a limited amount of branching, and rarely to

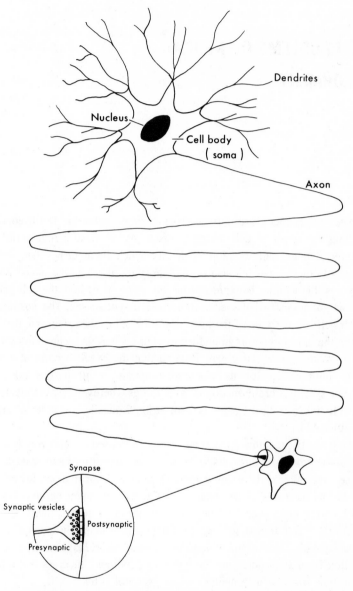

**Figure 1-1.** Component parts of a typical neuron, giving some indication of the relative lengths of axons and dendrites. No attempt has been made to signify axonal branches.

extend more than 1 mm from the cell body. Together with the cell body, these dendrites are the principal receiving surface for inputs from other neurons. A single slender process leaves the cell body. This is the axon which serves at its junction with the cell body as the site for the integration of inputs to the cell and, along the rest of its extent, as the means whereby this information is transmitted to other neurons. It is usually longer than the dendrites (as much as 1 meter) and generally shows a greater complexity of branching. Along its course and at terminations of branches there are specialized junctions with other neurons. These are called synapses and they closely resemble nerve–muscle junctions. They are the sites where a wave of membrane depolarization transmitted down the axon (termed the presynaptic process) causes the release of a chemical transmitter substance (stored in vesicles in the presynaptic terminal) which affects the membrane conductance of the other partner at the junction—the postsynaptic cell. This is the common way that information is transferred between neurons in the vertebrate brain.

In order to achieve a transfer of highly specific information, a neuron's dendrites, its cell body, and its axon branches must be distributed in a particular spatial fashion. If these features (the position of the cell body, the disposition of the dendrites, the disposition of the axon and its synapses) are used to classify nerve cells, it becomes clear that there is a highly predictable correlation between them. A neuron with a cell body at a particular location and with a certain dendritic distribution will tend to have an axon with a particular distribution of synapses: it differs from axons of all other neurons and can be recognized from one animal to another of a certain species or even between species. This predictability of neuronal form and connections is termed *neural specificity*; exceptions to this predictable pattern in particular circumstances represent *neural plasticity*. It should be emphasized first that these terms—neural specificity and neural plasticity—are descriptive and in no way serve as explanations of the respective phenomena and, secondly, that they are always relative to the techniques used to demonstrate them. The terms raise a number of questions, and a large part of the book will be devoted to a discussion of these problems.

### A. How specific is specific?

There are three aspects to this question: first, how exact is specificity of form and connections; second, is it technically possible to determine the degree of specificity of a neuron, particularly its connections; and third, how precise do the interconnections need to be for information transfer without loss of significant detail? The first two questions have been addressed successfully only in

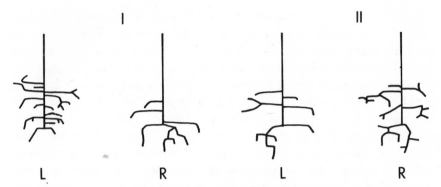

**Figure 1-2.** Drawing of computer reconstructions of the branching pattern of the same retinal axon on left (L) and right (R) sides of the brain of two Daphnia specimens (I and II) to show the considerable variability experienced (After Macagno et al., 1973).

simple invertebrate systems, such as the initial connections in the visual system of the water flea *Daphnia* (Macagno et al., 1973). That study involved computer reconstruction of electron micrographs of the axon of an identified retinal receptor cell. The pattern of distribution could then be compared from one side of the brain to the other and also from one animal to another, where each was an isogenic member of a single parthenogenetic clone. Despite the simplicity of the system and the genetic homogeneity of the population, there were obvious differences in the axonal branching patterns from one cell to another (Figure 1-2) and the variation in synapses formed on particular postsynaptic cells was not insignificant (Table 1-1). It should be noted, however, that even in this relatively

**Table 1-1.** Numbers of synapses formed by the same retinal axon (D2) with laminar cells on the left (L) and right (R) sides of the brain of four *Daphnia* specimens. Physical separation precludes the formation of synapses with three other laminar cells (2, 3, and 5) (from Macagno et al., 1973).

|          |      | Postsynaptic cell | | |
| :------: | :--: | :-: | :-: | :-: |
| Specimen | Side | 1 | 4 | Unidentified |
| I        | L    | 10 | 54 | 42 |
|          | R    | 8  | 50 | 37 |
| II       | L    | 9  | 65 | 54 |
|          | R    | 4  | 36 | 38 |
| III      | L    | 3  | 20 | 18 |
|          | R    | 3  | 22 | 20 |
| IV       | L    | 6  | 41 | 54 |
|          | R    | 5  | 44 | 45 |

simple system, a large number of synapses were unaccounted for simply because the anatomical methods were inadequate to trace them. These results indicate that even in a simple system specificity is not absolute and technical limitations still preclude a full analysis. What chance does one ever have, then, of addressing these questions in the vertebrate brain?

Given the enormous complexity of the vertebrate brain, it might be expected that the degree of variation in connections from the mean would be proportionately that much greater. Unfortunately, the difficulties in demonstrating the degrees of variation are also greater and as yet we have no more than a rough indication as to how specific connections really are. This fact is often overlooked when the matter of neural specificity is discussed.

The third point, concerning the exactness of connections necessary for functional specificity, has been largely unexplored by developmental neurobiologists. It is nevertheless a question of considerable importance in terms of how many bits of information are necessary to produce a satisfactory set of connections. For example, cell X may fire an action potential if cell A is stimulated but not if cell B is stimulated. Does this mean that A is connected to X, and that B is not? This is possible. However, a number of recent studies (reviewed in Chaps. 15 and 16) have shown that if A is cooled, damaged, or otherwise made ineffective, X may now respond to stimulation of B. This implies that B does indeed make connections with X but for some reason is unable to drive cell X in the presence of a normal input from A. The presence of such "silent" synapses and the mechanisms whereby they are suppressed are obviously important in turning what may be a quite diffuse anatomical connection into a quite specific functional one. The corollary of this, of course, is that physiological methods (particularly those for monitoring action potentials in postsynaptic cells) will not demonstrate degrees of specificity of anatomical connections.

### B. How does specificity arise?

It is obviously of great interest to know to what extent the predictable shape and connections of a neuron are determined by its genetic program independent of its physical environment, and then what role the environment (other neurons, neuroglial cells, chemical gradient phenomena, and other factors) plays in molding the neuron's structure and function and by what means it has this effect. These questions can be investigated in two ways—either by observing interrelations in normal development or by observing patterns of variation and invariance which may arise in abnormal circumstances as a result of particular mutations or chemical or surgical insults. The principal experimental approach to studying specificity of connections has perhaps focused undue attention on the spatial interrelation of two interconnected regions where the connections are distributed in an orderly fashion so that a topographic map of one region is relayed to the next. As a result, much of what is referred to as "specificity" concerns topographic specificity. This is obviously too restrictive a view since throughout their maturation a neuron and its processes are continually demonstrating specificities of behavior in relation to their environment, and the final position they occupy and connections they make are the end product of this continuous process. There are a number of phases in this process, and they can be outlined as follows:

   a. The neuron must leave the cell cycle at a set time.
   b. It must migrate to the appropriate region.
   c. It may develop a spatial identity with respect to its neighbors.
   d. Dendrites must develop in a characteristic shape and orientation.
   e. The axon must leave the cell body and grow in the right direction toward its region of termination.
   f. The axon must direct its branches to the appropriate side of the brain.
   g. The axon must direct branches to the right region or regions.
   h. Within a region, the axon must ramify in the right subdivision or layer.
   i. The terminal field of the axon must be ordered in a particular topographic relationship with the cell bodies in the regions of its origin and termination.
   j. The axon terminals may end only on certain cell types within the terminal distribution area.
   k. The axon terminals may end only on certain parts of these cells (parts of the dendritic surface, for example).

### C. What degree of plasticity is there in special circumstances?

Many examples are provided in later chapters to demonstrate plasticity at all the stages of development outlined in B above. There are three particular reasons for documenting the various situations in which plasticity occurs.

   a. Aberrations from the normal pattern give insight into the factors that control normal development.
   b. Human mental retardation is sometimes the outcome of aberrant patterns of development. Some forms of mental retardation, such as Down's syndrome, are the result of genetic abnormalities; some are the result of changes in response to viral infections or perhaps exposure to harmful chemicals at certain stages in development; some are probably due to lesions resulting from vascular problems; others are of unknown etiology. In assessing the mechanisms of plasticity in controlled experimental conditions, we may be able to identify some of the factors underlying these clinical disorders.

c. It is commonly suggested, although not proven, that learning and behavioral changes may be the product of changing patterns of neural inter-connections occurring throughout life. Is this likely? What other possibilities are there? Are such connections also susceptible to the difficulties which will be described in the ensuing chapters? Do the rules governing synapse formation in the relatively simple systems discussed in this book apply to the somewhat more nebulous systems involved in learning and behavior? These questions lead to two others:

### D. What role does function play in defining patterns of connections?

This question has been addressed most prominently by research in two areas: the imprinting behavior of many animals, particularly birds, and the role of visual input in the maturation of the visual system. Attention here will be given mainly to visual deprivation studies because they have provided more detail concerning the neural mechanisms involved. The degree to which these mechanisms may be common to other sensory systems and to the development of the brain and behavior in general is of obvious interest.

### E. How stable are the neuronal patterns in the adult animal?

There is considerable evidence for anatomical and functional modification of connections in the adult animal. In addition to its relevance as a possible substrate for learning, such modification is particularly significant in relation to the behavior of animals following brain injury, especially with regard to the matter of neural regeneration. In certain vertebrates, such as fish, amphibians, and lizards, the central nervous system shows some capacity for regrowth of damaged axons and even for restitution of extirpated brain regions by further cell division. Over the past 35 years the behavior of the regenerating axons in these animals has been a major source of concepts of developmental neurobiology in vertebrates. Accordingly, this work will be discussed mainly in the parts of the book devoted to development rather than in the sections on the adult nervous system. However, it might be supposed that an axon regrowing along a course already defined by a previous pathway to cells which could still have "labels" related to the previous axon terminals might bear no similarity to an axon growing for the first time into a region which itself is still forming. For this reason several authors (e.g., Gaze, 1974) have cautioned that to derive rules concerning development from studies of regeneration may be presumptuous. Still, we will see that in those cases where regeneration and developmental studies have been conducted on the same system, they usually parallel one another closely.

Although there is evidence of regrowth after lesions in young mammals and birds, regeneration of the adult brain is of very limited scope. The growth in young animals may be due to a continuation of normal developmental growth, but most studies have not provided data on the timing of this process. Why substantial growth after injury does not occur in the adult mammalian central nervous system is unclear, and this topic will be discussed in Chapter 17.

### F. May some simple basic mechanisms be defined which underlie the development of the intricate patterns making up the central nervous system?

In the succeeding chapters it will become clear that while different regions of the brain show elaborate and distinct patterns of organization, there are nevertheless a limited number of general mechanisms which may be hypothesized as underlying their development. In fact, these same mechanisms play a part in the development of other tissues, although the complexity of the nervous system gives them greater scope.

# 2

# AN INTRODUCTION TO THE GENERAL ORGANIZATION
# OF THE BRAIN AND ITS DEVELOPMENT

In order to appreciate the significance of the studies described in this book, one needs to have at least a general understanding of the logic behind the anatomical organization of the central nervous system. For this purpose, this chapter provides a brief discussion of some general principles of organization, the development and order of the basic regions of the brain, and, in slightly more detail, the regions which receive the most attention throughout the rest of the book. No attempt has been made to present the functional organization of the brain, since this is covered so well elsewhere (e.g., by Shepherd, 1974; and Kuffler & Nicholls, 1976).

## 1. General Principles

A number of very basic comments may be made about the pattern of organization of the vertebrate central nervous system:

    1. The central nervous system is composed of a large number of distinct, recognizable regions. These regions can be predicted from one animal to another within a species, between animals of different species, and even between animals of different vertebrate orders.

    2. Neurons with a cell body in any one region will send an axon to a limited and predictable number of other regions: connections between regions are not random, nor, unfortunately, are they made only between adjacent regions. This last point has been a major barrier to elucidating the organization of the brain, since axons may follow an intricate course connecting two widely separated regions, which is often extremely difficult to follow.

    3. Groups of axons running in a common direction will usually do so in a compact bundle called, variously, a nerve, tract, peduncle (stalk), or brachium (arm). A single bundle may contain as many as 1 million axons.

4. Cell bodies of neurons tend to be segregated into groups. These may form sheets of cells called layers or laminae or more compact groups called nuclei (not to be confused with the nucleus within each cell). A nucleus is generally the smallest regional subdivision of the brain which can be made. Its borders are often defined by tracts of axons that separate it from other nuclei. In general, nuclei defined on anatomical grounds have characteristic functional responses and patterns of connections which distinguish them from neighboring nuclei. Similarly coherent groups of nerve cell bodies also occur outside the central nervous system, where they are called ganglia.

5. Many of the regional interconnections in the brain are between regions which share functional affinities. This is perhaps self-evident, since clearly a region derives many of its functional properties from its inputs.

## 2. Basic Terminology

Before considering the patterns of development and organization of the brain, a few simple descriptive terms must be introduced. The central nervous system is arranged as an enlarged mass at the front, the brain, from which runs the spinal cord. For descriptive purposes (Fig. 2-1), those regions closer to the face of the animal are more *rostral* while those closer to the tail are more *caudal*; those parts closest to the top of the head or the back are *dorsal* while those on the opposite

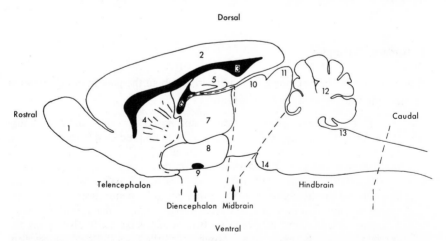

**Figure 2-1.** Sagittal section through a rat brain to show descriptive terms used for orientation of brain and major brain parts. (1) Olfactory bulb, (2) neocortex, (3) corpus callosum, (4) basal ganglia (corpus striatum), (5) hippocampus, (6) fornix, (7) thalamus, (8) hypothalamus, (9) optic tract, (10) superior colliculus, (11) inferior colliculus, (12) cerebellum, (13) dorsal column nuclei, (14) pons.

side are *ventral*. Use of the terms anterior and posterior, while frequent, becomes confusing in animals such as humans that stand upright and have a spinal cord positioned at 90° to the rest of the brain.

Among the most commonly used terms in descriptions of the nervous system are the following:

> **Neuropil.** A mass of neural tissue.
>
> **Grey matter.** An area of neural tissue containing mainly cell bodies and relatively few axons that possess an insulating sheath (myelin); it has a grey appearance in fresh tissue.
>
> **White matter.** An area of neural tissue in which the predominant component is axons invested with myelin. The myelin gives the tissue a white appearance in fresh preparations.
>
> **Lesion.** In experimental work, this usually means an injury deliberately made at a selected site.
>
> **Nerve fiber.** Synonymous with axon.
>
> **Afferent connections.** These are made by axons projecting into the region or onto the cell under consideration from elsewhere.
>
> **Efferent connections.** These are made outside the region under consideration by neurons whose cell bodies lie within the region.
>
> **Extrinsic connections.** Afferent connections arising from outside the region being considered, or efferent connections being made outside the region.
>
> **Intrinsic connections.** These are made within the region being considered, by neurons already present there.

## 3. Development of the Basic Brain Regions in Mammals

The nervous system first appears as part of the surface ectoderm in the neurula stage of the developing embryo. This region, termed the neural plate, continues to develop into neural tissue when transplanted to other parts of the embryo. Although some degree of regulation is possible within the neural plate, local areas are already predestined to develop into certain brain regions. The maps of the presumptive brain regions and their behavior when transplanted are beyond the scope of this book, but a good basic introduction to the topic is provided by Jacobson (1970). The next step in development is for the cells to migrate towards the midline, and for each cell to develop a constriction on one surface. The cells on the edge of the neural plate become narrower on their inner margin, while those more centrally located become narrower on their outer surface. As a result the neural plate folds in to form a groove—the neural groove (Fig. 2-2). It gets deeper and eventually zips up along its outer edge to form a tube of cells around a central fluid-filled space. This is the neural tube.

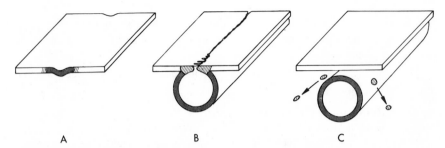

**Figure 2-2.** Three stages in development of neural tube. (A) Neural groove, (B) closing up to form tube, (C) separation of tube from surface ectoderm and migration of neural crest derivatives. Neural tube is denoted by cross-hatched area, neural crest by lined area.

Initially the tube is open rostrally and caudally but it closes off in time; if it fails to do so one of a series of anomalies, such as anencephaly (in which the forebrain fails to develop), results. If the neural tube does not close properly, the result is a condition known as spina bifida, in which the spinal cord is not confined by the vertebral column. Adjacent to the line of closure of the neural tube is a specialized region, the neural crest, from which cells migrate to form groups of neurons (ganglia) in various parts of the body. Such cells innervate glands and smooth muscle. In time the ganglia are themselves innervated by axons growing out from cells in the central nervous system. Two of these ganglia—the ciliary ganglion and the superior cervical ganglion—will be mentioned a number of times in this book. Another derivative of the neural crest is the dorsal root ganglion, whose cells send dendrites (sensory processes) to the skin and an axon to the central nervous system.

Once the neural tube is formed, two events occur: (1) cells in the wall of the tube undergo frequent cell divisions, and (2) at the rostral end the wall balloons out to form the three vesicles that define the major divisions of the brain—the forebrain (prosencephalon), the midbrain (mesencephalon), and the hindbrain (rhombencephalon). These are shown in Figure 2-3. The caudal part of the neural tube becomes the spinal cord.

As the primary divisions develop, a bulge appears on the ventrolateral aspect of the forebrain on each side. This is the optic vesicle. Its wall apposes the surface epithelium with which it maintains contact and eventually forms the eye (Fig. 2-4). The connection with the main part of the forebrain becomes attenuated as the optic stalk. The cavity of the optic vesicle becomes thinner and the stalk itself eventually acts as a guide for the optic axons as they grow into the brain. The retina and an associated layer (the pigment epithelium) form from the terminal part of the optic evagination, and as these layers appose one another, the cavity of the optic vesicle is obliterated.

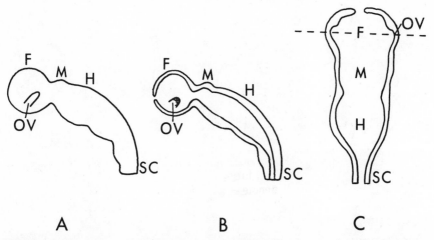

**Figure 2-3.** Three views of rostral expansion of neural tube early in development. (A) From side, (B) sagittal section, (C) dorsal view. (F) Forebrain, (M) midbrain, (H) hindbrain, (SC) spinal cord, (OV) optic vesicle.

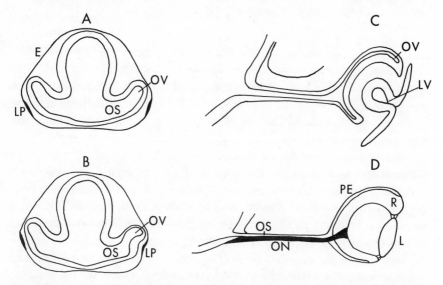

**Figure 2-4.** Cross sections showing four stages in development of retina. (A) shows the early outpouching of the optic vesicle (OV) and the associated optic stalk (OS). There is a thickening of the surface ectoderm (E) adjacent to the outpouching where the lens placode (LP) forms. This section is taken through the dotted line in Fig 2-3c. (B) shows the further thickening of the wall of the optic vesicle. (C) shows the invagination of the wall and of the overlying ectoderm which forms the lens vesicle (LV). (D) shows the differentiation of the wall of the optic vesicle into retina (R) and pigment epithelium (PE) and the formation of the optic nerve (ON) and lens (L).

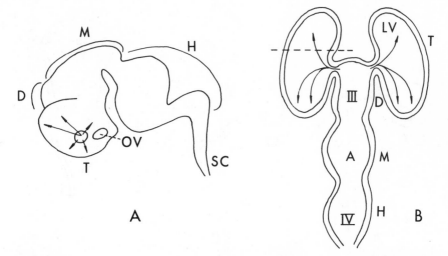

**Figure 2-5.** Further development of brain showing outpouching of telencephalon (T) of forebrain in the direction indicated by the arrows. (A) Lateral view, (B) dorsal section. (D) diencephalon, (M) midbrain, (H) hindbrain, (SC) spinal cord, (OV) optic vesicle, (LV) lateral ventricle, (III) third ventricle, (A) aqueduct, (IV) fourth ventricle.

A further outpouching on each side of the forebrain forms a separate division called the telencephalon (Fig. 2-5) which surrounds its own cavity, the lateral ventricle. The wall of each telencephalic vesicle differentiates into a number of regions as shown in Figure 2-6 and summarized in Table 2-1. The cerebral cortex is characterized by its development into a set of laminae; the hippocampus by its elaborate pattern of infolding; and the basal ganglia by the fact that they come to lie deep in the forebrain adjacent to the lateral ventricle. As the cerebral cortex develops, a major tract grows between it and subcortical structures. This is termed the internal capsule and it grows through the basal ganglia, either as a major bundle or as a series of smaller bundles (giving a striated appearance to part of the basal ganglia—the corpus striatum). Another major pathway is the corpus callosum, which grows across the midline connecting both sides of the cerebral cortex. The hippocampus has a major fiber tract which connects it with the septum and hypothalamus. This is the fimbria.

The central part of the forebrain is termed the diencephalon. The cavity of the diencephalon becomes elongated and has a thin roof through which blood vessels pass. The neural tissue on either side of the ventricle increases in bulk, especially in the more dorsal part which forms the thalamus. The thalamus relays information from various parts of the central nervous system to the neocortex. It is divided into a series of nuclei, each of which is concerned with a particular set of

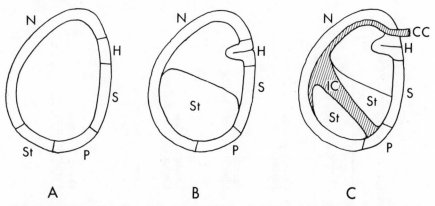

A                     B                     C

**Figure 2-6.** Diagrammatic sections of progressive development of forebrain telencephalon taken at the level of the dotted line in Fig. 2-5. (A) early tube, (B) internal development of basal ganglia and infolding of hippocampus, (C) development of fiber pathways. (N) Neocortex, (H) hippocampus, (S) septum, (P) pyriform (olfactory) cortex, (St) striatum, (CC) corpus callosum, (IC) internal capsule.

functions or sensory inputs. Most parts of the thalamus receive feedback from the region of cortex to which they project. Two nuclei of the thalamus that will be mentioned in later chapters are the nucleus ventralis posterior (associated with sensory input from the body) and the lateral geniculate nucleus (associated with the visual input). They are indicated in Figure 2-7. In many animals the lateral geniculate nucleus is located on the dorsolateral aspect of the thalamus (Figure 2-7C). While in higher primates it occupies this position early in development, eventually it comes to lie more ventrally as shown in Figure 2-7D, largely because of the massive development of an adjacent nucleus—the pulvinar. Coursing round the ventrolateral border of the thalamus is the internal capsule, which contains axons

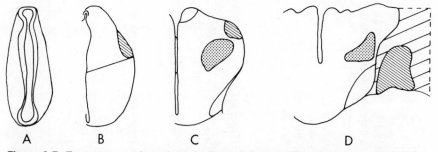

A          B             C                    D

**Figure 2-7.** Transverse sections through diencephalon at different stages to show progressive development of thalamus, in particular the relative positions of nucleus ventralis posterior (dotted area) and the lateral geniculate nucleus (lined area).

**Table 2-1.** Summary of major divisions of the mammalian central nervous system and the principal structures associated with each division. The diencephalon, midbrain, and hindbrain are collectively termed the brainstem.

| Major division | Neural tube derivative | Major regions and nuclei | Major fiber tracts |
|---|---|---|---|
| 1. Forebrain | | | |
|   Telencephalon | Lateral ventricle | Cerebral cortex (neocortex) Hippocampal formation Septal nuclei Basal ganglia (striatum) | Corpus callosum Fimbria |
|   Diencephalon | Third ventricle | Thalamus Hypothalamus | Internal capsule Medial forebrain bundle |
|   Eye | | Neural retina Pigment epithelium | |
| 2. Midbrain | Aqueduct | Tectum (superior and inferior colliculi) Red nucleus Oculomotor & trochlear nuclei | Cerebral peduncle Cerebello-rubro-thalamic tract |
| 3. Hindbrain | Fourth ventricle | Cerebellum Pontine nuclei Inferior olive Dorsal column nuclei Trigeminal nucleus | Cerebellar penduncles Pyramidal tract |
| 4. Spinal cord | Central canal | Dorsal and ventral horns | Dorsal columns Pyramidal tract |

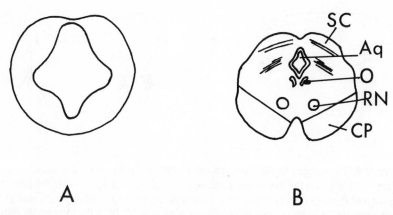

**Figure 2-8.** Transverse sections of midbrain. (A) Early stage of development. (B) Differentiated midbrain at level of superior colliculus (SC), showing aqueduct (Aq), oculomotor nucleus (O), red nucleus (RN), cerebral peduncle (CP).

connecting the thalamus and cortex as well as various groups of axons of cortical cells ending throughout the brainstem and spinal cord.

In the midbrain (Fig. 2-8), the ventricular space (the aqueduct) becomes relatively small compared with the neural tissue in its wall. The dorsal surface, forming a roof (or tectum) to the midbrain, enlarges into two pairs of protuberances in mammals called the colliculi, of which the more rostral (the superior colliculi) assume considerable importance in the following chapters. The homologue of the superior colliculus in nonmammalian vertebrates is the optic tectum, a region of major significance in developmental studies. Several nuclear groups develop within the tissue of the midbrain. Among these are the red nucleus and two nuclei involved in controlling the eye muscles, the oculomotor (shown in Fig. 2-8) and the trochlear. The axons traveling from cortical to subcortical regions in the internal capsule continue their course through the midbrain in two large fiber tracts—the cerebral peduncles.

The cavity of the hindbrain (Fig. 2-9) is maintained throughout development as a large space, the fourth ventricle. The neural tissue around it develops unevenly, forming a roof and a floor which comprise two regions—the pons more anteriorly and the medulla more posteriorly. Within the pons and medulla are a large number of nuclei; mention will be made in this book of those concerned with the relay of sensory information from the head (trigeminal nucleus), and upper and lower limbs and trunk (nucleus cuneatus and gracilis, respectively) and of nuclei that are connected with the cerebellum—the pontine nuclei and the inferior olive. Many fiber tracts run through this region, the most significant

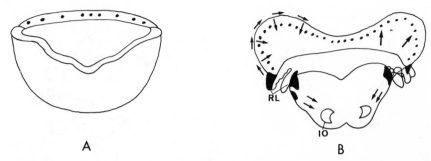

A                                                              B

**Figure 2-9.** Two stages in the development of the hindbrain. (A) Early stage showing Purkinje cells (dots) in the roof of the fourth ventricle. (B) Later stage showing the pattern of migration of Purkinje cells (arrows, right) from their region of generation on the roof of the fourth ventricle. In addition, migrating cells coursing ventrally from the rhombic lip (RL) to form the inferior olive (IO) are shown, as is the dorsal migration on the surface of the cerebellum. Cells from this external granule layer migrate inward (arrows, left) to form the granule cell layer lying below the Purkinje cells.

being those which connect the cerebellum with the rest of the brain (the cerebellar peduncles) and the continuing pathway from the cortex to the spinal cord (the corticospinal or pyramidal tract).

The spinal cord (Fig. 2-10) maintains a small ventricle—the central canal. In the cord the neural tissue develops into two main bodies of nerve cells—the dorsal and ventral horns. Peripheral to these areas are the columns of axons running rostrally and caudally in the cord. Of these the only ones that will be mentioned here are the dorsal columns, which are composed of sensory axons destined for the gracile and cuneate nuclei (described collectively as the dorsal column nuclei). The dorsal horn is concerned with sensory events and receives axons from the dorsal root ganglion. Other axons or axon branches from the

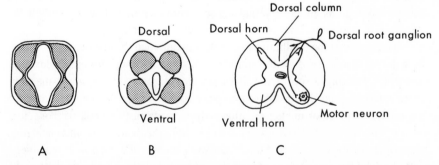

A                    B                           C

**Figure 2-10.** Transverse sections of spinal cord. (A) Early development, (B) development of fiber columns around grey matter (shaded), (C) mature cord showing main components.

dorsal root ganglion run directly into the dorsal columns. The ventral horn is concerned with motor events and contains the cell bodies of axons which innervate muscles. These cells are termed ventral horn cells.

## 4. Selected Regions

Most of the work on development and plasticity of the central nervous system has been conducted in a limited number of regions. These include the regions processing visual information, those handling body sensory inputs, the surface layer of the cerebellum (the cerebellar cortex), and the hippocampal formation, in particular the dentate gyrus. In this section, a brief introduction to the organization of each of these systems and regions will be given.

### Visual system (Fig. 2-11)

The visual system is of major importance in developmental studies of the brain for a number of reasons: (1) the various parts are well circumscribed; (2) a map of the visual world is maintained in the visual areas of the brain, and this can be modified experimentally; (3) there is a considerable degree of dynamic interaction between its various components; (4) the retina is uniquely accessible to experimental manipulation that does not directly involve the rest of the central nervous system; and (5) the visual image can be more readily controlled than almost any other sensory modality.

The visual image is received in the retina by receptors on its posterior surface, adjacent to the pigment epithelium. It is relayed through a bipolar cell to the ganglion cells; two sets of interneurons, horizontal cells and amacrine cells, modify the input at the relay points (Dowling, 1970). The axons of ganglion cells run across the surface of the retina and leave the eye at the optic disk, where they enter the optic nerve. Depending on the animal, some or all of the optic axons from each eye cross the midline and intermingle with fibers from the other eye in the optic chiasm on the ventral surface of the forebrain. The axons continue around the side of the brainstem and finally end in the superior colliculus (or optic tectum). On their way they innervate a number of nuclei, the most significant being the dorsal lateral geniculate nucleus, referred to in this book for simplicity as the lateral geniculate nucleus (see Kaas et al., 1972). Cells in this nucleus send axons to the primary visual cortex (area 17) which end predominantly in the middle layers of the cortex.

The cortex is divided into a series of layers (numbered I to VI from the

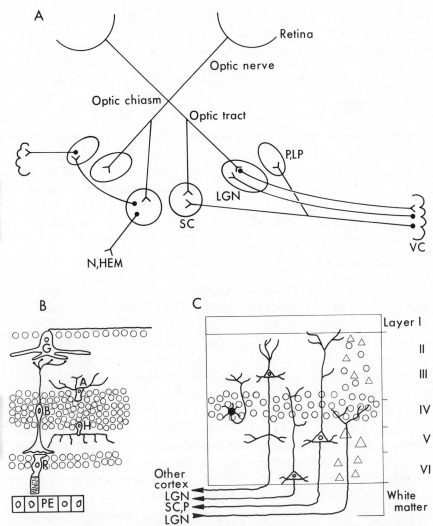

**Figure 2-11.** Summary of visual system organization. (A) Plan of principal connections. (LGN) Lateral geniculate nucleus, (N,HEM) nuclei serving head and eye movements, (P,LP) pulvinar, nucleus lateralis posterior of thalamus, (SC) superior colliculus, (VC) visual cortex. The left side of the diagram emphasizes the connections of the superior colliculus; the right side, the connections of the visual cortex. (B) Section through retina showing major cell types. (PE) pigment epithelium, (R) receptor, (B) bipolar cell, (H) horizontal cell, (A) amacrine cell, (G) ganglion cell. (C) Section through primary visual cortex (area 17) showing lamination and the specific connections formed by cells of certain layers. Abbreviations as in (A).

surface) characterized by the grouping of neuronal cell bodies. The dendrites of the cells may traverse several layers even though the cell bodies of a particular cell type are restricted to a single lamina. Depending on the laminar position of the cell bodies, certain cells have characteristic projections to other regions. This is indicated in Figure 2-11, which presents in simplified form data on the visual cortex of monkeys (Lund et al., 1975).

The superior colliculus is also organized as a series of layers, but these are defined by the presence of bundles of fibers interspersed with layers rich in cell bodies (Lund, 1972). The superficial layers are concerned with visual input, the deeper layers with auditory and somatosensory inputs. Of the other regions receiving optic input, mention will be made of the suprachiasmatic nucleus (in the hypothalamus adjacent to the optic chiasm), ventral lateral geniculate nucleus, and pretectum.

### Somatosensory system (Fig. 2-12)

Sensory information from the body is first received by the peripheral processes of the dorsal root ganglion and then transmitted into the spinal cord. Some axons end in the dorsal and ventral horns of the spinal cord at their level of entry or several segments more rostral or caudal. Other axons and branches run rostrally in the dorsal white matter of the cord (the dorsal columns) to end in the dorsal column nuclei (nucleus gracilis and cuneatus) of the medulla. Neurons in these nuclei send their axons across the midline to end in the thalamus at the nucleus ventralis posterior. From here, axons travel through the internal capsule to the cerebral cortex. Sensory afferents from the head end in the trigeminal nuclei, which fulfill the same role for the head as do the dorsal column nuclei and dorsal horn for the rest of the body. The trigeminal nuclei send axons to the nucleus ventralis posterior, which in turn sends axons to the somatosensory cortex. At each relay a "map" of the body surface is represented in the region. The maintenance of this map and the convergence of sensory pathways with other afferents are matters of special interest.

### Cerebellar cortex (Fig. 2-13)

The surface of the cerebellum is folded into a series of folia. Its uniform structure has attracted a good deal of attention because of the relatively few cell types involved, their developmental history, and the geometric configuration of their processes. There are two principal neuron types: Purkinje and granule cells. The Purkinje cell has an elaborate dendritic field oriented like a fan across

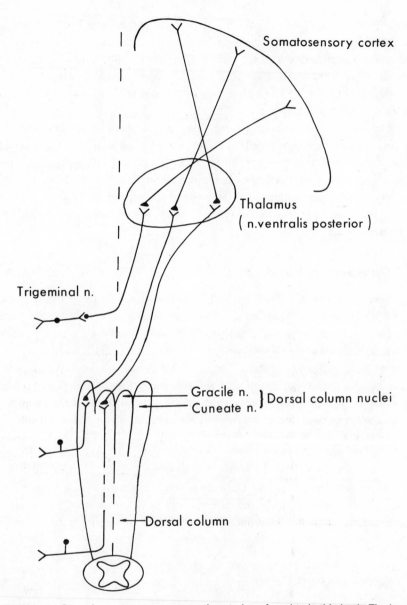

**Figure 2-12.** Plan of somatosensory connections to be referred to in this book. The lowest input on the left comes from the lower limb; the middle, from the upper limb; and the highest from the head. The dotted line signifies the midline of the body.

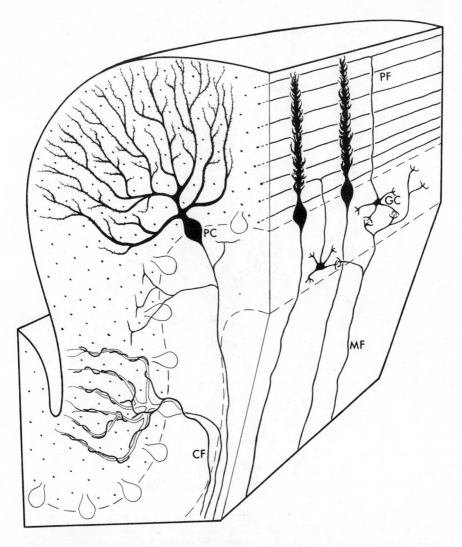

**Figure 2-13.** Three-dimensional view of a folium of the cerebellum showing two cell types and their associated axons. Purkinje cells (PC) have dendrites extending across the folium, oriented perpendicular to the parallel fibers (PF) of granule cells (GC). Climbing fibers (CF) and mossy fibers (MF) are shown (courtesy E. Elekessy).

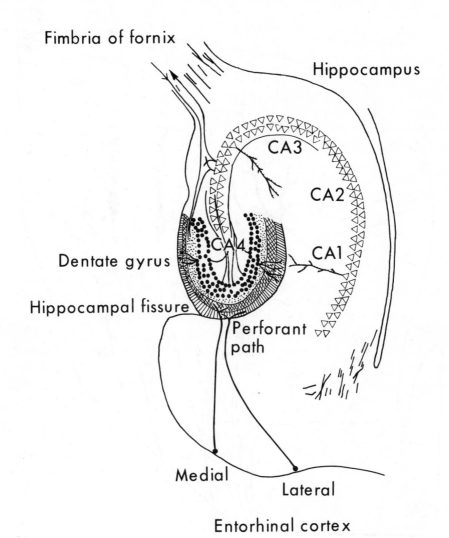

Fimbria of fornix

Hippocampus

CA3

CA2

CA4

Dentate gyrus

CA1

Hippocampal fissure

Perforant path

Medial

Lateral

Entorhinal cortex

**Figure 2-14.** Cross section of hippocampal formation showing the main components. Note the laminar order of inputs to the dentate gyrus from the lateral entorhinal area (lined area), medial entorhinal area (cross-hatched area), and commissural and association afferents from CA3 (dots). Because the hippocampal formation curves round the forebrain, horizontal sections appear quite similar, although the adjacent structures may differ.

the folium. Its axon runs deep into cerebellar white matter, although some branches return to the cortex. The granule cell has a collection of small dendrites and an axon which runs toward the surface and then branches along the plane of the folium. In this disposition the axons of granule cells are called parallel fibers. They end on the dendrites of Purkinje cells, one parallel fiber making synaptic contact with many Purkinje cells. There are two main inputs from the brainstem into the cerebellum: one from the inferior olive wraps around the dendrites of the Purkinje cells and is termed a climbing fiber; the other (arising from the pontine and other nuclei) ends mainly on granule cells and is called a mossy fiber. Of the other cells in the cerebellar cortex, mention should be made of the stellate neuron of the molecular layer and a nonneuronal cell type, the Bergmann glial cell. Detailed discussion of cerebellar organization may be found in Llinas (1969) and Palay and Chan-Palay (1974).

### Hippocampal formation (Fig. 2-14)

This is a complex of several regions, the main part of which is infolded in an elaborate pattern. It is composed of the hippocampus proper (sometimes termed *cornu ammonis,* or ram's horn, and subdivided as CA1-4), the dentate gyrus, and some adjacent cortical regions. What makes the region, and in particular the dentate gyrus, interesting is that it receives a number of inputs which are spatially segregated along the length of the dendrites. The dentate gyrus receives separate inputs from the medial and lateral parts of the entorhinal cortex (adjacent to the hippocampus), from part of the hippocampus (CA3), and from the septal nuclei. The entorhinal afferents, which are predominantly but not exclusively from the ipsilateral cortex, cross the hippocampal fissure (for this reason they are sometimes called perforant pathways) and terminate along the distal part of the dendrites of the granule cells in the dentate gyrus (see Fig. 2-14). The afferents from CA3 arise partly from the ipsilateral hippocampus (association fibers) and partly from the contralateral hippocampus (commissural fibers). The two are intermixed and end proximally on the dendrites, close to the cell body. Septal afferents appear to end partly distally and partly very close to the cell body. The axons of the granule cells of the dentate gyrus are distributed among the cell bodies and proximal dendrites of neurons in regions CA3 and 4 of the hippocampus, and there are also recurrent branches which run back to end among the granule cell dendrites. Local neurons have axons that are distributed within the granule cell dendritic field. A full description of hippocampal organization can be found in Brodal (1969) Isaacson and Pribram (1975), and in the relevant papers discussed in Chapter 13.

# II
# NERVE CELL BIOLOGY

Fundamental to any consideration of development and plasticity of the nervous system is an understanding of the organization and cell biology of neurons and other cells of the nervous system which are important for the orderly maturation and survival of the neurons. In particular it is important to know something of how neurons differ from one another and what schema are available for sorting out the differences. We need to know what the significant subcellular components of nerve cells are, how they are distributed, what role they play in the cell's overall organization; and what interactions occur within the nervous system to maintain it in a state of order. Very much tied up with these questions is the availability of techniques for studying nerve cells, since it is the techniques that both direct and limit progress.

In this section these basic matters are reviewed in three chapters. The first deals with the classification and components of nerve cells and the descriptive events following injury. The second considers the dynamic properties of nerve cells. The third reviews the ways in which neural diversity may be demonstrated and points out some of the important limitations of each methodological approach.

# 3

# MORPHOLOGY OF NERVE CELLS

In this chapter we will cover the basic identifying features of neurons and neuroglia, giving special attention to the ways in which cells are classified and the subcellular components that are defined. This will provide a basis for understanding the variations that occur under particular conditions and the dynamic interactions between neurons.

We will first consider the morphology of neurons and then of neuroglial cells.

## Neurons: Classification

Neurons, as mentioned in Chapter 1, usually have a number of processes arising from the cell body. These are the dendrites and the axon. Their configuration relates to the requirements of the cell with respect to its input–output characteristics. The variety of shapes has led to a series of cell classification schema and three will be considered here.

*I*. Classification according to number and disposition of processes (Fig. 3-1).

a. *Multipolar cells*. Like the characteristic cell of Chapter 1, these have one axon and more than one dendrite arising from the cell body. Many neurons in the mammalian brain are of this type. Various subclassifications or descriptive terms are applied to particular varieties of multipolar cells. For example, whereas large cells with a symmetric spread of dendrites around the cell body are usually called multipolar cells, similarly arranged smaller cells are usually called "stellate" or granule cells. Other multipolar cells have a skewed dendritic orientation and may be called fusiform (a bush of dendrites above and below the cell body), pyriform (dendrites above cell bodies), pyramidal (a pyramidally shaped cell body and one large dendrite—the apical dendrite—extending toward the surface of the brain), Purkinje (a very elaborate pyriform cell), and so on. Some types are unique

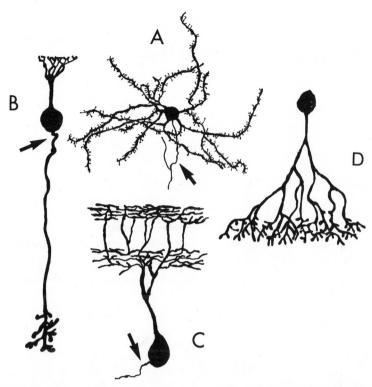

**Figure 3-1.** Examples of different classes of cell. (A) Multipolar (stellate) cell from cerebral cortex with spiny dendrites, (B) bipolar cell of retina, (C) ganglion cell of retina, also bipolar, (d) amacrine cell of retina—unipolar and anaxonic. In A-C, the axon is indicated by an arrow (drawing courtesy J.S. Lund).

to particular regions. Thus, pyramidal cells are peculiar to the telencephalic derivatives, while Purkinje cells are found only in the cerebellar cortex.

b. *Bipolar cells.* These have two processes, usually an axon and a dendritic trunk, arising from the cell body. Both may branch, however, beyond their point of origin. The bipolar cells of the retina are a good example of this type.

c. *Unipolar cells.* These have one process arising from the cell body which may divide to form a traditional axon and dendrite. Most cells in invertebrate nervous systems fall within this category. In the vertebrate brain, the cells of the dorsal root ganglion and some amacrine cells are examples.

d. *Apolar cells.* Interneurons without processes have been identified in the superior cervical ganglion; they are both presynaptic and postsynaptic at the cell body. So far, such cells have not been demonstrated in the central nervous system.

*II*. Classification according to axon. The two basic categories are cells with axons and those without. Cells with axons are commonly classified as Golgi type I or Golgi type II. Golgi type I neurons have axons projecting away from the region of the cell body, while Golgi type II neurons have axons that ramify locally. These divisions however, are not particularly helpful for two reasons. First, the axons of many cells do both, and second, what constitutes a "local region" becomes a matter of definition. For example, cells projecting from one layer to another in the cerebral cortex, while technically Golgi type II, could be classed as type I on the grounds that each cortical layer functions as a separate nuclear group.

Clearcut cases of cells without axons, in addition to the apolar cells mentioned above, are the horizontal and amacrine cells of the retina. The dendrites of these cells are both presynaptic and postsynaptic to other neurons. In a number of other central nervous system regions, cells have been identified whose dendrites are both presynaptic and postsynaptic to other elements, but in some regions at least these cells certainly have axons.

*III*. Classification according to cell size. Cell body size is used to classify neurons in many experimental studies, yet its significance is obscure at present. Since most of the protein synthesis in a neuron takes place in the cell body, its size probably reflects the amount of cytoplasm the neuron has to support, particularly the volume of the axon. However, there is no good evidence to support this idea. Attempts have been made to subdivide neurons according to somal size and to correlate size with a number of other properties (Altman, 1970; Jacobson, 1974). Basically two classes of cell are recognized. Macroneurons have large cell bodies and are generated early in development. They are typically Golgi type I cells and are considered to be more rigidly programmed in development than are microneurons (Jacobson, 1974). By contrast, microneurons have small cell bodies, are generated later in development, and are of Golgi type II. They are postulated to be more plastic in development. Although the two types can be distinguished according to generation time in such regions as the cerebellum and the hippocampus, that criterion is clearly not generally applicable. In the cerebral cortex, for example, large and small cells are generated at the same time. The separation of Golgi types I and II has already been criticized and the correlation is not necessarily related to cell size or cell birthday. The suggestion that one group is rigid and the other plastic in development has no foundation. In Chapters 9 and 10 we will see that one of the classic "macroneurons," the retinal ganglion cell, has a very plastic development; current technical limitations prevent us from saying anything about the plastic properties of microneurons. Thus, while the basic distinction holds up for a few regions, it does not apply to the brain in general, nor should implications regarding plasticity be drawn from its use.

Two general problems with cell classifications are first that they are subjective, and second that defining a series of classes leads to a pigeonholing such that cells having intermediate characteristics are not recognized. Even subjective classification is not a useless exercise, however. It provides, at the very least, a framework for recognizing neuronal heterogeneity. Further, it is a first step toward determining how cell shape, particularly dendrite shape, arises. Do cells derive from a common pool and gradually acquire the characteristic shape of one or another sort, or is each cell type determined at an early stage long before the adult shape is expressed? These two alternatives offer the extreme possibilities, and some intermediate condition would seem more likely. However, without rather elaborate mathematical descriptions of cell shape instead of the qualitative descriptions used at present, such problems of cell taxonomy cannot be easily approached. Only with mathematical methods will it be possible to chart the range of variation for a particular cell type. Such methods may indicate how tightly its developmental program is controlled and possibly which elements of its shape are basic to the function of the cell and which are less important.

A second problem with cell classification based on detailed criteria of cell shape is the natural expectation that if two cells look the same, they must function in a similar manner. Some doubt has been cast on this by West and Dowling (1972), who found that one identifiable class of ganglion cell in the retina (defined in Golgi preparations) may receive one or the other of two different configurations of synaptic input (when examined with the electron microscope), which should result in quite different physiological properties. Light microscopy gives no indication as to which configuration the cell receives.

Thus, cell classification, while basic to our need for defining cell differences, must be treated with some caution. It is obviously important to define cells in more objective terms. It is important to rely not only on the dendrite configuration, but also on the axonal ramifications and, if possible, on the patterns of synapses at the cell surface as well as the physiological response properties of the cell. In some clearly ordered tissues such as the hippocampus and cerebellum, where there are sheets of identical cells, such a comprehensive analysis is possible. Elsewhere we are in a less fortunate position.

## Cytological details (Fig. 3-2)

Neurons possess all the characteristics of eukaryotic cells. The nucleus situated in the cell body (or soma) has a variable outline but is frequently folded. There is a distinct nucleolus, clearly distinguishable from the nuclear chromatin. Most cells of the mammalian brain are diploid, although there is evidence that some are polyploid, including the large Betz cells of the motor cortex, pyramidal cells of the hippocampus, the Purkinje cells of the cerebellum, and the larger motor neurons of the spinal cord, all of which are tetraploid (Herman and Lapham,

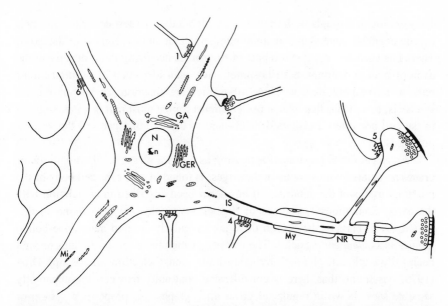

**Figure 3-2.** Diagram of neuron to show major cytoplasmic components and a variety of synaptic relationships. For clarity, intracytoplasmic filaments have been omitted. (N) nucleus, (n) nucleolus, (GA) Golgi apparatus, (GER) granular endoplasmic reticulum, (IS) initial segment of axon, (Mi) mitochondrion, (My) myelin, (NR) node of Ranvier. Synapses are signified by numbers: (1) axodendritic, (2) axospinous, (3) axosomatic, (4) axoaxonal (initial segment), (5) axoaxonal (terminal), (6) dendrodendritic.

1969). There is no clear correlation between ploidy and cell body size as in other tissues; some medium-size cells in the spinal cord for instance, are octoploid. A nuclear membrane surrounds the nucleus. Pores in it allow continuity between nucleoplasm and cytoplasm. The cytoplasm, studied most extensively in the axon, appears to be an elastic solid which can be liquefied by calcium ions and high salt concentrations (Gilbert, 1975a,b). Various subcellular components are embedded within the cytoplasm. These are often distributed preferentially in certain parts of the cell.

Ribosomes in polyribosomal aggregates associated with membrane (granular endoplasmic reticulum—GER) are found in quantity only in the cell-body and proximal dendrites. In large neurons the region of the cell body immediately adjacent to the origin of the axon is devoid of granular endoplasmic reticulum. This is termed the axon hillock. Frequently the endoplasmic reticulum is delineated into clumps. Termed Nissl substance, these clumps can be demonstrated by light microscopy using basophilic stains. In addition to the endoplasmic-reticulum-associated ribosomes, others lie free in the cytoplasm.

They are found throughout the cell body in all but the smallest dendrites (rarely if ever in dendritic spines) and in small numbers in the initial segment of the axon. There has been an occasional report of ribosomes occurring more distally in the axon of mature neurons, but ribosomes have been identified in axon terminals only in developing tissues. Since ribosomes are the primary site of protein synthesis, this means that axonal proteins must for the most part be manufactured in the cell body and transported from there to the terminal parts of the various processes.

A typical Golgi apparatus, appearing as a series of sheets of smooth membranes associated with vesicles at the edges, is found in either the cell body or the proximal trunks of the dendrites. It has never been described in association with the axon. Sometimes lysosomes or digestive vacuoles are also found in this region. In all parts of the neuron there are irregularly ordered membrane-bound tubules, saccules, and vesicles. The extent to which these are continuous has not been fully explored, although by using high-voltage electron microscopy Droz (1975) has found that there is considerable continuity between the apparently separate sacs shown in regular electron micrographs. He proposes a system of internal membrane-bounded channels connecting the cell body with distal parts of the neural processes.

Three fibrillar proteins are identified in neurons—microtubules (sometimes called neurotubules), neurofilaments, and microfilaments.

### Microtubules (Fig. 3-3)

These appear as unbranched tubes 25nm in diameter and a pale core 15 nm in diameter. Within the wall, 6 protofilaments have been identified (Wuerker and Kirkpatrick, 1972). Microtubules are found in most parts of a neuron except dendritic spines and (in normal preparations) axon terminals. They are quite unstable and are best demonstrated with primary fixatives containing gluteraldehyde; they show up less impressively with primary osmium or paraformaldehyde fixation. Using hypotonic treatments prior to fixation, Gray (1975) has found that microtubules may also exist in axon terminals. Whether they are present in normal living tissue or were condensed from cytoplasmic subunits by the unconventional treatment is not altogether clear. Within large dendrites microtubules are notably evenly spaced from one another, and they are frequently absent from the smallest dendritic branches. In the initial segment of the axon they show a curious pattern of aggregation in which groups of 2–10 microtubules come to lie within 10 nm of one another and are connected by cross bridges. In the rest of the axon, the microtubules are somewhat more haphazardly arranged than in the dendrites.

Detailed cytological and chemical studies show microtubules in nerve cells to be identical with those in cells of other tissues (see review by Stephens and Edds, 1976). A major component of microtubules is the protein tubulin, which shares a number of properties with actin and belongs to the same class of structural proteins (tektins). Tubulin is notable in binding with colchicine, vinblastine, and a number of other related drugs; as a result investigators have used these drugs to inactivate microtubules in an attempt to investigate their function. A note of caution must be added here. None of these chemicals is absolutely specific in its binding properties; colchicine, for example, also binds to some membrane proteins. In many tissues, microtubules seem to provide a cytoskeleton for cells. Evidence based on the effects of colchicine on nerve outgrowth in vitro (Wessells et al., 1971, 1973) would tend to support that hypothesis for the nervous system (with the proviso about the specificity of colchicine). Thus, it is possible that microtubules may play a significant role in the maintenance of the characteristic shape of individual cell types. Whether they actually have any role in determining the shape of a cell is uncertain.

A second possible role for microtubules concerns their relation to the mechanisms by which materials are transferred along axons and dendrites, both away from and toward the cell body. These transport mechanisms, orthograde and retrograde, which will be discussed in more detail in the next chapter, are blocked by the application of colchicine or vinblastine. Reduction in transport can occur without any loss of microtubules; blockage is accompanied by a reduction in numbers of microtubules, while dosages sufficient to remove all microtubules results in cell death. Thus, although there is some sort of correlation between obstruction of transport and microtubule loss, it is not a direct one. The exact transport mechanism is unclear. On the one hand is the sliding vesicle hypothesis of Schmitt (1968); on the other is a more elaborate model involving an interaction of intra-axonic channels with a microtubule-neurofilament system (see Droz, 1975).

## Neurofilaments

These are spiral protein threads having a diameter of about 10 nm, a central core of 3 nm and small side chains (Wuerker and Kirkpatrick, 1972). In Figure 3-3 they can be seen in comparison with microtubules. They are found in most parts of a neuron, including, in a few cases, dendritic spines, where they form a ring. Ring formations are also found in some normal terminals, and, as will be seen later, become more common in degenerating terminals (see Fig. 3-7A). One curious fact is that not all neurons have neurofilaments: in others they appear to be concentrated in the myelinated segment of the axon, and in still others large

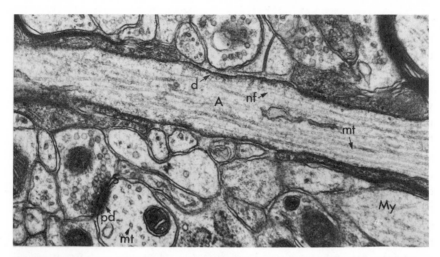

**Figure 3-3.** Electron micrograph of neural tissue. (A) axon at node of Ranvier, with surface density (d), microtubules (mt), and neurofilaments (nf). Note synaptic specializations at S, including postsynaptic density (pd) and microtubules cut in cross-section. My denotes myelin.

numbers are found in all parts of the cell. Their capricious appearance is difficult to account for, although it is true that larger axons tend to have more neurofilaments. In a number of studies the physical and chemical properties of these filaments have been investigated: the most complete studies are those of Huneeus and Davison (1970) and of Gilbert (Gilbert et al., 1975; Gilbert, 1975a, b) of filaments taken from axons of the marine worm, *Myxicola*. An individual neurofilament is made of a fundamental α-helix which is further coiled. There are indications that some neurofilaments may coil one with another. Gilbert suggests that this coiling may be responsible for the frequent observation that the cut end of an injured axon adopts a corkscrew configuration; he proposes that neurofilaments contribute to the skeleton of neural processes and may be concerned with various rotational movements (as of the nucleus) observed in neurons. Chemical studies show that neurofilaments have nothing in common with microtubules.

## Microfilaments

These appear in many types of cells as 6–9 nm strands. They may occur in bundles, possibly concerned with cell locomotion; or they may appear in a fine network close to the surface, where they are apparently involved in local surface movements. Microfilaments appear to bind heavy meromyosin, which would imply that they are made up of actin or an actin-like substance. They differ from the other filamentous arrays in neurons in being disrupted by the drug cytochalasin and in being associated with intracellular contractile processes in development. They have no obvious role in the adult neuron, but they are found in growing tips during development and in that context they will be discussed further in Chapter 9.

Other cytoplasmic components include mitochondria and lysosomes. The former are concerned with oxidative phosphorylation and appear as sets of disks (cristae) in an elongated membrane-bound profile about 0.5 $\mu$m wide and as much as 20 $\mu$m long. The mitochondria in terminals seem to have different staining characteristics from those situated elsewhere in adult neurons, but this difference is not obvious in the immature nervous system. In addition, mitochondria show a differential packing of cristae in terminals of certain axons (in particular, optic nerve terminals where they are loosely packed), which has proven useful for identifying terminals of certain origin in normal material. The lysosome system is concerned with degradation of intracellular debris. It is located in the cell body and has high acid phosphatase activity. The reader is referred to papers by Novikoff et al. (1971) and Holtzman et al. (1973) for a detailed discussion of lysosomes, Golgi apparatuses and associated intracellular membrane systems, and how such systems may interrelate.

### Surface membrane and associated structures

Chemical, electron microscopic, and X-ray diffraction studies have described cell membranes as somewhat rigid layers of lipoprotein. More recent studies (summarized and organized into a new theory by Singer and Nicolson, 1972) have modified that view of membranes in two significant ways. They indicate first that the phospholipid in the membrane is organized largely as a fluid bilayer and second that the proteins are embedded in this lipid to a greater or lesser degree as relatively independent globular molecules with ionic and highly polar groups protruding from the bilayer structure (Fig. 3-4). These discoveries have several implications. The membrane can no longer be considered a rigid structure; indeed, a number of studies using plant agglutinins and fluorescent antibody labels in a variety of cells have shown that the proteins they tag are capable of considerable movement across the sheet of membrane. Further, the actual membrane clearly extends beyond the bilayer leaflet seen in electron micrographs.

The surface proteins include antigens, enzymes, and glycoproteins, are

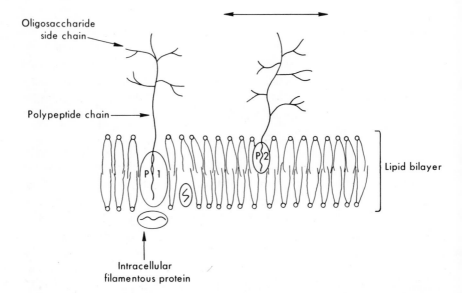

**Figure 3-4.** Diagram of current model of membrane structure. The external surface is uppermost. Membrane proteins not held by intracytoplasmic proteins have the capacity to move laterally, as indicated by the arrows. (redrawn from Nicolson, 1974).

thought to be important in cell recognition behavior. The antigens can be recognized by using labeled antibodies; the enzymes by adding the appropriate labeled substrates; and the glycoproteins by using plant agglutinins or lectins, such as concanavalin A or wheat germ agglutinin (Nicolson, 1974), which bind specifically with certain carbohydrate groupings. Another property of the surface proteins of membranes is that active groups can be masked by molecules which can be removed by specific enzyme treatments. As will be seen at various points in this book, the configurations and behavior of the proteins of the cell surface are most important in determining the selective interaction a neuron exhibits towards other cells, both neuronal and nonneuronal. As yet, however, work on the surface proteins of neural cells is just beginning. On the cytoplasmic surface of the membrane bilayer of neurons, there is an accumulation of protein material in three places: at the initial segment of the axon, at nodes of Ranvier (see Fig. 3-3) (in both cases, the protein appears to be organized in a helical array), and at the postsynaptic membrane of synaptic junctions (see Fig. 3-5). It is significant perhaps that at each site the membrane is physiologically very active, and that the density associated with the postsynaptic membrane at least is a region of high concentration of certain enzymes.

## Synaptic junctions

Synapses have four basic features (Fig. 3-5):

1. *A site of adhesion between two membranes.* There is stainable material in the gap of 10–20 nm between the two membranes and this can sometimes be resolved as an array of filaments. The adhesion withstands centrifugation and a variety of drastic treatments, although it can be broken by treatment with detergents (see Jones, 1975).

2. *An array of vesicles in the presynaptic processes.* These contain chemicals, termed transmitter substances, which when released into the synaptic cleft change the conductance of the postsynaptic membrane to certain ions. Conventional electron micrographs show that a terminal contains a preponderance of either agranular or granular vesicles. The latter usually contain aminergic transmitters such as norepinephrine. The agranular vesicles can be divided into subclasses of different morphology if hyper-

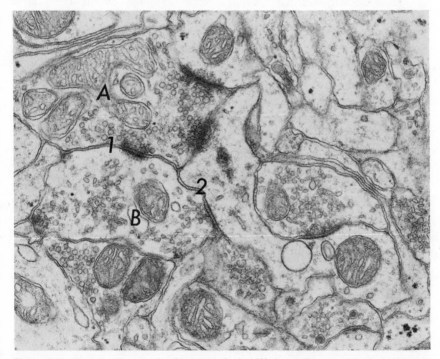

**Figure 3-5.** Electron micrograph of two terminals. (A) (an optic nerve terminal with pale mitochondria) contains round vesicles and makes a type I junction at 1; (B) contains flattened vesicles and makes a type II junction at 2. This is an example of one terminal synapsing on another × 32,060.

tonic fixatives are used. The two basic categories are round and flattened (or pleomorphic), but several investigators have demonstrated that some flattened vesicles are flatter than others and that by adjusting the osmolarity certain subpopulations can be differentiated (e.g., Valdivia, 1971; Lund, 1972). The question of vesicle morphology is of interest in that two inhibitory transmitters, glycine and γ-aminobutyric acid (GABA), are both associated with terminals containing flattened vesicles, while acetylcholine, an excitatory transmitter, is associated with round vesicle terminals. It is still perhaps too early to accept the generalization of Uchizono (1965) that round vesicle terminals are excitatory and flat vesicle terminals are inhibitory, but so far this hypothesis is consistent with most available data.

3. *Receptor molecules.* These are the postsynaptic sites with which molecules of transmitter (released from the presynaptic process) bind. Most of the work on receptors has been done at cholinergic synapses outside the central nervous system, since investigators can make use of one of the neurotoxic components of some snake venoms, α-bungarotoxin, which binds specially with acetylcholine receptor molecules. The substance can be conjugated with fluorescein or tritiated so that the nature and behavior of the receptors can be identified. It is presumed that the receptors reside within the postsynaptic membrane and not in the associated postsynaptic density.

4. *A postsynaptic density.* This is a densely staining region on the cytoplasmic surface of the postsynaptic membrane. Its specific staining properties and components have been characterized in some detail (see Banker et al., 1974; Matus et al., 1975). It is made up largely of protein, including a significant amount of tubulin, and contains high concentrations of certain enzymes such as $3'5'$-phosphodiesterase. Banker and colleagues suggested that the density may function to stabilize the overlying membrane or may serve as a binding site for the enzymes in this region. Two distinct classes of postsynaptic density are recognized (see Fig. 3-5), a very pronounced one (Gray type I or asymmetric) and a relatively insignificant density (Gray type II or symmetric).

It has been suggested that the magnitude of the density may affect the strength of the adhesion between pre- and postsynaptic membranes and hence the capacity of the apposed membranes to break contact. Somewhat surprising is the fact that most terminals containing flattened vesicles are associated with type II densities and most round-vesicle terminals have type I densities.

Apart from the traditional synapses, there are zones of close apposition of neural membranes, so-called gap junctions. These allow transport of small molecules from one cell to another and function as channels of low resistance. In the mammalian brain, they have been found in certain specific regions; elsewhere they occur infrequently.

## Classification of synaptic patterns (see Fig. 3-2)

Most synapses in the central nervous system are from axons to dendrites; these axodendritic synapses constitute 90% or more of the contacts in most regions. Some are on dendritic shafts, others are on dendritic spines (axospinous). Synapses from axons to cell bodies (axosomatic synapses) usually represent less than 1% of the synapses in a region. Synapses from axon terminals to other axons (axoaxonic synapses) take two forms. In one type the axon terminal contacts the initial segment of another axon; in the other it contacts the terminal region. As was indicated in the discussion of cell classification, some cells have dendrites which themselves contain synaptic vesicles and are presynaptic to other cells. These form dendrodendritic or dendrosomatic synapses. Dendroaxonal synapses have not been described, although it is quite possible that they do sometimes occur. Dendrodendritic synapses tend to occur in certain regions, such as the thalamus, the retina, and the superior colliculus, and to be very rare or absent in other regions (e.g., the cerebellum and all derivatives of the telencephalon). The rare apolar cells as well as occasional neurons in the CNS having processes (reported in the optic tectum, superior colliculus, lateral geniculate nucleus) have presynaptic specializations in the somal region, forming somadendritic synapses.

Studies of various regions show that certain classes of axons will be predictably involved in particular sets of synaptic patterns. For example, optic terminals have never been described postsynaptic to other terminals, but a certain (and where studied, quite constant) proportion are presynaptic to terminals containing flattened synaptic vesicles (Fig. 3-5). Comparable examples can be cited for many other regions. Although statistics are often unavailable, the impression gained is that in any region certain terminals are involved only in particular synaptic patterns, and this suggests a measure of selectivity at the synaptic level.

### Degenerative reactions

We will consider the events following transection of an axon at a point X, in Figure 3-6. Those degenerative changes occurring distal to the cut (i.e., distant from the cell body) are called orthograde, anterograde, descending, or Wallerian degeneration. (These terms are interchangeable.) The changes occurring proximal to the cut are generally termed retrograde degeneration. If one neuron degenerates, the neurons with which it is connected may themselves undergo atrophy or degeneration. Such an effect is termed transneuronal degeneration and

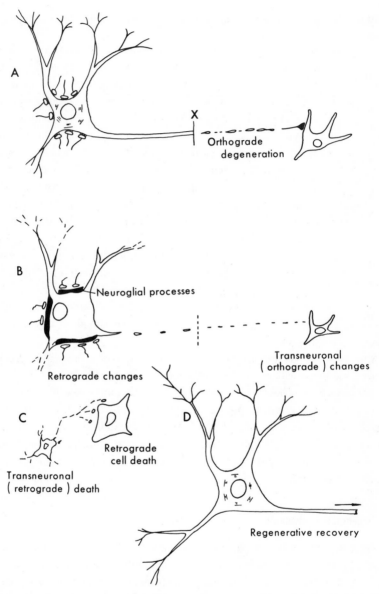

**Figure 3-6.** The sequence of events which may follow axon injury at X. From B, the cell (and even those connecting to it) may either die as in C or recover as in D.

may be either orthograde or retrograde, depending on whether the subsequent or the preceding neuron in the chain of cells is affected.

### Orthograde changes

Simple Orthograde Degeneration. The first sign of degeneration distal to a cut is the appearance of local swellings along the axon. It is suggested that this may be caused by failure of active ion pumping mechanisms in the axon membrane followed by the loss of selective permeability of the membrane (Mire et al., 1970). The process can be delayed by increasing extracellular calcium. Study with the electron microscope has revealed three sets of degenerative events. In the first, the terminals become pale within a day or so of the lesion. The subsequent behavior has not been described. In the second, the terminals become pale and fill with neurofilaments that form a ring around a central core of vesicles and mitochondria (Fig. 3-7A). This ring can be stained by neurofibrillar light microscopic methods (Fig. 3-7B). This reaction first appears a day or so after the lesion and lasts for only a few days; the neurofilamentous terminals then become electron dense.

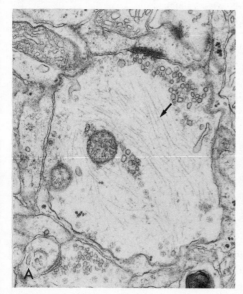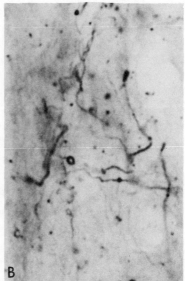

**Figure 3-7.** Changes in optic nerve terminals in the superior colliculus of an adult rat three days after eye removal. (A) Hyperplasia of neurofilaments (arrow) in electron micrograph × 25,850. (B) Associated terminal rings seen in light micrograph, stained by a neurofibrillar method × 2070.

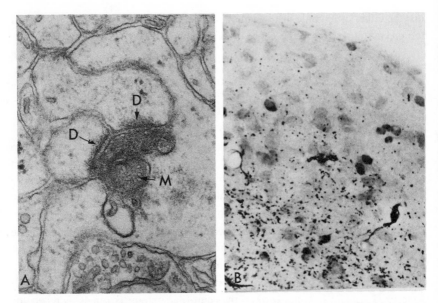

**Figure 3-8.** Part of a fragment of superior colliculus which had been transplanted from one rat to another early in development to show degeneration due to removal of the host eye. (A) Dense terminal degeneration in electron micrograph. Note postsynaptic density (D) associated with dead terminal and mitochondrion (M) in terminal × 47,300. (B) Light micrograph of adjacent region stained with Fink-Heimer stain. The small black granules correspond in most cases to terminals like the one in (A) × 210. (See Chap. 13 for description of the specific experiment.)

The transitory terminal neurofilamentous hyperplasia is difficult to account for, especially since this part of the nerve cell has been separated from the region of major protein synthesis by the lesion. Two of a number of possibilities are:(1) that the changing microenvironment provokes a condensation of axonal proteins into neurofilaments (but calcium ions accelerate the breakdown of neurofilaments—Schlaepfer 1974), or (2) that axonal filaments are being slowly transported into the terminal where they are normally broken down. This breakdown process may be blocked if the axon is cut. In the third and most commonly described reaction, the axon and terminal become electron dense and tend to shrink (Fig. 3-8A). This reaction appears within a day or so after the lesion, and terminals may persist in this state for as little as one to two days and as much as nine months. The reaction is stainable by the Nauta methods of light microscopy and derivatives thereof, the most important being the Fink-Heimer method (Fig. 3-8B; see Chap. 5). Why different groups of terminals should undergo fundamentally different degenerative reactions is not altogether clear. Do the reasons relate to physiological or metabolic activity, to axon diameter, or to other factors? A further question is that of degeneration rate. It appears that small axons degenerate before large axons, although there are contrary views in the literature, some of which are based on less sensitive or totally inadequate techniques. This matter has been discussed in some detail elsewhere (Lund et al., 1976).

Transneuronal Orthograde Changes. If the afferents to a region are removed in an adult animal, the cells on which those axons ended will generally show no obvious anatomical change, although significant electrophysiological events may occur (see Kjerulf et al., 1973). Sometimes, however, a profound and rapid shrinkage of the cell-body region accompanied by degeneration of dendrites, with presumed retraction of the overall dendritic field, will result from deafferentation. The best studied example is the lateral geniculate nucleus of primates (including man): in monkeys changes in cell body size are detectable within four days after eye removal (Matthews et al., 1960). In young and developing animals, similar transneuronal changes are much more dramatic and provide an important parameter in the regulation of cell numbers.

Retrograde changes

Simple Retrograde Degeneration. After axon section, the cell body shows one of three reactions—no detectable change, a series of pronounced changes followed by regeneration of the axon and recovery by the cell body of its normal appearance, or a set of changes followed by degeneration of the neuron proximal to the lesion. The series of initial changes in the last two cases appears to be the same. There is first a transitory but significant increase in RNA synthesis. This is followed by dispersion of the granular endoplasmic reticulum and of its light microscope correlate, the Nissl substance. This response is termed chromatolysis. In addition many more free ribosomes are found, occurring singly rather than in polyribosomal clusters; the cell body is swollen; the nucleus is displaced to one side of the soma; and the toal dendritic field may shrink (Sumner and Watson, 1971). Retrograde degenerative changes are more likely if the axon transection is close to the cell body, and are also more common in young animals. A full discussion of the retrograde changes associated with axotomy is given by Cragg (1970), Lieberman (1971), and Grafstein (1975). If the axon regenerates, there is a dramatic increase in ribosomal RNA; eventually the cell returns to normal and the dendrites expand to their original size. If the cell dies, dendrites and cell body show an electron-dense reaction. Reasons for the cellular responses of axotomy will be discussed in Chapter 4.

Transneuronal Retrograde Changes. These changes were first identified in the visual system of primates (Van Buren, 1963). If the visual cortex was lesioned, simple retrograde degeneration of cells in the lateral geniculate nucleus resulted. With longer survival time, however, loss of ganglion cells in the retina was also noticed. Since the axons of these cells end on cells in the lateral geniculate nucleus, this effect is considered to represent a transneuronal retrograde effect. A similar phenomenon has been reported in the limbic connections of young rats (Hart, 1969).

In addition to cell death, another, more subtle transneuronal retrograde change has been identified: the terminals ending on a neuron undergoing chromatolysis disappear and are replaced by microglia. This change has been shown in facial (Blinzinger and Kreutzberg, 1968) and hypoglossal neurons (Hamberger et al., 1970), motor neurons (Kerns and Hinsman, 1973), and in neurons of the superior cervical ganglion (Matthews and Nelson, 1975). Associated with the loss of terminals is a failure of conduction across the synapses (Kuno and Llinas, 1970; Mendell et al., 1974). Since each of these neurons is capable of regeneration, it is assumed that the terminals reestablish contact at some point, although exactly when this occurs is not known. The position formerly occupied by the terminals is taken by microglial cells. Whether the glial cells remove the terminals or secondarily occupy the synaptic sites is not clear and the primary stimulus responsible for the event is not known. Curiously, "synapse stripping" has not been reported for chromatolytic cells projecting within the central nervous system. Thalamic cells may undergo retrograde degeneration and still have normal synapses ending upon them (see Barron, 1975). Possibly, as Hamberger and his coauthors imply, the response is related more to regenerative phenomena than to degenerative events.

The degenerative reactions described in this section have been used as experimental tools to define patterns of interconnections within the central nervous system. They have a further importance which has been emphasized only in recent years, and that is in pointing to some of the dynamic interactions which occur between various parts of a neuron and between different nerve cells. Such interactions have been termed "*trophic*" and many recent studies have been directed toward defining more specifically what is meant by "*trophic effects.*" These will be considered in detail in Chapter 4.


## Neuroglia

Neuroglia are the interstitial cells of the nervous system. They are not themselves directly concerned with the conduction of specific information, but are nevertheless essential for the integrity and successful functioning of the neurons. Three separate classes are recognized: astrocytes, oligodendrocytes, and microglial cells.

### Astrocytes and similar cells

Astrocytes are characterized as having many processes radiating from their cell bodies in light microscopic preparations. They are perhaps best demonstrated for light microscopy with a fluorescent-conjugated antibody to the sclerotic plaques of multiple sclerosis (GFA - glial fibrillar antibody), which is specific for this class of neuroglia and has the added advantage of staining their processes. In electron microscopic preparations, they appear to have processes that form no characteristic shape of their own but rather interpolate round and between other cell profiles. Their cytoplasm contains few organelles and has a watery appearance. Some cells have tightly packed bundles of microfilaments, which tend to be more obvious in higher animals and when the cells are in reactive states. Of somewhat similar appearance to astrocytes are the radial glial cells in developing brains. This cell type has a single radial process from the ventricle to the surface, with small side processes. In later development it may mature into an astrocyte (Schmechel and Rakic, 1973). In addition, a number of specialized cell types resemble these two. Among these are the Müller cells of the retina, the Bergmann glia of the cerebellum, and the pituicytes of the pituitary. (The last are exceptional in that they receive synaptic contacts, and perhaps should not be considered as glia at all.)

The possible functions of this class of neuroglia are:

1. They may act as a reservoir for extraneuronal potassium ions.

2. They may serve to insulate unmyelinated neural processes from each other, thus preventing cross talk between them.

3. They appear to isolate and remove degenerative debris, but do not digest it.

4. They may act as a channel for moving metabolites from blood vessels to neurons; however, evidence for this role is lacking (see Kuffler and Nicholls, 1966).

5. During development the radial glial cells and others like them may act as guides for neuronal migration. This will be discussed in Chapter 7.

### Oligodendrocytes (literally "cells with few branches")

These cells have very little dendritic spread and are identified with the electron microscope by their blotchy nucleus and dense cytoplasm. They are commonly found in fiber bundles, where individual processes form the myelin sheath of axon segments, and adjacent to neurons where they function as satellite cells. Large neurons of the lateral vestibular nucleus have as many as 35–40 satellite cells per neuron (Hyden and Pigon, 1960). Similar cells surround the neurons of the dorsal root ganglion, and here they appear to have an essential supportive role, which will be discussed in the next chapter. Whether the satellite oligodendrocytes of the central nervous system function similarly remains to be seen.

### Microglial cells

This is a class of small cells which become reactive and increase in number in areas of focal degeneration. They appear to be involved in phagocytosis of degenerative material. Various studies have indicated that they are derived from ectodermal undifferentiated cells; others have identified them as mesodermal elements, possibly leucocytes (see Vaughn and Skoff, 1972; Matthews, 1974). Labeling studies in the hippocampus have shown that focal degeneration causes microglia to divide over a broad area and then to migrate towards the focus of degeneration (Lynch et al., 1973).

# 4

# DYNAMIC INTERACTIONS

The static forms of neuroanatomical preparations tend to lull one into thinking of the brain much as a telephone system or a computer network; a set of physical interconnections which once laid down function more or less successfully with a minimum of maintenance. While analogies of this kind may help one grasp the patterns of information transfer through the brain, they are, in fact, misleading for understanding the nervous system as a network of living cells. Individual neurons depend for their successful functioning or even survival on their interactions with other neurons and with neuroglia and individual parts of any neuron depend on continued interaction with other parts. A consideration of these interactions is fundamental to understanding many of the events occurring in development, the reasons for some of the degenerative reactions described in Chapter 3, and the problems associated with regeneration and maintenance of stable organization of the mature nervous system. In this chapter we will review the various interactions that occur between the various parts of a neuron and between individual nerve cells.

## Events Associated with Synaptic Transmission

When a wave of depolarization of the surface membrane of an axon invades its terminal, a series of events leads to the release of transmitter sequestered in synaptic vesicles at a specific region of the presynaptic membrane. The predominant mechanism by which the contents of synaptic vesicles are released appears to be exocytosis. The vesicles join with the specific region of the presynaptic surface membrane and lose their contents. In order to maintain a continuous supply of vesicles and to prevent excessive expansion of the surface membrane (caused by addition of membrane from newly released vesicles), it is obviously desirable to have a mechanism in which surface membrane is taken back into the terminal (see Fig. 4-1). Several lines of evidence support the concept of membrane recycling. It has been demonstrated that prolonged stimulation causes a reduction in vesicle numbers in certain terminals. Major support also comes from studies using horseradish peroxidase to identify intracellular membrane systems that open to the surface (Ceccarelli et al., 1972; Heuser and Reese, 1973). If put into a bath around an active neuromuscular junction, the horseradish peroxidase is taken into the terminal, first in coated vesicles, then into membrane saccules, and finally into synaptic vesicles. This implies a cycle of membrane movement from synaptic vesicle to surface membrane, to coated vesicles, to reticulum and back to the synaptic vesicle again. Since several studies have shown that the membrane of synaptic vesicles is chemically different from plasma membrane, it might be expected that the surface membrane at the synapse would also be of different composition, and this indeed seems to be the case (see Livett, 1976).

One consequence of such a recycling is that if the nerve is stimulated at very high rates, there should be a buildup of membrane at the rate-limiting step in the cycle. This would appear to be the uptake process since stimulation at high rates has been shown to increase the surface area of axon terminals (Pysh and Wiley, 1974). A more dramatic increase in surface membrane has been shown after treatment with black widow spider venom (Clark et al., 1972). This causes massive depletion of synaptic vesicles and an elaborate folding of the presynaptic membrane.

A                                                    B

**Figure 4-1.** (A) Pattern of recycling of synaptic membranes (after Heuser and Reese, 1973). ⇅ signifies release of synaptic vesicle contents to affect the postsynaptic membrane. Uptake of membrane into the terminal is by coated vesicles. (B) Diagrammatic representation of effect of excessive stimulation and inadequate recycling (as after application of black widow spider venom). There are few vesicles in terminal and the surface membrane is increased in area and folded.

## Orthograde Effects

### Orthograde transport of materials within the axon

The cell body has been considered the trophic center of the neuron. This conclusion is amply supported by evidence of the concentration of ribosomes in this region and minimal protein synthesis in axons. Measurements have indicated that the dendrites and soma of a pyramidal cell of the cerebral cortex account for 26,000 $\mu m^2$ of the surface area and 6,600 $\mu m^3$ of the volume of the cell, while the axon has a surface area of 250,000 $\mu m^2$ and a volume of 250,000 $\mu m^3$ (Sidman, 1974). In order to ensure that material manufactured in the soma is transported down the axon in sufficient amount to replace that which is used in normal metabolism as well as in export from the axonal terminal, an efficient transport mechanism must operate in the axon.

The first direct evidence of transport of material in axons was provided by Weiss and Hiscoe (1948), who showed that there was a damming up of material proximal to a ligature placed around the axon and that this material was released when the ligature was removed. Droz and Leblond (1963), using tritiated amino acids which are incorporated into proteins in the cell body and then transported along the axon, were able to study this movement more directly, without the danger of surgical trauma. They and a number of succeeding workers estimated the rate of movement of materials at 1–3 mm/day. However, such a rate would mean that it would take one to three years for material to reach the terminals of an axon one meter long. In the late 1960s, a number of investigators showed a faster rate of as much as 400 or more mm/day. Intermediate rates of about 60 mm/day have also been demonstrated. In addition to identifying a fast rate of 240 mm/day and slow rates of 2–4 mm/day and 4–8 mm/day, Willard et al. (1974) also described an intermediate rate of 34–68 mm/day. A discussion of all aspects of axoplasmic transport is not possible within the scope of this book, and the reader is referred to several excellent reviews of the subject (Droz, 1975; Grafstein, 1975; Lubinska, 1975; Livett, 1976).

A number of general comments may be made, however. Fast axoplasmic transport (i.e., 40 mm/day and faster) carries many materials destined for synaptic vesicles, the presynaptic membrane, and (to a lesser extent) the mitochondria, as well as materials which will leave the axon terminal in one form or another to be taken up by other cells. Among the materials transported are phospholipids, RNA, adenosine, and a variety of proteins (including glycoproteins and various enzymes). Droz (1975) has proposed that after synthesis in the endoplasmic reticulum, proteins then pass through the Golgi apparatus (where glycoproteins are formed) into a network of agranular reticulum which extends along the axons and dendrites. Whether different materials may follow different channels is not altogether clear. Fast transport is unaffected, at least in the short term, by separation of the axon from the cell body and by inhibition of protein synthesis. It is blocked by chemicals disturbing oxidative metabolism and by colchicine. The effect of colchicine indicates that fast transport is in some way related to microtubular systems; this was discussed in Chapter 3. The rate of transport may vary according to the material being transported, but all materials must leave the cell body at the same time after injection of the precursor chemical for this observation to be accurate. Within limits, the rate varies directly with temperature. During development, it increases with maturity. Whether the rate increases in regenerating nerves is not altogether clear, although it is certain that under these conditions more materials are transported. Similarly, prolonged stimulation does not apparently alter the rate of transport, but it does increase the amount of material transported.

Materials involved in the growth and maintenance of neural processes tend to be transported slowly. Microtubules and neurofilaments move in this manner, as apparently do mitochondria and a whole group of so-called "soluble" proteins. The mechanism of slow transport is not altogether clear. Weiss has suggested that peristalic movements of the axoplasm may be involved. Because of difficulties of measurement, it is not so easy to identify factors that influence slow transport rates. One study has indicated that the slow transport rate in the optic nerve declines as development progresses, reaching its adult level at about the time that axonal growth is completed (Hendrickson and Cowan, 1971). Studies of other systems have not shown this trend, but much depends on the maturity of the system at the time of investigation, which is not usually known.

### Transneuronal influences

We have seen that the cell body of a neuron may atrophy if a major source of afferents is removed. This leads to the idea that axons may exert some sort of maintenance effect or trophic control over postsynaptic cells. Is this trophic effect related to the mechanism of chemical transmission at the synapse or to the transport of a sustaining material made in the presynaptic cell to the postsynaptic cell? The latter proposal would seem to be supported by the finding that colchicine (which blocks axoplasmic flow), if placed on a nerve, will produce much the same effects in the postsynaptic cell as axon section does (e.g. Albuquerque et al., 1972; Hofmann and Thesleff, 1972). However, apart from blocking the flow of materials down the axon, colchicine also depresses synaptic transmission

(Pilar and Landmesser, 1972; Perisic and Cuénod, 1972). We clearly need a more direct test.

### Evidence of transneuronal transfer of materials.

Korr et al. (1967) produced evidence of transfer of the tritiated amino acids across a neuromuscular junction. Since then, studies of the visual system of mice, monkeys, and other animals (see Schubert, 1976) have shown that if $^3$H-proline or $^3$H-fucose is injected into an eye, it will be detected not only in the lateral geniculate nucleus and superior colliculus, to which the optic axons project, but also in structures to which neurons in these regions project. This is demonstrated by specific labeling of an intermediate layer of the visual cortex (to which the lateral geniculate nucleus projects) and of the parabigeminal nucleus (to which the superior colliculus projects). The question arises: what is being transported? We know that the amino acids and sugars are incorporated into proteins to be transported to the terminals. Whether they are transferred from one neuron to another as protein, or whether they are broken down at the terminal and transferred as amino acid and sugar, is not clear.

Using puromycin to block protein synthesis in the postsynaptic cell, Droz et al. (1973) found a reduction of 90% in incorporated label, suggesting that all but 10% of the transneuronally transported material labeled in their studies was being broken down and resynthesized. Transfer of materials may be specific, occurring only between synaptically interconnected neurons, or more general in that material could be released into the extracellular space, there to be taken up by any cell that lies in the area. As a result, this method cannot be used to indicate that, for example, the cells of the superior colliculus which project to the parabigeminal nucleus necessarily make direct optic connections. It may be true, but it remains unproved. Another possibility must also be considered—that protein exocytosed by the terminal may serve to bind the transmitter within synaptic vesicles, rather than exert a trophic influence (Geffen and Livett, 1971; Musick and Hubbard, 1972).

Several studies have indicated that adenosine is transported in an orthograde direction and transferred from one neuron to another. The first of these experiments involved the pathway from the visual cortex to the lateral geniculate nucleus (Schubert and Kreutzberg, 1974), but did not satisfactorily eliminate the possibility that the adenosine might in fact have been transported in a retrograde direction by geniculocortical axons. Since, as we will see, adenosine can be transported in a retrograde manner, this remains a valid interpretation of the results. Subsequent studies (Schubert et al., 1976; Wise and Jones, 1976) have shown what appears to be transneuronal orthograde transfer of adenosine in situations where it would be difficult to account for the experiments otherwise. However, one recent study (Kruger and Saporta, 1977) failed to find trans-neuronal transfer from optic nerve terminals to lateral geniculate neurons in monkeys. Further clarification of this issue is needed.

### Stimulation-related effects

1. *Increased Adenosine Levels*. These have attracted attention for two rea-sons. First, increases in available adenosine stimulate production of cyclic AMP (adenosine 3',5'-monophosphate), which raises cell enzyme levels; and second, adenosine does appear to cross from one neuron to another in an unincorporated state and could serve as a transneuronal trophic agent. McIlwain (1973) showed that stimulating brain slices caused adenosine to be released into the medium. Other studies on peripheral neurons have confirmed the release of adenosine derivatives after stimulation of cholinergic and monoaminergic nerves. Schubert et al. (1976) observed a local increase in the release of [3]H-adenosine from entorhinal afferents (which had been labeled by injection of the entorhinal cortex) to granule cells of the dentate gyrus after stimulating the entorhinal cortex with an implanted electrode.

2. *Increased Cyclic AMP Levels*. A number of studies reviewed by Costa et al. (1974) showed that cyclic AMP levels in adrenal medulla and superior cervi-cal ganglion can be increased by a variety of methods of stimulation, and that such stimulation is ineffective if the presynaptic nerve is cut or agents blocking synaptic transmission (such as hexamethonium or mecamylamine) are applied. The results suggest that transsynaptic activation of one class of cholinergic recep-tors on the postsynaptic membrane (the nicotinic receptors) leads to the increase in cyclic AMP.

3. *Changes in Enzyme Levels*. A considerable amount of evidence has been gathered to show that enzyme levels in a neuron can be influenced by afferent activity. The systems most frequently studied are, again, the superior cervical ganglion and adrenal medulla. Stimulation by reserpine, cold, or swimming stress increases the amounts of two enzymes involved in norepinephrine synthesis—tyrosine hydroxylase and dopamine-$\beta$-hydroxylase. A third enzyme, DOPA-decarboxylase, is unaffected. The effect is prevented by cutting the pre-synaptic nerve (described in the literature as decentralization) and by ganglionic blockers. Actinomycin D (which blocks formation of messenger RNA) and cy-clohexamide (which blocks the translational stage of protein formation) are both effective in preventing stimulation-induced increases in enzyme levels. The in-duction effect is initiated by a relatively short period of stimulation (i.e., two

hours of swimming stress). An increase in RNA occurs at 10–24 hours after stimulation and an increase in tyrosine hydroxylase can be recorded within 16–20 hours and may last for four days. Since cyclic AMP shows an increase for 90 minutes after stimulation and is blocked by the same substances that block an increase in tyrosine hydroxylase levels, Costa et al. (1974) suggest that cyclic AMP is in fact a "second messenger" in the chain of events. Thoenen (1974) is more cautious, noting that dibutyryl cyclic AMP, when administered in high doses, causes increases in the levels of each of the three enzymes studied, although stimulation does not affect DOPA-decarboxylase. Obviously, this issue is not yet resolved.

In summary, it is becoming clear that: (1) materials are transported from one neuron to another; (2) the transneuronal transfer of adenosine, at least, appears to be increased by nerve stimulation; and (3) nerve stimulation causes metabolic changes in the postsynaptic cell.

It is not yet possible to say that transported materials cause postsynaptic metabolic changes. It may be that different mechanisms—transport (both diffuse and discrete), stimulation, a combination of transport and stimulation—all have trophic effects. To eliminate any one at the moment would seem premature. The role of such trophic effects is important in neural development and in the maintenance of normal neuronal connections. We have already discussed transneuronal atrophy and degeneration as an important mechanism by which cell numbers are regulated. Beyond this, some of the same mechanisms of enzyme-level control outlined above are also at work in the early development of innervation. These issues will be explored further in Chapter 12.

## Retrograde Effects

The reasons why a reaction to axotomy may result in death of the cell are not altogether clear. In fact, it is surprising that such changes should occur since the actual lesion should not isolate the neuron from any essential part of its synthetic machinery. Cragg (1970) reviewed a number of possible explanations. The most likely of these involves some form of regulation being exerted at the soma by material flowing back along the axon from the terminal region. This requires a mechanism for retrograde axonal transport.

### Retrograde axonal transport

While it has been known for many years that virus particles, tetanus toxin, and several other materials could be transported from a terminal region to the cell body of a neuron, it was never absolutely clear that this transport occurred within the axon or outside it. The earlier studies are well reviewed by Lubinska (1975). More recently, it has been shown that a number of substances may be taken up by axon terminals and transported back to the cell body region in significant amounts within intracellular membrane systems. Among these are horseradish peroxidase (see LaVail, 1975), adenosine (Wise and Jones, 1976), nerve growth factor (see Hendry, 1976), and tetanus toxin.

Horseradish peroxidase, a glycoprotein isolated from horseradish roots, appears to be transported in a retrograde direction by most neurons (although a few negative cases which deserve further study have been recorded). The active molecule is one of a series of isoenzymes (isoenzyme C) that is distinguished from the others by its high isoelectric point (Bunt et al., 1976). What role transport of this molecule may have is uncertain. At the very least, the work shows that some glycoproteins are capable of being taken up by nerve terminals and axons and of being transported back to the cell body region.

Derivatives of adenosine stimulate the metabolism of cells in a variety of ways. As such, the retrograde transport of adenosine is clearly of significance for its possible trophic role.

Nerve growth factor (NGF) is particularly interesting as a chemical which can be transported from terminal to cell body. Several recent reviews have been devoted to it (Levi-Montalcini et al., 1972; Varon, 1975; Hendry 1976). NGF is produced by a wide variety of mesenchymal cells and possibly even by neuroglial cells. Only a limited number of cell types, however, are capable of taking it up and transporting it. These are the cells of the superior cervical ganglion and dorsal root ganglion (both derivatives of the neural crest) and certain groups of cells in the central nervous system which may be central adrenergic cells (Ebbott and Hendry, 1978). The reason for the specificity of uptake and transport is unknown; presumably it relates to some feature of the surface membrane. NGF has a specific trophic effect on those cells which take it up, promoting axonal sprouting, maintaining levels of certain enzymes (Hendry and Thoenen, 1974), and reversing the effects of axotomy (Purves and Nja, 1976). The presumption, therefore, is that these neurons require NGF to maintain normal function and that a significant amount reaches the cell body by transport from the axon terminals. Axotomy deprives the cell body of this source and leads to the characteristic retrograde events. One crucial experiment to test this hypothesis—placing NGF at the end of the cut axon—has not yet been reported for adult animals. The reversal effect shown has so far been produced by placing a plug of NGF on the axotomized ganglion.

NGF is present in sufficiently large amounts in certain places such as mouse salivary gland and the venom of some snakes, to allow its purification and the identification of active components. The mouse salivary gland NGF is a high-molecular-weight protein composed of three subunits—$\alpha$, $\beta$, and $\gamma$. The $\beta$ subunit, a highly basic molecule, is the neurotrophic element. The $\alpha$ subunit is an acidic polypeptide that appears to protect cell membranes from damage, while the $\gamma$ subunit is an arginine esterase that binds to the $\beta$ subunit. Studies of the molecular structure show that NGF has some striking similarities with insulin. On the basis of these parallels, and of certain reactions of NGF, Frazier et al. (1973) have suggested that NGF and insulin may have a common evolutionary origin.

Of some interest is the possibility that NGF is one of a number of comparable substances which function as trophic agents and serve to stabilize neuronal connections.

Varon and his colleagues have addressed the further question of whether neuroglia which exist as satellite cells might have a trophic effect on neurons that is comparable or complementary to the trophic effect of NGF (see Varon, 1975). Studying the glial satellite cells that surround dorsal root ganglion cells, these authors have been able to isolate preparations which either contain mainly glial cells or which are rich in neurons and have low percentages of glial cells. Using such preparations, they have defined two roles for the glial satellite cells. First, these cells are capable of replacing NGF in supporting neuronal adhesion, neuronal survival, and neurite outgrowth in culture; and secondly, they are essential elements in permitting NGF to function at all. Thus, when neurons are grown in culture with less than a critical number of satellite cells per neuron, however much NGF is subsequently added, none of the normal supportive functions characteristic of NGF can be demonstrated. But once that critical number of glial cells is present in the culture there is then an inverse relationship between the amount of NGF needed to produce an optimum effect and the number of glial cells in the culture i.e., the more glial cells there are, the less NGF is needed to provide optimum growth. NGF has no additional effect when a level of three nonneurons/neuron (for attachment) or four nonneurons/neuron (for long-term survival) is achieved. From these figures, it would appear that the effect of a single satellite cell is equivalent to that of 0.1–0.2 biological units of NGF. The question then arises as to whether the satellite cells are making NGF or an NGF-like protein. This possibility is strengthened by the finding that antibody to mouse submandibular-gland NGF eliminates the supportive effect of glial cells. This experiment, however, is open to alternative interpretations, as outlined by Varon et al. (1974), who suggest that the antibody may be disturbing surface binding by the glia of an NGF precursor from the serum or perhaps disrupting a transfer mechanism from glia to neurons.

At the very least, these results indicate that there is a specific relation between neurons and a particular glial cell population which has to do with NGF activity in maintaining the neurons' integrity and studies of the relationships may provide a more specific definition of the supporting function of neuroglia. To what extent these results from dorsal root ganglion cells in culture are generally applicable to other regions and to in vivo situations remains to be seen.

## Summary

There is clear evidence from the experiments discussed in this chapter that neurons are far from static elements. Synaptic events require a considerable amount of membrane mobilization: there is continued transfer of material both between the various compartments of a neuron and between one neuron and another. It is apparent that neurons depend on their connections with other neurons for the maintenance of their normal metabolic health. This trophic inter-dependency appears to depend in part on stimulation-related phenomena and in part on the transport of sustaining materials like NGF, which bear little relation to physiological activity. It is apparent that neuroglia, particularly those classed as satellite cells, may also perform some sort of maintenance function which parallels or even complements that of NGF.

It will be seen in the later chapters that while these dynamic interactions are undoubtedly important in maintaining the integrity of the adult nervous system, they are even more important in development, when interdependency between cells is much more sensitively attuned.

# 5

# TECHNIQUES IN NEUROBIOLOGY

The course of developmental neurobiology, like that of all sciences, depends on the development of experimental approaches which serve to answer particular questions and promote new hypotheses. This book is concerned largely with a discussion of these approaches and of the ideas deriving from their implementation. Of extreme importance is the adequacy of the methods used to gather and analyze the data. In recent years, major advances have been made in neurobiological techniques, particularly those used to define morphological features of cells; and these advances have substantially expanded the range of questions about developing neural systems that can be profitably asked.

In this chapter we will review some of the principal methods used in experiments on the development and plasticity of the brain, with special emphasis on their intrinsic advantages and disadvantages. The chapter may thus serve as a basis for qualifying results discussed throughout the rest of the book. Further discussion of the techniques reviewed here, with particular reference to their practical application, can be found in three symposia (Nauta and Ebbeson, 1970; Kater and Nicholson, 1973; and Cowan and Cuénod, 1975). The reader is referred to these for additional information and for access to details of the methodology. Other references are provided here only for more recent technical advances and for further discussion of interpretive difficulties associated with the use of particular methods. Since physiological methods have played a major part in this field only in analyzing the mapping of retina on optic tectum and in tracing the effects of visual deprivation, their discussion will be reserved for the relevant chapters (14 and 15).

In studying the nervous system with the light microscope, it is clearly valueless to stain everything since with a section of average thickness, the tissue would appear black. Three approaches are possible: (1) to stain selected cells in their entirety, neglecting the rest; (2) to stain particular subcellular components of all neurons; (3) to label certain metabolic features of active neurons.

**Table 5-1.** Summary of some of the basic neurohistological methods, their fine structural counterparts, and their uses. See text for further discussion.

| Technique | Structure stained or labeled | Fine structural correlate | Parts of cell stained | Use |
|---|---|---|---|---|
| Golgi, intracellular injection | Whole neurons. neuroglia (especially with Golgi) | Cytoplasm | All but nucleus | Visualization of total ramification of processes of single cells |
| Thionine, cresyl violet | Nissl substance | Clumps of granular endoplasmic reticulum | Cell body and proximal dendrite | Cytoarchitectonics/ degenerative change of cell body/ gliosis |
| Cajal, Ranson, Holmes, Bodian, Bielschowsky, Glees, etc. | Neurofibrils | Bundles of neurofilaments | Variable, according to cell/ axon: sometimes dendrites and cell bodies (e.g., ventral horn cells) some terminals | Have been used for pathway tracing (unreliable)/ fiberarchitectonics/ show neurofilamentous degenerative changes |
| Nauta, Fink-Heimer | Degeneration granules | Electron dense degeneration | Degenerating axons and terminals | Orthograde and (in some cases) retrograde degenerative changes |
| Rasmussen, Armstrong et al. | Terminals | Mitochondria | Terminals containing number of mitochondria | Light microscope visualization of some terminals |
| Weigert, Weil Luxol fast blue | Myelin | Membrane lamellae | Myelinated part of axon | Major fiber pathways and brain subdivisions/ long-standing tract degeneration |
| Tritiated amino acids | Incorporated proteins | Various | Any part of neuron distal to incorporation site | Pathway tracing—orthograde |
| Horseradish peroxidase | Incorporated enzyme | Membrane-bound granules | Cell bodies whose processes lie in injection site | Pathway tracing—retrograde |
| Catecholamine fluorescence | Monoaminergic transmitters | Membrane-bound granules | All, especially axons | Distribution of monoaminergic axons and cell bodies |
| Deoxyglucose | Areas of active glucose uptake | | Active regions (low resolution precludes cytological localization) | Active regions during selective stimulation |

These techniques will be considered separately: a summary is provided in Table 5-1.

## Staining Selected Cells

There are two principal methods. The first approach is to use stains which for some unknown reason selectively impregnate only a small percentage of the total cell population. The Golgi methods are now most commonly used for this purpose. There are three varieties:

a. **Golgi Cox.** Unlike the other methods this stains a good selection of neurons in a relatively even manner. Its drawbacks are that it fails to impregnate terminal parts of dendrites in many cases, dendritic spines tend to stain poorly, and only the initial part of the axon is usually impregnated.

b. **Golgi Kopsch.** This usually stains irregular groups of cells and often has a tendency to favor stellate and granule cells. However, it shows dendritic spines well, appears to stain the whole dendritic ramification, and often stains axons for some distance. It is often successful where the others may fail in nonmammalian vertebrates.

c. **Golgi Rapid.** This again stains erratically, tending to favor pyramidal cells. However, it impregnates processes completely, including dendritic spines and the total axon plexus. For this reason it is the method of choice for quantitative studies of dendrites, and for studying local axon plexuses. An example is shown in Figure 5-1A.

The second approach is to inject the cell with a substance that diffuses along its processes but stays localized within the cell. The value of this approach is that it can be used to record from a cell, characterize its response properties, and then define its morphology. Three substances have been used in this way with some success in the mammalian central nervous system. They are procion dyes, horseradish peroxidase, and tritiated amino acids. In addition, cobalt nitrate may be taken up by a nerve process such as an axon and thus put to similar use.

Although they have yielded useful information, procion dyes suffer from poor resolution, since they do not impregnate the whole cell. Also, because they rely on fluorescence methods for demonstration, they do not allow counterstaining of the rest of the tissue. Since amino acids tend to move quite freely out of injected cells, their use in intracellular labeling is limited. Horseradish peroxidase (Fig. 5-1B), while difficult to expel from small pipettes, seems to be

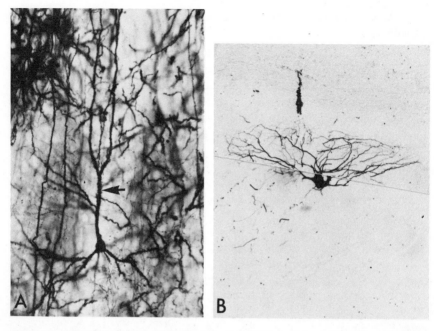

**Figure 5-1.** Whole-cell stains. (A) Golgi stain of pyramidal cell in cat visual cortex. Apical dendrite signified by arrow × 310. (B) Adjacent sections of cell in dorsal horn of spinal cord of cat which has been characterized physiologically and injected intracellularly with horseradish peroxidase × 50 (Courtesy of P. J. Snow, P. K. Rose, and A. G. Brown).

the material of choice for intracellular injection since it fills cells and their processes, including the axon, rather well. Furthermore, it can be used for electron microscopy, thereby allowing identification of synapses on identified cells. Cobalt has not been used so commonly in mammalian brains, but it does seem valuable in certain circumstances (e.g., Cunningham, 1976).

Though it is clearly important to define the shape of neurons, neither Golgi nor intracellular injection methods are ideal. The capriciousness of the Golgi methods is their principal drawback; no systematic protocol has emerged for getting entirely reliable staining. Such unreliability is obviously important when using the method to compare a normal and aberrant cell population. The extent to which differences between two populations is artifactual or real is sometimes difficult to assess.

The intracellular injection methods also have drawbacks. When recording from vertebrate brains, it is often difficult to hold a cell long enough to inject it, and the yield of successfully impregnated cells is small. Further, with the exception of some especially large cells, impregnation is rarely complete. It is valuable, if not essential, to couple any intracellular injection study with a parallel Golgi study to define the total pattern of the cells studied.

Since few cells are impregnated with both Golgi and intracellular injection methods, it is possible to cut thick sections (e.g., 100 $\mu$m) to allow visualization of a large part of one cell on a single section. Sections of this thickness are beyond the depth of focus of the commonly used lenses and for this reason it is customary to represent the results in drawings rather than photographs.

## Staining selected subcellular components of all neurons

a. Nissl Stains. Stains such as cresyl violet and thionine stain RNA, particularly in the nucleus and granular endoplasmic reticulum, and were exploited in the nineteenth century by Nissl, by whose name they are collectively referred. It should be emphasized here that while individual ribosomes may be stained, these are below the resolution of the light microscope and only ribosomes in sufficient number to be resolvable will be shown. This principle also applies to other stains with a specific substrate. Since the granular endoplasmic reticulum is distributed in the perikaryal region and proximal dendrites, the method is a useful indicator of the general disposition, cytoarchitecture, and size of cell bodies and as such is often correlated with other studies (Fig. 5-2A). Use of the method to compare cell body size between experimental and control animals can be criticized because it is possible that the distribution of ribosomal aggregrates could vary without any actual change in the size of the soma. Nevertheless, such comparisons are widely made.

b. Reduced Silver Stains. A whole series of techniques is based on two methods—those of Bielschowsky and Cajal. All these methods compare closely with a photographic process. The stages are outlined in Table 5-2. In basic form the methods stain groups of neurofilaments more or less specifically. The Bielschowsky method, together with its modifications and derivatives such as the Glees method, tends to be more specific, while the Cajal method and its varieties and derivatives (e.g., Holmes, Ranson, Bodian) may stain other components as well, perhaps including microtubules. With a pretreatment stage, using chromium salts a mitochondrial stain results (e.g., Rasmussen, 1957). Neurofibrillar staining is blocked by pretreating with phosphomolybdic acid, uranyl nitrate, or phosphotungstic acid. These chemicals are, or can be, used as the first step in the Nauta-Gygax stain for demonstrating degenerating axons (Nauta and Gygax, 1954) and may also be used in staining such components as microtubules and neurofilaments for electron microscopy. Thus, the principle of suppression

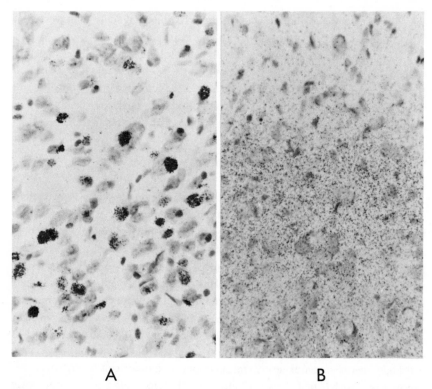

A                                              B

**Figure 5-2.** (A) Autoradiogram of adult neural tissue which had been injected with ³H-thymidine early in development. Those cells in S-phase of division at the time of injection have a heavy concentration of silver grains over their nucleus. Nissl counterstain × 170. (B) Autoradiogram showing terminal field of axon plexus. ³H-proline had been injected in the region of the cell bodies of the axons and transported down the axon. The area of heavy dispersed grains is the region of dense distribution of axons. Nissl counterstain × 170.

applied is in binding materials to active sites on neurofilaments to prevent subsequent staining. A light oxidative step in the Nauta methods appears to make the products of degeneration in damaged neurons more susceptible to staining. Thus one basic staining method (the Bielschowsky method) has been the source for some important histological procedures.

Neurofibrillar methods. These are useful techniques for showing the disposition of axons and the behavior of neurofilaments, particularly in degeneration. They are not very useful for tracing the detailed course of pathways, especially since neurofilaments may often be found in myelinated but not unmyelinated segments of axons. Determining exactly where axons end or indeed which is the beginning and which is the end of a stained segment is often impossible.

**Table 5-2.** Outline of basic neurofibrillar methods and some of their derivatives

| Stain | Pretreatment | Silver bath | Reducer | Component stained |
|---|---|---|---|---|
| Cajal | — | Pyridine AgNO₃ | (a) Pyrogallol | Neurofilaments |
| | | | (b) Hydroquinone | Neurofilaments |
| Bielschowsky | — | AgNO₃ + ammoniated AgNO₃ | Formalin | Neurofilaments |
| Nauta-Gygax (1954) | (a) Uranyl nitrate <br> (b) Phosphomolybolic acid } + KMnO₄ | ʺ | Formalin | Dense degeneration |
| Fink-Heimer I | KMnO₄ + uranyl nitrate/AgNO₃ | ʺ | Formalin | Dense degeneration |
| Armstrong et al. | Potassium dichromate + chromium fluoride | ʺ | Formalin | Terminal mitochondria |

**Mitochondrial methods.** Two essentially similar methods by Armstrong et al. (1956) and by Rasmussen (1957) have been used to stain mitochondria. In immature animals they appear to stain mitochondria throughout the cell, but in older animals they stain only mitochondria in the terminal region. Whether they also stain synaptic vesicles has never been determined, nor has the question of whether they stain all mitochondria in terminals or indeed whether they stain all terminals. This is a rather important point since the methods have been used for counting synapses on cells and for studying their behavior in particular degenerative states.

**Nauta-derived methods.** Several methods—Nauta-Gygax (both 1951 and 1954), Nauta-Laidlaw (Chambers et al., 1956), Wiitanen (1969), and Fink-Heimer (1967)—have been developed over the years for staining the products of degeneration as black granules against a golden brown background (see Fig. 3-8B). The initial Nauta-Gygax method (1951) stains degenerating elements as well as normal axons and is little used nowadays because of the interference of normal staining. The second method (1954) suppresses the staining of normal axons, but has a tendency to suppress some of the finest terminal degeneration; this tendency is much more marked in the Nauta-Laidlaw variety. The Fink-Heimer methods (Fig. 3-8B) overcome this problem and stain both degenerating axons and fine terminals. In some systems, such as some pathways in the monkey brain, the Fink-Heimer methods are not very efffective. For this reason a number of modifications have arisen, the best of which is perhaps the Wiitanen method.

It should be noted that these stains are specific for all degeneration—not only terminal degeneration but cell death and retrograde axonal and dendritic changes. Dissociation of orthograde and retrograde degeneration often requires careful adjustment of survival time: the shorter the time the less likely that the degeneration is due to retrograde changes. Survival time is also critical with orthograde systems, especially in young animals, since it is apparent that different pathways degenerate at different rates. Although there is some controversy over this, it would appear that small-diameter axons probably degenerate more quickly. In some systems they may totally disappear within two days—a survival time shorter than that used in most degeneration studies. This problem has only been recognized quite recently, and a number of studies have overlooked it.

### Labeling of Neuronal Activity

This approach depends on the use of materials that reflect the activity of a neuron or that can be used to follow the progress of certain molecules through the various compartments of the neuron. Such methods have the advantage over traditional neurohistology of providing more information on the activity of the neuron at the time of labeling. On the other hand, they have the drawback of depending on metabolic events for their success, so that a failure of staining may mean that the neuron or individual processes were either inactive or absent. It has been common to assume that only the latter explanation holds.

Three different approaches will be considered: (a) the labeling of cells in division, (b) the labeling of materials being transported along axons, and (c) the labeling of general cell metabolism.

a. Labeling of Cells in Division. Before mitosis a neuron synthesizes new DNA and for this process takes in thymidine, which will remain in the nucleus with little degradation. If the cell divides again the thymidine will be diluted by new thymidine taken in, and so on through each successive division. If tritiated thymidine is injected into an animal, it will be taken up by all cells in the process of DNA synthesis. It clears from the animal quite quickly—somewhere between 30 minutes and one hour (see Sidman, 1970; Nowakowski and Rakic, 1974)—and therefore cells dividing at the time of injection can be identified at later times by finding which ones are emitting $\beta$-radiation (from the tritium) from their nuclei. Emissions are identified by cutting thin sections of the region in question, coating them with an emulsion, leaving them in the dark, usually for 3–4 weeks, and then developing the emulsion (Fig. 5-2A). Since tritium penetrates only small distances, it is preferable to use section thicknesses of around 5 $\mu$m; at greater thicknesses, labeled cells lying deep in the section may not be identified.

Heavily labeled cells will have been in their final cycle of division at the time of injection. Those with progressively lighter label are likely to have divided one or more times further, but some may not have been in the DNA synthetic phase for the whole time that the labeled thymidine was available. For this and other reasons, it is unlikely that the density of label (measured by the number of developed silver grains) on the emulsion will be a step-like progression of $N$ for a terminally dividing cell, $N/2$ for a cell that undergoes one further division, and so on. Since the number of grains also depends on the time during which the emulsion was exposed to the sections and on other technical features, it becomes extremely difficult sometimes to decide what a terminally dividing cell is and what cut-off point in grain numbers should be taken to indicate that one or no further divisions has occurred. Although this question has been raised (see Sidman, 1970), it usually has not been applied to relevant studies.

One limitation of the method is that thymidine is usually given to the whole animal, so that all cells dividing at the time of injection are labeled. It would often be desirable to label only small groups of neurons to follow their subsequent migration in the brain; this has not been achieved so far.

**b. Labeling Components of Axon Transport.** *Orthograde techniques.* These methods rely on the fact that isotopically labeled compounds are transported along axons as indicated in Chapter 4, and therefore offer a good way of testing the distribution of axons from the region injected. If, for example, tritiated leucine is injected into a brain region, it will be taken into cell bodies in that region, incorporated into protein, and then transported along the cells' axons to their terminal regions, generally within 24 hours. Tissue treated in this way can be sectioned, coated with a photographic emulsion, and then kept in the dark for a number of weeks exactly as in $^3$H-thymidine labeling studies. The emissions from the tritium will expose the emulsion. Final development and fixing will make them permanent against a relatively clean background and the section may be counterstained with a Nissl or similar stain (Fig. 5-2B). The method has several major advantages over degeneration techniques:

1. A lesion damages not only cell bodies in the region but also axons running through it. Fortunately, however, amino acids are only taken up by cell bodies in the area for transport—axons do not appear to transport free amino acids in significant amounts.
2. Generally autoradiographic grain is heavier in an area of terminal distribution than is stained degeneration of the same pathway, and certain subtleties of a terminal field not readily seen in degeneration studies are obvious with autoradiography.
3. Survival times are not so much of a problem.
4. Sections can often be viewed at low power using dark-field illumination that permits easier resolution of the relation of a projection area to the whole tissue (see Fig. 15-2).

There are, however, a number of reasons for caution in the use of this method:

1. Different amino acids may be incorporated into proteins that are distributed differently.
2. At any survival time, autoradiographic grain is never completely localized at terminals but will also be found in axons. Further, there is no proof that all terminals will necessarily contain the labeled protein, nor that terminals making more synaptic connections have more label in them. Thus, the practice of providing counts of grains as an indication of synaptic density is of questionable validity.
3. The amount of material transported depends on the amount available for uptake. Because it is almost impossible to control from one animal to another how much gets taken up, comparison of counts made between two

groups of animals is of limited value unless elaborate statistical surveys are made and there is an independent control.

4. Material may be transferred transneuronally. This should be no problem with short survival times and the appropriate amino acids, but with longer survival and the use of materials such as proline or fucose, there is substantial transneuronal transfer.

Advantage can be taken of transneuronal transfer in order to study transneuronal projections. The degree to which this property is specific to interconnected neurons or is more broadly spread has been discussed already. The fact that material may be transported transneuronally, for instance from the eye through the lateral geniculate nucleus to the visual cortex, has proven extremely valuable in recent years. One example of its application is in testing the relative influence of the two eyes on the visual cortex in normal and experimental conditions (see Chap. 15). The major difficulty associated with the use of this technique is that with the high dosages of tritiated compounds and long survivals and exposure times necessary for this technique, labeled proteins are detected beyond the areas to which they are transported intra-axonally. As a result a high background level of autoradiographic grain is encountered over the whole tissue, which obscures lightly labeled pathways.

*Retrograde transport.* As indicated earlier, a number of materials are transported back to the cell body after they are injected into the terminal region of an axon. Using this approach, it is possible to inject a substance into a region and identify cell bodies which send axons to the area of the injection from another region. Horseradish peroxidase (HRP) is widely used for this purpose (Fig. 5-3), but a number of considerations have to be taken into account when using it:

1. Apart from the specific retrograde transport, there is also a nonspecific filling—both orthograde and retrograde—of damaged axons, and there are a few indications of true orthograde transport of HRP. Fortunately, neither of these can be confused with the typical perikaryal granules of true retrograde transport.

2. The area from which material is taken up for transport does not usually correspond to the whole brown area around the injection site but rather to a more localized zone immediately adjacent to the injection.

3. Axons of passage take up HRP for retrograde transport. This means that one must take care to account for axons running through the injection area.

Other problems are discussed elsewhere (Bunt et al., 1975). As we have already seen, other substances such as nerve growth factor are also transported in

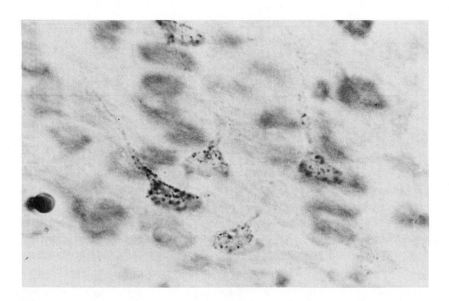

**Figure 5-3.** Cells in visual cortex of cat. The region of distribution of their axons has been injected with horseradish peroxidase which transports in a retrograde direction along the axon and appears in the cell body as dark granules. Nissl counterstain × 408.

a retrograde direction by some neurons. The movement of NGF has been followed by labeling the molecule with [125]I.

**c. Transmitter-related Substances.** Some transmitter substances, together with their precursors or associated enzymes, can be stained specifically to identify certain pathways. One approach of some significance here relies on the fluorescence of catecholamines to demonstrate the presence of aminergic pathways. The most commonly used method involves freeze drying tissue immediately after removal from the animal, cutting sections of the block after paraffin infiltration, and exposing these to paraformaldehyde. A temporary fluorescence of aminergic pathways results. The method has been used extensively to show plasticity in aminergic pathways, but recently a more sensitive method using glyoxylic acid (Lindvall et al., 1974) has been introduced. A number of qualifying comments on the use of the technique are particularly relevant to the matter of plasticity and regeneration in adult brains:

1. There is no proof that varicosities seen along the course of axons represent terminals, though this assumption is often made.

2. The method shows the amount of transmitter precursor: if this were depleted, the axon would appear to be no longer present.

3. We know little about the absolute resolution of the method. Thus, if a fluorescent axon appears in one preparation and not in another, it is possible that the axon is still present in the latter but, due to turnover and other factors, contains too little transmitter precursor to be stainable. This problem has been highlighted by the glyoxylic acid technique, which shows more axons than the traditional method used in most of the plasticity studies.

4. It is not possible to counterstain the tissue directly, and so the relation of fluorescent fibers to general tissue features is often difficult to define precisely.

d. Activity-related Histology. A radioactive analogue of glucose, $(1\text{-}^{14}C)\text{-}2\text{-}$deoxy-D-glucose, has been used recently to label active brain regions. Taken in like glucose by active cells, this compound binds irreversibly to glucose receptors. As a result, it gives a clear picture of certain regions of the nervous system which are more active than others. It is particularly valuable for showing trans-neuronal effects of local lesions and of functional deprivation (Kennedy et al., 1975, 1976). At present its resolution is poor, since all cells are likely to take up glucose whether stimulated or not, but with time some way round this problem may be found. So far the method has had little application in development studies, but many possibilities can be envisioned and for this reason it is mentioned here. In a recent review the method's applications and potential are discussed (Plum et al., 1976).

A number of other histological methods have been used in development and plasticity studies, but they will not be included here since they seem less fundamental than those mentioned above. Minor comment, however, should be made of light microscopic identification of neuroglia. One significant approach to specific staining of neuroglia derives from the isolation of a specific protein (glial fibrillar acidic protein—GFA) from scar tissue in the central nervous system (see Bignami et al., 1972; Bignami and Dahl, 1973). Antibodies developed against this protein and conjugated with fluorescein allow visualization of all cells containing the protein. These are the astrocytes and astrocyte-like cells described in Chapter 3. Most neuroglial cell types can be demonstrated by Golgi methods (especially when staining for neurons is less than optimal) and the cell bodies are clearly shown by Nissl methods. When the latter are used, the differentiation of neurons and neuroglia assumes importance in counts of neurons under normal and experimental conditions. It is easy to distinguish large neurons from neuroglia but quite difficult to distinguish small neurons from neuroglia. Some investigators have accepted the erroneous view that neurons have nucleoli and neuro-glia do not. This is clearly unsatisfactory; while the nucleus of oligodendrocytes is frequently difficult to distinguish from the nuclear chromatin, such is not the case for astrocytes.


## Discussion

Given this battery of techniques we should point out which would be chosen to study particular problems.

1. General cell shape and local axon ramification are best traced with Golgi methods and if possible with intracellular labeling methods.

2. The general health of the cell is usually studied with Nissl methods. The question of cell shrinkage in Nissl stains relative to the real size of the cell body and to the extent of the dendritic tree has not been adequately surveyed.

3. The course of an axon and its distribution can be studied with autoradiographic and degenerative methods. Preparing alternate sections stained by one or the other method after injecting one region with a tritiated amino acid and lesioning another allows an assessment of the interaction of two pathways at once. Staining with horseradish peroxidase and physiological techniques are also useful for tracing axons.

4. The only satisfactory way of studying the distribution of synapses anatomically is by using electron microscopy. With the light microscope there is no absolute guarantee that a terminal impregnated with a Golgi or mitochondrial stain and lying adjacent to a neuron actually forms a synapse with that neuron rather than with another adjacent, unimpregnated or small process. In Golgi preparations axons may sometimes be seen extending endfeet to dendritic surfaces, which would strongly suggest synaptic contact, but one should be extremely cautious even in these cases. In mitochondrial stains of large neurons (e.g., motor neurons of spinal cord) terminals are seen lining the large dendrites and these move with the dendrites in cases where tissue shrinkage has occurred. It is generally implied that these terminals do indeed end on the dendrites, but they may not be the only terminals. Intracellular physiological studies should give indication of synaptic connection between two neurons; extracellular recording, on the other hand, will only show a connection if a particular afferent drives the cells or if it modifies an ongoing activity.

5. The important matter of determining the numbers of synaptic inputs into a region or onto a neuron cannot be easily undertaken with any techniques currently available. Attempts have been made with mitochondrial methods, autoradiography, and degeneration. Their limitations have already been mentioned. Electron microscopy should be a satisfactory method, but this too presents many problems, especially when used to compare normal

and abnormal populations. In such comparisons one must account for volume changes and for changes in synaptic length. In this respect, developmental studies showing rates of increase in the absolute numbers of synapses in a brain region often do not take into account the change in total volume of the tissue. If the whole tissue is enlarging, the figures will always underestimate the total number of synapses, and various discontinuous trends shown in numbers of synapses relative to unit volume could equally be accounted for by nonlinear changes in the total volume of the region under investigation. Perhaps the best current approach is that of Cragg (1975a) who normalized synaptic counts with respect to the total number of neurons per unit volume. Of course, this approach presumes that neuronal numbers do not change significantly. Labeling individual pathways to get an estimate of the percentage of terminals from one input is precarious. Degenerative methods are of limited value, since not all terminals degenerate at the same time, and further, there is considerable plasticity of residual connections after lesions. Each of these connections would upset any count of the proportion of terminals from one pathway. Autoradiography has the added problem that the number of terminals over which grains are found increases with exposure time, and one can never guarantee that all terminals contain the labeling precursor.

Despite these difficulties, the various methods have provided some useful quantitative data. Their limitations should be kept in mind however when one is extracting biological principles from the experiments.

# NEURAL GENERATION AND MIGRATION

During development, cells undergo mitoses either in a position close to the lumen of the neural tube or ventricle or in a secondary subventricular zone. Once they leave the mitotic cycle and become postmitotic, they migrate to their appropriate location, there to differentiate into a mature neuron. The whole subject of cell division and migration offers many interesting questions. What controls the rate of the mitotic cycle? Do individual stem cells make several neural types? What determines that a cell should stop migrating once it reaches a certain position? Are there mechanisms whereby abnormal patterns of migration are corrected as, for example, by death of inappropriately placed cells or by modification of subsequent connection patterns? What directs the postmitotic migration patterns?

Unfortunately, we have few answers. Yet studies of a variety of genetic mutant disorders have revealed cell generation and migration patterns that are highly abnormal, and, in time, they may provide answers to some of the questions asked above. Most of the relevant work has been done on the mouse, since the various inbreedings necessary to isolate and develop a mutant variety are easier to handle in this animal. A summary handbook of some of the mutant varieties of the mouse has been published (Sidman et al., 1965). Beyond this, the literature on major human brain disorders which can be attributed to aberrations of neural migration is growing, and in the next few years may become important in experimental work.

# 6

# CELL GENERATION

## Normal Cell Cycle

Neurons, like any other cells, go through a series of well-defined stages in the course of cell division termed, in aggregate, the cell cycle. There are four stages: the actual mitotic stage (M phase of Howard and Pelc, 1953), the postmitotic gap (G1), the DNA synthetic phase (S phase), and the premitotic gap (G2). The mitotic stage is divided into prophase and metaphase, during which the chromosomes align at the equator of the cell and the spindle appears; anaphase, during which the chromosomes are pulled out (a process blocked by colchicine); and telophase, during which cytokinesis occurs (blocked by cytochalasin). Most neurons become postmitotic at G1; tetraploid cells become postmitotic at G2.

During early neural development, all mitoses occur in cells that have nuclei lying close to the ventricle and that have rounded cytoplasm. This led to some misinterpretations until Sauer (1936), citing the position of the mitoses and the fact that nuclei farther away from the ventricle are larger (and therefore likely to be synthesizing DNA), suggested that nuclear migration occurs during the cell cycle, the nucleus moving to the ventricle for division and away from it for the S phase. This theory, though much disputed, has been supported by three lines of evidence: (1) Colchicine administration, arresting cells in anaphase, leaves all the nuclei close to the ventricle (Watterson et al., 1956). (2) Microspectrophotometric studies show that nuclei farther from the ventricle have twice the amount of DNA of those close to the ventricle (Sauer and Chittenden, 1959). (3) Studies using pulse label of $^3$H-thymidine show that nuclei labeled in S phase will then undergo mitosis, and can be followed through the whole sequence of migration (Sauer and Walker, 1959).

Changes in neuronal shape associated with nuclear movement in the cell cycle have been difficult to sort out—it is easy to see a nucleus, but very difficult to define the complete cytoplasmic boundaries of a cell when it has long thin processes. Nonetheless, studies involving electron microscopy of serial sections (Hinds and Ruffet, 1971; Hinds and Hinds, 1974) and scanning electron microscopy of fractured cerebral vesicles correlated with some transmission electron microscopy (Seymour and Berry, 1975) have provided a complete picture of changes in cell shape relative to the stage in the cell cycle.

From these studies it appears that at the early stages of neural development, all cells of the pseudostratified columnar neuroepithelium are tightly attached at the ventricular surface by junctional complexes, including gap junctions. No prominent junctions are found between the cell processes at the outer surface. The sequence of cell shape changes and the pattern of nuclear migration are sketched in Figure 6-1. Depending on the region, the nuclei of prophase cells either have reached the ventricular surface or are moving towards it. When the nucleus arrives at the ventricular surface, the whole cell becomes pear-shaped, with fine processes extending from its apical surface and microvilli from the rest of the cell surface. Whether the apical processes actually lose contact with the external surface, as indicated in Figure 6-2, or whether there continue to be very fine threads which cannot be detected by the methods used so far, is not altogether clear, but the available evidence suggests the former explanation. By anaphase, the microvilli on the cell surface have disappeared and are replaced by

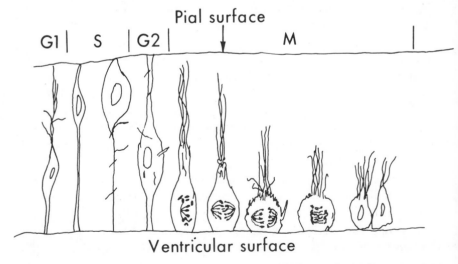

Figure 6-1. Summary of stages of cell cycle, nuclear configuration, and pattern of nuclear migration in neuroblasts of early cortical plate (redrawn from Jacobson, 1970; and Seymour and Berry, 1975).

bumps. The cytokinesis occurring in telophase is usually, though not invariably, in the plane indicated in Figure 6-1. In G1, the cells reestablish major cytoplasmic contact with the outer surface. If the cell becomes postmitotic at this stage, it loses cytoplasmic contact with the ventricular surface and migrates as a whole away from it.

## Timing of Cell Cycle and its Subcomponents

There are a number of ways of determining the duration of the cell cycle. The most satisfactory involves pulse labeling with ³H-thymidine. This substance is taken up only by cells in the S phase, and in mammals unincorporated thymidine is removed within 30 minutes (Nowakowski and Rakic, 1974). (In birds, the thymidine is not cleared and a cumulative labeling procedure must be used—see for example Fujita, 1962). The labeled cells will subsequently go through the cell

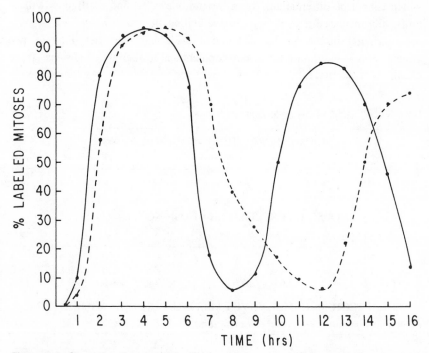

**Figure 6-2.** Generation time in midbrain of normal mouse (continuous line) and splotch mutant (*Sp/Sp*) at 10 days gestation. Abscissa: hours after injection of ³H-thymidine (from Wilson, 1974, with permission).

cycle, undergoing mitosis. A graph can be prepared that expresses the number of labeled cells in mitosis as a percentage of all cells in mitosis (see Fig. 6-2). The distance between corresponding points on the two curves is the generation time—the length of one whole cell cycle. To deduce the timing of parts of the cell cycle, one must also calculate the mitotic index (M.I.). This is the number of cells (labeled and unlabeled) in mitosis expressed as a percentage of all cells capable of division. From these figures, the following time may be calculated:

M phase = M.I./100 × generation time.
G2 + 1/2M = 50% point on ascending curve (hence G2).
S phase = interval between 50% points on ascending and descending curves.
G1 = total generation time minus (M + G2 + S).

One important factor to consider when interpreting these figures is that the mitotic index is expressed as a percentage of all cells *capable of division*. The figures will be artificially reduced if there are postmitotic cells in the population which cannot be differentiated from intermitotic cells, and the figures will be artificially increased if there is excessive cell death.

Variation in the length of the cell cycle has been described in two circumstances—with progress of development and in aberrant conditions. Three examples of the first kind of variation are given in Table 6-1. As can be seen in

**Table 6-1.**  Change of cell cycle timing with age

| MOUSE NEURAL TUBE—KAUFFMAN (1968) | | | | | |
|---|---|---|---|---|---|
| *Fetal age (days)* | *Total cell cycle (hrs)* | *S* | *G2* | *M* | *G1* |
| 10 | 8.4 | 4.0 | 0.8 | 1.3 | 2.3 |
| 11 | 10.5 | 5.4 | 1.2 | 1.2 | 2.7 |

| MOUSE TELENCEPHALON—HOSHINO et al.  (1973) | | | | | |
|---|---|---|---|---|---|
| *Fetal age (days)* | *Total cell cycle (hrs)* | *S* | *G2* | *M* | *G1* |
| 10 | 7.0 | 5.1 | 1.0 | 0.8 | 0.1 |
| 13 | 15.5 | 6.9 | 1.0 | 0.8 | 6.8 |
| 17 | 26.0 | 10.4 | 1.0 | 0.8 | 13.8 |

| CHICK TECTUM (CUMULATIVE LABELING METHOD)—WILSON (1973) | | | | | |
|---|---|---|---|---|---|
| *Embryonic age (days)* | *Total cell cycle (hrs)* | *S* | *G2* | *M* | *G1* |
| 3 | 8 | 4 | 1.5 | .3 | 2.2 |
| 4 | 9 | 5 | 1.5 | .4 | 2.1 |
| 5 | 13 | 4 | 1.5 | .8 | 6.7 |
| 6 | 15 | 5 | 1.5 | 1.4 | 7.1 |

**Table 6-2.** Cell cycle timing in the midbrain of normal and splotch mutant mouse embryos, from Wilson (1974)

| Age (days) | Total cell cycle (hrs) | S | G2 | M | G1 |
|---|---|---|---|---|---|
| 10 normal | 8.5 | 5.0 | 1.0 | 1.0 | 1.5 |
| 10 mutant | 12.0 | 5.5 | 1.0 | 1.9 | 3.6 |
| 11 normal | 11.0 | 6.0 | .9 | 1.2 | 2.9 |
| 11 mutant | 15.0 | 8.0 | .9 | 1.7 | 4.4 |

each case, the cell cycle time increases with age. In the mammalian studies, G2 and M change very little, while S and G1 increase with time; in the chick study, G2 and S stay stable. The differences are not easy to explain. In the study of Hoshino et al. (1973), there must have been contamination from postmitotic cells. The cumulative labeling method was used in the chick study, which could account for the differences observed. The mechanism of the increased cell cycle time is not altogether clear. It could be that all cells in division cycle more slowly with time or that there is a spectrum of cell cycle times in the population: those cycling fastest leave the cycle first and slower cells remain. The significance of the increase is even more obscure.

However, there have been some reports of alteration in cell cycles that resulted in major, usually lethal, changes in brain development. Several genetic mutant varieties of mouse—ocular retardation (or), fidget (fi) (Konyukhov and Sázhina, 1971), splotch (Sp) (Wilson, 1974), and looptail (Lp) (Wilson and Center, 1974)—have been identified in which the cell cycle is lengthened (Fig. 6-2). In each case, there is a major increase in G1. This is shown for the splotch mutant in Table 6-2, where only G2 remains stable. As can be seen from Table 6-2, a mutant with a prolonged cell cycle still shows an increase in the length of the cycle with age. Whether the prolonged cell cycle in the mutant is the cause of the neurological defect, or whether it reflects some other problem is not clear at present. A number of other studies have suggested that X irradiation (Fujita et al., 1964) and urethane (Kauffman, 1969) affect the cell cycle time, but this seems to be secondary to general toxic effects.

## Cell Lineage

One matter of obvious importance concerns the time at which a neuron acquires the information necessary to develop its characterizing features. As will be seen in Chapter 14, parameters associated with the establishment of spatial identity in

a sheet of cells become fixed at approximately the time the first neurons contributing to the map become postmitotic. Other properties might be decided much
earlier. Among these may be the regional designation (i.e., forebrain compared
with hindbrain, thalamus compared with hypothalamus; or lateral geniculate
nucleus compared with nucleus ventralis posterior), cell type (i.e., pyramidal as
opposed to stellate), and functional division (i.e., visual system and other tightly
interconnected regions such as the cerebellum and its associated nuclei). Are
there, in fact, individual cells that are the single source of groups of neurons
sharing common attributes? For example, do all the cells of the thalamus or of the
visual system arise from a single progenitor? Do the pyramidal cells of the cortex
arise from one class of dividing cells while the stellate cells arise from another?
How early in development do such differences become established? At present,
this important area of cell lineage has yet to be studied in the vertebrate brain,
although there are examples of invertebrate systems in which single cells give
rise to a group of daughter cells which share a particular characteristic that
distinguishes them from other groups of cells (see Jacobson, 1970, p. 18, for a
number of examples).

## Summary

The cell cycle is accompanied by a complex pattern of nuclear migration and
associated changes in cell shape. The cycle slows with age and is prolonged in
certain mutant cases. Although these events seem important in establishing the
basic histogenetic patterns in development, we know virtually nothing about why
they happen. The related matter of when distinguishing features are impressed on
developing nerve cells has received little attention. The question of whether there
are identifiable patterns of cell lineage throughout development is also unanswered for the vertebrate brain.

# 7

# CELL MIGRATION

As the wall of the neural tube thickens, five main events occur:

1. The generative cells no longer follow a pattern of making and breaking contact with the external limiting layer.

2. A separate zone of cell generation involving cells farther away from the ventricle may develop. This is termed the subventricular zone, as opposed to the ventricular zone which comprises those cells lying immediately adjacent to the ventricle.

3. A class of nonneuronal cells with processes from the ventricle to the outer limiting layer becomes recognizable. They are termed radial glial cells.

4. Postmitotic cells migrate in an orderly and predictable fashion from their region of generation to the region where they develop and express their individual characterizing features.

5. Associated with the migration of cells away from their site of generation is the development of new zones of postmitotic cells and their axons situated outside the generative layers (see Fig. 7-1).

The details of these events, the factors underlying them, and aberrations which occur in cell migration will be described for three areas: the cerebral cortex, the midbrain tectum, and the hindbrain, in particular the cerebellum. An interesting study of ectopic cells in the chick brain is discussed in addition because it raises the matter of how specific cell migration really is. Two further issues are introduced: gradients of cell generation and in vitro studies of the tissue organization that occurs after dissociated cells are placed in culture.

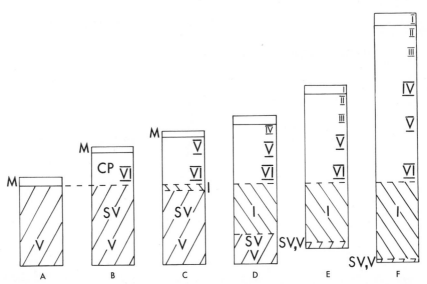

**Figure 7-1.** Sequence of development of neocortical layers. Cells are generated in ventricular (V) and subventricular (SV) zones and migrate upward to come to lie just below a zone containing very few cell bodies (the marginal zone, M, which eventually becomes layer 1). The postmitotic neurons stop migrating in the cortical plate, the inside-out sequence of development of layers (VI–II) being formed as indicated. As development proceeds an intermediate zone (I) containing axons coursing to and from the cortex interpolates between generative layers and cortical plate. In the rat, stage A is found at fetal day 16; (B) fetal day 17; (C) fetal day 18; (D) fetal day 20; (E) birth; (F) postnatal day 4.

## Cerebral Neocortex

### Normal development (Fig. 7-1)

The cerebral neocortex develops from the wall of the telencephalic forebrain vesicle. Peripheral to the generative cells of the ventricular zone lies a cell-sparse region, the marginal layer. Gradually cells migrate from the ventricular zone and collect beneath the marginal zone. This accumulation of cells develops into the cortical plate, the region which will mature to form the adult neocortex. Fibers entering and leaving the cortex form an intermediate zone which develops between the generative layer and cortical plate, and a secondary generative zone— the subventricular zone—appears. In early work it was proposed that the cells formed first are situated most superficially while the last to be generated lie deepest in the cortex (Tilney, 1933). This view, which seemed eminently reasonable, was accepted until a study was published by Hicks et al. (1959) in which rat fetuses were irradiated in the uterus at later gestational times and the cortex was studied postnatally. The irradiation selectively destroyed the last dividing neurons to contribute to the cortical plate. Instead of the deeper cortical layers being disturbed, defects were found in the upper layers. This led to the suggestion that, in fact, the cells formed last come to lie most superficially. Autoradiographic studies (Angevine and Sidman, 1961) confirmed this and showed that the first cells to be generated come to lie deepest in the cortex; cells generated later migrate between them to occupy progressively more superficial positions. It is clearly of interest to know what guides this migratory behavior, what stops each cell's migration, and whether situations exist in which these migratory patterns are disturbed.

Berry and Rogers (1965) suggested that the generative cells maintain contact at all stages of development with both ventricular and pial surfaces. They proposed that nuclear division is not accompanied by cytoplasmic division: one nucleus migrates to the appropriate position, buds off from the parent cell together with some cytoplasm, and differentiates into a mature neuron. Morest (1970) offered a slightly different interpretation of Golgi images of developing forebrain. Like Berry and Rogers, he proposed that a developing neuron maintains contact with ventricular and pial surfaces. In contrast he thought that cell division is accompanied by cytokinesis, the nucleus of the new daughter cell migrating away from the zone of cell division to a position where the cell begins differentiation. In the process the cell loses cytoplasmic contact with the pial surface.

Against these proposals are the findings of transmission and scanning electron microscopy (Hinds and Ruffett, 1971; Seymour and Berry, 1975) that cells round up at division, losing cytoplasmic contact with the pial surface. In addition, no binucleate cells were found with the electron microscope. It should be added too that since the light microscope is unable to resolve the small distances between adjacent cells, it can never be used to define binucleate cells, nor is it possible to separate adjacent stained profiles in Golgi preparations.

A further suggestion which appears frequently in the literature is that the axons from cells in the thalamus arrive in the marginal zone before the cortical plate begins to form and then act as guides for the migrating cells. The evidence for early arrival of thalamic axons is poor and recent studies on the rat (Lund and Mustari 1977) have indicated that, in fact, the axons arrive in the intermediate zone four days after the first cells migrate into the cortical plate and may actually enter the cortical plate *after* the cells on which they are to end have reached their appropriate position. Work on the monkey is consistent with this (Rakic, 1977).

Rakic (1974) has put forward a more plausible theory of cell migration, based on the interpretation of electron micrographs and on Golgi-stained sections of developing monkey cortex. This theory reemphasizes the importance of an ephemeral class of neuroglial cells, radial glia, which have processes extending

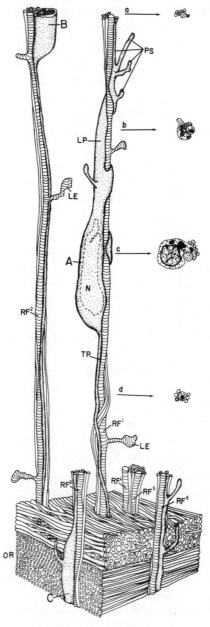

**Figure 7-2.** Drawing of migrating neuron (A) in visual cortex of monkey showing its close association to a radial glial process (RF$^1$). To the right are shown cross sections at levels a through d. (OR) white matter (from Rakic, 1972a, with permission).

from ventricular to pial surfaces. In electron microscopic preparations, Rakic has found that migrating neurons are closely associated with radial glia. Although there are no obvious specialized junctions between the two, it is common to find the processes of neurons closely applied to and even enveloping one or more radial glial processes (Fig. 7-2). Computer reconstructions of serial sections (Rakic et al. 1974) confirm this impression and provide an interesting additional observation—that many of the leading processes of the migrating cells appear to be spread over several glial profiles (Fig. 7-3). For the cells to achieve their final position, several of their leading processes would have to be retracted.

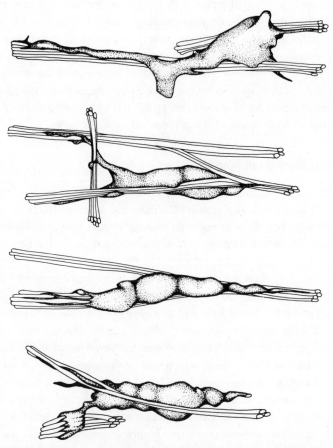

**Figure 7-3.** Reconstructions from serial electron micrographs of migrating neurons from monkey cortex, showing the complex patterns of leading processes associated with radial glia. Neurons are migrating from left to right (from Rakic et al., 1974, with permission).

The idea that radial glia may be the guide for neuronal migration is certainly attractive; definite proof is still lacking, though studies of the cerebellum of mutant mice, to be discussed below, support the idea. Several questions remain unanswered, not the least of which is where and when the radial glia are generated and what relation they have with the generative cells that give rise to the neurons. It would be expected that they should be among the first postmitotic cells since if they guide neuronal migration, they should be present as the first migrating neurons leave the generative zone. Thymidine studies do not support this idea and the possibility remains that the glia continue dividing. Comparable cells of the cerebellum (Bergmann glia) eventually become postmitotic quite late in development. We will return to this question. Another problem concerns the nature of the proposed interaction between neurons and glia. Clearly, if the radial glia are to act as guides, they cannot simply be passive elements; they are not the only processes present and indeed most neurons migrate through a layer of tangentially oriented axons in the intermediate zone without apparently following them. Some sort of specific interaction between the neurons and glia has to be proposed. Sidman has discussed this problem a number of times (e.g., Sidman, 1974) but so far it has not been approached experimentally.

### Aberrations of cell migration

The major experimental study of disturbed migration in the cerebral cortex involves the reeler mouse mutant (Caviness and Sidman, 1973). In this animal cells are more broadly distributed through the depth of the cortex with respect to their time of generation than in normal animals, and, more significantly, the first generated cells lie near the surface while the last generated are deepest in the cortex (Fig. 7-4). In other words, the cortex is inverted compared with normal. Despite their aberrant position, the cells express themselves by virtue of their connections as they would have had they been normally situated. But many of the pyramidal cells are inappropriately oriented, a phenomenon which is found only rarely in normal animals. Associated with the inversion of the layers are some highly abnormal fiber patterns which will be discussed in Chapter 9.

In a number of cases cells have failed to migrate through the intermediate layer in the human cortex. (Stewart et al., 1975). In such cases it appears that the cells may develop in an abnormal position adjacent to the ventricle (Fig. 7-5), and in Golgi preparations (Fig. 7-6) highly differentiated cells typical of the normal superficial cortical layers are found mixed together (Williams et al., 1975). It is not known why migration should fail in these cases nor why perhaps cells should adopt anomalous stopping points. Does it reflect a failure of specific

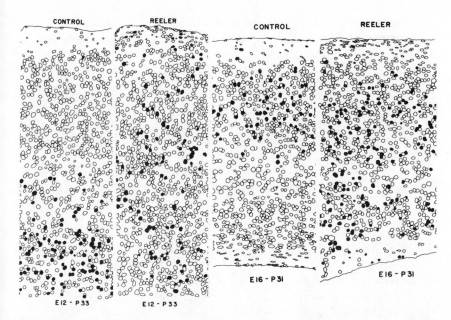

**Figure 7-4.** Sections through the neocortex of control and reeler mice injected with ³H-thymidine earlier (embryonic day, E12) and later (E16) in cortical development and fixed on postnatal days 31 and 33. Labeled cells appear dark. As can be seen, in normal animals, earlier generated cells come to occupy deeper cortical layers and later generated cells occupy more superficial layers. In the reeler, this is not so: note also the absence of a cell-sparse layer at the surface (From Sidman 1968, with permission).

recognition between neurons and radial glial cells, or is there some other fundamental problem, such as a temporary metabolic disturbance?

### Comparative studies

A related point concerns comparisons of mammalian and bird forebrains. Karten and colleagues have compared thalamic projections to the telencephalic forebrain in the bird with those in mammals (Karten, 1969, Nauta and Karten, 1970). Previously a large mass of tissue deep in the avian forebrain had been considered to represent an expansion of the mammalian basal ganglia, and the various parts were named accordingly: neostriatum, ectostriatum, paleostriatum, etc. However, Karten's work showed that only the paleostriatum compares with the mammalian basal ganglia: the other regions compare closely with various parts of the mammalian neocortex.

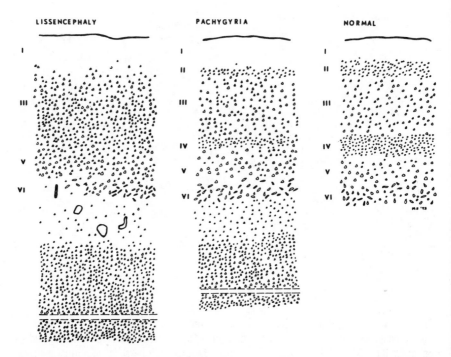

**Figure 7-5.** Cell patterns in normal human cortex and in two situations in which there has been a failure of migration of cells and there is an accumulation of neurons deep to the white matter (from Stewart et al., 1975, with permission).

It is known that a special zone of cell proliferation exists between the basal ganglia and neocortex. The proposal was made that in mammals the cells generated here invade the neocortex and provide the intermediate layer in which thalamic axons have their main termination; in birds, these same cells migrate over the striatum and form a cap over it (Fig. 7-7). Despite an "anomalous" position, they still receive input from the same thalamic axon. This interesting idea has no direct support, since at present there is no easy way of following the migration patterns of small groups of cells. We are only able to characterize cells in terms of when (but not precisely where) they are generated. The idea nevertheless is worth consideration because it suggests a reasonable means whereby subtle changes in cell migration patterns could cause some of the variations in brain structure found between different groups of animals and such variation could be one of the factors that has influenced the evolution of the brain.

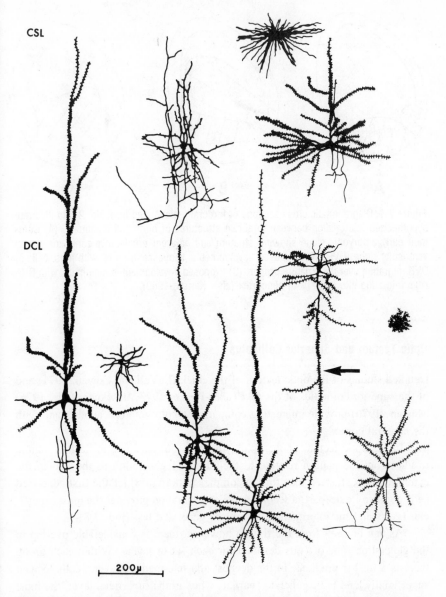

**Figure 7-6.** Drawing from Golgi preparations of unmigrated cells of the lissencephalic cortex of Fig. 7-5 showing a variety of cell types; some, such as the inverted pyramidal cell (arrow), are oriented abnormally (from Williams et al., 1975, courtesy of V. Caviness).

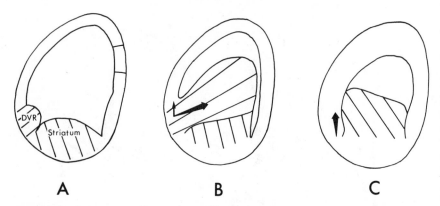

**Figure 7-7.** Diagrammatic cross sections of forebrain, elaborated from Fig. 2-6 to illustrate hypothesized interrelation between forebrain structures of birds and mammals. (A) Common early embryonic stage showing striatum and adjacent proliferative zone, the dorsal ventricular ridge (DVR) (B) Proposed developmental sequence in birds, with most cells of DVR migrating over the striatum proper. (C) Proposed development in mammals, with DVR cells migrating into layer IV of neocortex (after Karten, 1969).

## Optic Tectum and Superior Colliculus

Detailed studies of the optic tectum of the chick (LaVail and Cowan, 1971) and of the superior colliculus of the rat (Taber-Pierce, 1973, Bruckner et al., 1976, Mustari, 1976) provide interesting comparisons both with one another and with the cerebral cortex.

In the optic tectum of the chick, the cells do not come to lie in a simple inside-to-outside pattern relative to their time of generation as they do in the cerebral cortex. Instead they form a complex pattern in which the first generated cells come to lie deepest; a second wave of generation provides the most superficial layers and the intermediate layers are generated last (Fig. 7-8).

As can be seen from the figure, however, there is considerable overlap in the generation time of cells destined for each set of layers. Within each group there is a further gradient. In the deepest and intermediate layers, cells located superficially tend to be generated earlier, while in the superficial layer, the most superficial cells are generated late. This pattern is indicated by arrows in Figure 7-8. Beyond this, two more points are of interest. First, although the large cells of the deep layer are among the first to be generated, there is no overall pattern of large cells being generated before small cells. Second, in addition to a laminar

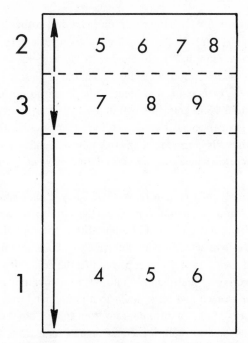

**Figure 7-8.** Order of maturation of layers in chick optic tectum (after LaVail and Cowan, 1971). As indicated on the left, the deepest layers mature first, the superficial next, and the intermediate last. Arrows signify gradients of maturation within each group. Numbers within each layer signify the embryonic days of generation of cells in each layer.

gradient of generation, there is a gradient from rostrolateral to caudomedial. The second gradient has also been found in the optic tectum of frogs (Straznicky and Gaze, 1972). The complexity of the laminar generation patterns observed by LaVail and Cowan (1971) has no parallel in any other study so far undertaken. There are two possible reasons for this. First, the laminar ordering of the chick tectum is in itself complex and very precise. It is possible that comparable patterns of cell generation in some other regions may be obscured by less precise cell body placement in development. Second, the autoradiographic technique employed (using rather thick sections and short exposure times) might tend to dilute the grain counts and reveal patterns which may otherwise be lost with techniques yielding large numbers of grains.

The rodent superior colliculus shows none of the complex patterning seen in the chick tectum, perhaps for the reasons mentioned above. In the rat (Mustari

1976), cells destined for each layer of the superior colliculus are generated between fetal days 13 and 18, the majority between 13 and 16. Cells generated on day 13 contribute more heavily to the deeper layers, and, as in the chick, the largest cells of the superior colliculus are among the earliest to be generated. By day 16, more of the cells generated come to lie in the more superficial layers. There are no obvious local gradients within individual layers. Again, an overall rostral-to-caudal gradient was observed. In the superior colliculus, therefore, as in the optic tectum, cells generated at any one time can end at a variety of depths, in contrast to the cortex where there is a correlation between laminar position and generation time.

How do the collicular cells achieve this? They do not simply move to the surface and then stop, as might happen in the cortex, where contact with a marginal zone might end a cell's migration. Brains of animals studied shortly after ³H-thymidine is injected provide information about the migrating behavior of neurons. From such material it is apparent that the collicular cells do not wait in the germinative layers and then migrate out in a coherent group. They may migrate up to the surface and back down to a particular level. Each cell could migrate for a set time but at a different rate from that of other cells generated at the same time; as a result cells generated at any one time would be spread throughout the depth of the midbrain. If this is not the case, it may be necessary to postulate that cells recognize the level at which they must stop and therefore are programmed at their time of generation for a stopping point.

## Hindbrain

### General development (see Fig. 2-9)

The cell migrations of the hindbrain and particularly of the cerebellum are of interest from a number of viewpoints: (1) not all cells migrate in a radial fashion; (2) certain cells migrate before they have become postmitotic; (3) different cell types migrate in opposite directions; and (4) aberrations in cell position occur in both normal and mutant varieties.

The hindbrain is made up of two main tissue masses dorsal and ventral to the fourth ventricle; the cerebellum (which forms the roof) and, depending on the level, the medulla or pons (which form the floor). Early development progresses as in most regions; cells are generated in a ventricular zone and migrate once they become postmitotic into the adjacent marginal zone which is surrounded by the cell-free mantle zone. Of special interest is the fact that an asymmetry develops such that the neural tube becomes separated into dorsal and ventral blocks, as indicated above. At the lateral borders, the ventricular and, more importantly, the subventricular zones become situated very close to the surface.

A distinct area of the subventricular zone, the rhombic lip, becomes a significant proliferative zone and contributes cells both ventrally into the brainstem and dorsally into the cerebellum. The ventrally directed cells form two cell groups—the pontine nuclei and the inferior olive. Both groups project into the cerebellum, the first providing part of the mossy fiber input and the second, the climbing fiber afferents.

In addition, cells also migrate from the region of the rhombic lip across the surface of the cerebellum to form the external granule cell layer. It is of some interest that while the cells deriving from the rhombic lip migrate in several different directions they eventually become interconnected in the cerebellum. This is one of the few known examples of cells in the vertebrate nervous system with a common origin that show an affinity with one another which is expressed in their interconnections. In this case, however, it is uncertain whether a single cell is progenitor of the whole group, or whether the pattern is indeed a common phenomenon which, owing to technical problems, has not been demonstrated more widely. With regard to the migration of cells contributing to the inferior olive, an interesting anomaly has been described in human cases in which the cerebral cortex is also marked by a failure of normal cell migration (e.g. Stewart et al., 1975). Instead of appearing as a single complex of cells grouped in the ventromedial brainstem, the inferior olive is distributed in several patches which extend from the position of the former rhombic lip to the normal position (Fig. 7-9). It can only be presumed that there is a failure in the migration of certain cells, including the precursors of the inferior olive, which as a result come to lie in an ectopic position.

A second point of interest regarding the cells of the rhombic lip concerns their relation to a brainstem tumor, termed a medulloblastoma, which is typically manifested as dividing cells around the brainstem and is found in young children (Kadin et al., 1970; Rubinstein, 1972). It is unusual among tumors in that it can be successfully controlled with X-irradiation, and the suggestion has been made that it represents cells of the rhombic lip which have failed to stop dividing.

The question of the control of cell division is also relevant to the contribution of the rhombic lip to the cerebellum. Unlike most migratory neurons, these cells continue to divide after having left the rhombic lip. Even if the cells are depleted by irradiation or low levels of drugs (Altman et al.,

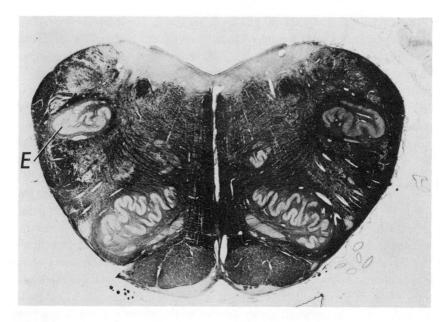

**Figure 7-9.** Section through the medulla of a lissencephalic human in which, besides the partial failure of migration of cortical cells as shown in Fig. 7-5, there is a disturbance of migration of cells originating in the rhombic lip destined for the inferior olive. As a result an ectopic inferior olive is found (E) in addition to the normal one. Myelin stain, ×6 (from Stewart et al., 1975 with permission).

1969; Shimada and Langman, 1970), there is a considerable recovery in the development of the external granule layer, apparently due to a compensatory prolongation of cell division. This would suggest a feedback that effectively limits cell division once a critical number of cells has been produced. Such a view is consistent with tissue culture studies of nonneural cells.

### Normal development of cerebellar cortex (see Fig. 2-9)

The development of the cerebellar cortex has attracted considerable interest (see Sidman and Rakic, 1973; Rakic, 1975a) largely because the two principal cell types, the Purkinje cells and granule cells, have such different developmental histories. The granule cells are generated postnatally in many mammals and are therefore selectively (along with stellate and basket cells generated at the same time) available for experimental manipulation by antimitotic agents. The Purkinje cells, by contrast, arise early in development and should be unaffected by postnatal application of such agents. The Purkinje cells are generated in the roof of the fourth ventricle and migrate to the more superficial part of the mantle layer. In time, with the development of the initial tracts in and out of the cerebellar cortex and with increased size of the cell body, the Purkinje cells become more clearly differentiated from other cells formed in the roof of the fourth ventricle. The progenitors of the granule cells lying within the external granule layer become postmitotic and emit tangentially directed axons while still close to their position of generation. The cell body then migrates inward, trailing a perpendicular axon, and eventually coming to rest below the Purkinje cell layer, at which point it develops dendrites (see Fig. 9-1). In humans the first granule cells do not lie immediately below the Purkinje cells; rather there is a gap between them, the lamina dissecans, that is eventually filled as more granule cells migrate down (Rakic and Sidman, 1970). Thymidine labeling experiments in rodents show that the granule cells do not adopt a laminar order (as in the cerebral cortex) which relates to their relative time of generation, but instead are randomly scattered. It appears, however, that the parallel fibers are ordered, the earliest formed being deepest in the molecular layer.

The main research interest in the granule cell in recent years has been focused on what guides it and its progenitors first in a tangential and then in a radial direction. One possible explanation is that the cell tends to move away from regions of active mitosis, first from the rhombic lip and then from the external granule layer enveloping the cerebellum. But even if this were the primary reason for the movement, it is not apparently the only one because under such circumstances the pattern of migration would presumably be less ordered than in fact it is. Attention has been restricted to the inward radial migration of granule cells, and in particular to the possible role played by Bergmann glia in the process. These glial cells have a soma lying close to the Purkinje cell layer and a series of beaded parallel processes running up to the surface. From studies of Golgi preparations of the developing monkey cerebellum (Rakic 1971), it appears that the Bergmann glia arise from the ventricular or subventricular zone and can be recognized by fetal day 80 as bipolar cells termed "immature granular epithelial cells." In the next month the cell bodies migrate into the Purkinje cell layer; they develop more apical processes and these processes acquire lateral thorns.

Since the appearance of these cells precedes the internal migration of granule cells, they serve as a potential guidance system for that migration. Earlier electron microscopic studies (Mugnaini and Forstrønen, 1967) have shown that migrating granule cells tend to lie adjacent to Bergmann glial processes, and this relation has been clearly demonstrated in more recent work (Rakic, 1971). Rakic

noted that the leading process of the descending cell has no growth cone specialization, unlike advancing processes in vitro. The descending limb of the granule
cell is apposed along its length to a radial glial process; sometimes several
descending cells lie adjacent to one glial process. At the region of apposition the
stainable membranes are separated by a gap of 20 nm, and the glial process is
notable in having no side spines or membrane irregularities along this interface.

While Golgi, electron microscopic (Del Cerro and Swarz, 1976), and immunofluorescence (Bignami and Dahl, 1974) studies attest to the presence of
substantial numbers of Bergmann glia prior to the migration of granule cells,
Bergmann glia are not identified in thymidine-labeled material (Das et al., 1974)
until late in the migratory sequence. This suggests that they are continually
dividing through the period when they may already be functioning as guides for
migration.

### Abnormal development of the cerebellar cortex

The idea of a relation between glial and neuronal migration both in the
cerebellum and (as already described) in the cerebral cortex although compelling,
is largely based on circumstantial evidence. Addressing the problem, Sidman and
Rakic turned to the use of mutant mice in which granule cell numbers are reduced
and patterns of distribution are abnormal. Two of these mutants, the weaver and
the reeler, have proven interesting.

The weaver mutant shows differences in both heterozygous (+/wv) and
homozygous (wv/wv) conditions. In the heterozygous condition the molecular
layer is thinner and some granule cells are found among the Purkinje cell bodies
(Rakic and Sidman, 1973b). Rakic and Sidman reported disorganization of some
Bergmann glia in Golgi preparations, but using an immunofluorescence stain
Bignami and Dahl (1974) found no change from normal in glial organization. It
is in the homozygous condition that the major disorganization occurs (Fig. 7-10).
There is little or no molecular layer, Purkinje cell bodies are distributed throughout the cortex, and their dendrites are oriented at random. Granule cells are rare
and again are not restricted to their usual layer below the Purkinje cell bodies.
Rakic and Sidman found also that the Bergmann glia are "very rare, although not
absent" (1973b) or "reduced in number" (1973c). Of particular significance is
their observation that the processes of those cells remaining are randomly distributed and very few are oriented in the usual radial pattern. So badly disorganized
are some of the cells that these authors could only tentatively identify them as
Bergmann glia. The reduction in number of granule cells results from a massive
degeneration in the external granular layer (Rakic and Sidman, 1973a, b; Sotelo

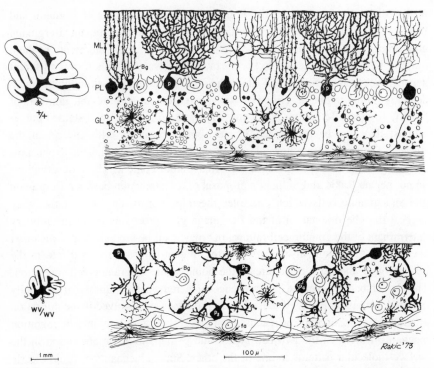

**Figure 7-10.** Cerebella of normal and homozygous weaver mutant mice. Note that the cerebellum is much smaller in the weaver mutant (diagram on the left) and that the cells of the cerebellar cortex (to the right) are disordered, in particular the Purkinje cells (P) and Bergmann glia (Bg). Note that there are few granule cells (g) and these are not distributed in a compact layer deep to the Purkinje cells as normal (from Rakic and Sidman, 1973b, with permission).

and Changeux 1974b) some five days after the cells become postmitotic. This cell death is considered secondary to a failure to migrate which occurs because available radial glial processes are no longer present (Rakic and Sidman 1973b; Rakic (1975a).

The arguments for a causal relation between the Bergmann glial disorder and the failure of granule cell migration, including the observation that the Bergmann glial aberration is seen before granule cell migration begins, are severely weakened by our ignorance of the site of the mutant gene's primary effect. In terms of logical relations, the orientation of Purkinje-cell dendrites is also severely abnormal; it could equally well be suggested that these and not the glia are the primary determinant of radial migration in the cerebellar cortex. A further difficulty is raised by the immunofluorescence studies of Bignami and Dahl (1974) which show that even in the homozygous weaver mutant, Bergmann glial processes develop normally and that at later stages there are plenty of them running in a radial direction to the surface. In a weaver mutant of a different genetic background, Sotelo and Changeux (1974b) found that although the main defect is similar to that described by Rakic and Sidman, considerable numbers of radially oriented Bergmann fibers are present in Golgi- and Cajal-stained preparations; they can be recognized in electron micrographs with endfeet at the surface. They differ from normal only in their extensive cytoplasmic expansions around Purkinje-cell spines and in their filamentous content. These observations do not negate Rakic and Sidman's proposal of an interaction between Bergmann glia and granule cells which promotes the migration of the latter; rather they suggest that the orientation of the Bergmann glial processes is not the primary determinant of the granule-cell failure, but may itself be secondary to some other aberration. At present, it appears that the mutant shows a reduced rate in the heterozygote and a major failure in the homozygote of granule-cell migration. The defect occurs even in the presence of a reasonable number of normally oriented Bergmann fibers. If indeed these processes are involved in the migration of granule cells, it may be necessary to postulate a failure in a recognition mechanism between the two cell types, due perhaps to some aberration in the surface molecular patterns in one or the other. Such a mechanism would clearly be more sophisticated than we can ascertain with current techniques, but certainly merits consideration.

The reeler mutant cerebellum has received less attention, but it too shows a number of interesting features (Sidman, 1972). The simplest way of describing it is to point out that the cell types have no laminar order but are randomly jumbled. Dendritic orientation of the larger cell types is quite abnormal and becomes more so the farther the cell body is from its normal position. Here it is relevant that Bergmann glial processes are randomly ordered both in Golgi (Sidman, 1972) and in immunofluorescent preparations (Bignami and Dahl, 1974). Nonetheless, there are appreciable numbers of granule cells, many of which have migrated away from the external granule layer. This finding tends to weaken further the correlation between the disarray of Bergmann glial processes and the pattern of granule-cell migration proposed for the weaver mouse. Granule-cell death in another mouse mutant, the staggerer, does not appear to be related to a cell-migration defect and will be discussed in Chapter 11.

In the reeler mutant, the heterozygote weaver, and to a lesser extent the homozygote weaver, granule cells are situated in anomolous positions. In apparently normal animals groups of ectopic granule cells have also been found in the molecular layer (e.g., Chan-Palay, 1972; Špaček et al., 1973). One point of note is that wherever they are located, granule cells receive a regular input from the mossy fibers; such fibers normally do not penetrate the Purkinje-cell layer. Indeed, it has been suggested (e.g., Rezai and Yoon, 1972) that it is the contact with the mossy fiber that stops granule-cell migration. Thus, a delay in cell migration would mean (since the mossy fibers are growing from deep in the cerebellum) that the granule cells would meet their input closer to the external granule layer. So far, no study has directly tested this idea. The other possible explanation as to why granule cells normally distribute in a discrete layer below the Purkinje cells is that the Bergmann glial somas are at the Purkinje-cell level and a migrating cell using these glia as guides would not be able to move any deeper.

## Isthmo-Optic Nucleus of Birds

In tracing the formation of the cerebellum, we have seen that if cells do not migrate away from their site of generation they degenerate. Is it also possible that if they migrate to the wrong place, they may also die? This is certainly not so in a number of cases where there is disorganized placement of cells (e.g., reeler mouse cortex) or a partial failure of migration (e.g., the cortex and hindbrain of lissencephalic humans). However, studies of the generation and death of cells contributing to a nucleus in the midbrain of birds suggest that cell death because of aberrant placment may sometimes happen (Clarke and Cowan, 1975, 1976, Clarke et al., 1976). In the course of development some cells lying well outside the confines of the isthmo-optic nucleus (more than 0.8 mm) share with the cells of the nucleus the property that their axons project to the retina. Later these ectopically located cells are no longer present; they have either died as the authors propose, or migrated secondarily into the nucleus.

These results are interesting since it is possible that some of the highly ordered arrays of brain cells may arise not simply by their precise migration to a terminal region but also by secondary elimination of inappropriately placed cells. Further discussion of this point will be left for Chapter 12.

## Spatial Gradients of Generation

In the examples given in this chapter we have seen that not all cells of any region are generated at the same time and that the final position a cell occupies in some cases relates to its time of generation. We have considered laminar order in relation to time of generation, but equally interesting are other gradients of generation. As mentioned above, there is a tendency for cells situated rostrolaterally in the cortex and the midbrain to undergo their final division earlier (on average) than those situated caudally and medially. A similar gradient has been identified in the thalamus (Angevine, 1970). In other words, there is a rostrocaudal gradient of maturation in each of the primary divisions of the central nervous system; and this can be seen shortly after these divisions first become clear. What role these patterns play in establishing the identity and autonomy of development of each region is not clear, nor indeed is the significance of the gradients themselves.

In addition to these gradients involving major divisions, individual areas also show local gradients in the timing of cell generation. One example will be mentioned. In the retina of fish (Hollyfield, 1972) and frogs (Straznicky and Gaze, 1971; Jacobson, 1976a) ganglion cells close to the optic disk are generated before those more peripherally located. Growth occurs by addition of new cells at the periphery of the retina, and this process carries on at low levels into adulthood. In rats, by contrast, ganglion cells destined for all parts of the retina are generated throughout the period of ganglion-cell formation, but at earlier times there tend to be more cells that occupy central retinal positions and at late times there tend to be more that are peripherally located. The relation of gradients of maturation to the course of axonal outgrowth will be discussed in Chapter 9.

## Cell Reaggregation Experiments

From early work on sponges and later studies of vertebrate tissues, it has been known that dissociated cells tend to reaggregate. (See Lilien, 1969 and Moscona, 1974 for reviews of this literature.) Furthermore, if cells from one source (one species of sponge or one tissue) are dissociated and mixed with cells from another source, they will sort out and reaggregate with cells of like origin.

Neural cells dissociated in culture also reaggregate (Lilien, 1968; DeLong, 1970; Garber and Moscona, 1972a). The amount of reaggregation varies from one region to another and is specially sensitive to the developmental stage of the tissue. Thus, optimal reaggregation is obtained early for medulla, later for cerebrum, and still later for cerebellum. Optimal reaggregation seems to occur in each region at a time when many cells are becoming postmitotic and beginning to migrate.

It is interesting that the size of reaggregates of cells from a particular source is increased by material taken from the supernatant of cultures of homotypic cells. In this way, one factor has been found which promotes retinal aggregation (retinal binding factor—Lilien, 1968; McClay and Moscona, 1974), and another which promotes aggregation of cerebral cells (Garber and Moscona, 1972b). The cerebral factor has no effect on the aggregation behavior of cells taken from a variety of other brain regions and from nonneural sources, although cerebral factor from mouse cells does have an effect on chick cerebral cells.

This work has led Moscona to hypothesize that there are components on the cell surface (termed "cell ligands") which function as receptor sites and serve to promote cell aggregation. The specificity of aggregation would depend on different ligands being present on different cells, and, since the antigens and glycoproteins on the surface membrane do appear to vary from one cell type to another (e.g. Goldschneider and Moscona, 1972; Pfenniger and Maylié-Pfenniger, 1976), this is not unreasonable. Work on purification of the retinal binding factor (McClay and Moscona, 1974) indicates that it is associated with a glycoprotein of molecular weight about 50,000. Its specific action requires that it be taken up by the surface membrane. Whether it functions by forming a bridge between two cells or by unmasking certain receptors on the membrane is not clear, nor is it certain whether the binding factor is a product of the cell membrane or derives directly from intracellular sources. Isolation from the supernatant does not resolve this matter since there are always sick or dying cells in any culture situation, especially after dissociation, and intracellular components from these cells may find their way into the medium.

It is obviously important to know what role cell aggregation mechanisms play in the developing brain, and, in particular, how they might relate to cell sorting behavior in the intact nervous system. In a number of reports the cell aggregation behaviors have been related to specificity at synapses. This would seem unlikely for two reasons: (1) affinities are demonstrated between cells of a region and not necessarily between cells which interconnect. Thus cerebral binding factor has no effect on either diencephalon or corpora quadrigemina despite the fact that these regions are elaborately interconnected with the cerebrum. (2) optimal binding occurs during the cell migratory phase in each region rather than at the time of maximum synapse formation.

It is more likely, therefore, that these cell reaggregation experiments should relate directly to the matters discussed in this chapter. To correlate them with the suggested role of radial glia in guiding neurons in a highly specific fashion, it would seem necessary to postulate that the particular binding factor is primarily concerned with maintaining specific glial–neuronal associations and perhaps that it is produced by the glia themselves. In this respect it is unfortunate that so far all the work on aggregation factors has been done with preparations containing all cell types from a region so that it has not yet been possible to isolate differential effects attributable to a particular population.

If aggregation behavior does indeed relate to the cell-sorting patterns of intact tissues, it might be expected that cells taken from mutant varieties in which cell migration patterns are abnormal would also show anomalous aggregation patterns. DeLong and Sidman (1970) tested this idea in the reeler mutant; and in aggregates obtained from both cerebrum and cerebellum they identified certain aberrations of organization. In the cerebrum, for example, there was no cell-free zone near the surface; and the cell processes were obviously ordered in a haphazard manner compared with those of normal tissue. While the experiments may be criticized because complete disaggregation of the cells was not achieved by the dissociation method, they do show that factors causing anomalies of cell sorting in intact tissues are still operative in culture. This clearly deserves further investigation.

## Summary

Cell migration is an orderly event, often but not always following a radial course from the site of generation. The work of Rakic and Sidman suggesting that the radial glial cells are an essential element in this migratory pattern is certainly reasonable, though at present it has to be admitted that neither the studies of normal tissue nor those of mutants constitute proof of a special relation between these cells and migrating elements.

We still know very little about why certain neurons migrate in such highly ordered ways and, in particular, why they come to a halt at a predictable point, which may bear some relation to their time of origin. Explanations of the cell aggregation behavior shown in vitro may be particularly relevant, especially the suggestion that the aggregation of like cells is mediated by glycoproteins on the surface membrane. In the intact brain, membrane recognition phenomena between migrating neurons and neuroglia could determine the specific course followed by each neuron as well as its stopping point. The latter could also be determined by some form of recognition between the neurons themselves. The failures of cell migration that occur in certain genetic mutants and in some human disorders might reflect a malfunction of surface molecules concerned in such recognition phenomena. It would seem important to attempt further parallel studies in intact brains and in culture, particularly by comparing normal animals and mutant varieties in which cell migration is disturbed. The existence of gradients of development in certain cell populations is of some interest, as it frequently provides a spatially distinguishing feature in an otherwise homogeneous population: a temporal gradient of neuronal maturation could be important in determining the specific connectivity of a neuron. We will return to this matter later in the book.

Finally, one issue not often considered is the extent to which degeneration of aberrantly placed cells may occur. There is evidence of this in the isthmo-optic nucleus of the avian midbrain; if of more general application it would raise the question of whether cell migration is as thoroughly directed as the usual experimental techniques indicate, suggesting the possibility that cells are distributed more broadly and inappropriately placed ones subsequently degenerate.

# IV

# THE DEVELOPMENT OF NEURAL PROCESSES

The neuron develops a single axon and one or more dendrites. The patterns of growth shown by these processes may largely determine the patterns of connectivity that characterize the adult nervous system. Although it is of obvious importance, we do not know why there should be only one axonal process per neuron and why its growth pattern and appearance should be so different from those of the dendrites. The actual outgrowth pattern of the axon depends to a large extent on the behavior of the growing axon tip. This growing tip must have a number of properties if the axon is to grow normally. First, it must be able to add new membrane; second, it must be able to interact with the substrate over which it is growing; third, it must interact with other neural processes growing in the same area; and fourth, it should have a mechanism for sending information back to the cell body about conditions of the environment at the growing tip. At the very least, the last requirement may be important to ensure that the cell body is providing the right amounts and types of precursors for the growth process, that the cell body will reduce their production once appropriate connections are made and the axon stops major growth, and that it will then make more of the materials associated with synaptic transmission. However, some of the interactions with the substrate and with other neural processes could be dictated from the cell body, in which case an efficient means of transporting "samples" back to the cell body from the area around the axon terminal would be essential. It might be expected that these same features would also characterize the growth of dendrites, even though they extend shorter distances and their shape is generally rather more ordered in comparison with axons.

In this part of the book we will first examine how neural processes grow in vitro: what substructural components are responsible for their growth and how the processes interact with the environment in which they grow. In many of the studies of neuronal growth in tissue culture, it is not clear whether the extending processes are axons or dendrites, even though their growth patterns are more characteristic of axons. For this reason it is perhaps better to use the term *neurite* rather than *axon*, although this convention is not always observed. Since it is obviously important to know to what extent observations from in vitro studies apply to in vivo situations, we will then examine how axons behave in the complex environment of the developing brain, how these patterns are modified by various circumstances, and what insight we can gain into the factors that regulate the growth of neural processes.

One major limitation in many studies of axonal growth patterns in the intact brain is that the experimental manipulations are performed at an early stage of development, whereas the effects of these manipulations are investigated when the animal is mature. Thus, the behavior of the axon during actual growth is unknown. Even in the normal brain, there have been very few detailed studies of growth patterns of axons. As a result we know very little about how much correction of inappropriate growth occurs and what is the nature of the normal interactions that occur between single axons and groups of axons. And the primary object of tissue culture studies, the process of neurite extension, has been given little attention in vivo.

This deficiency is largely the result of technical difficulties, but also, until recently, the issue had not seemed important. In the past few years, however, a number of experiments, mostly on mammals, have shown that after local injury in development, the distribution of undamaged pathways is expanded or significantly modified. An example is given in Figure IV-1. Regions X and Y normally send axons to adjacent zones x and y respectively (Fig. IV-1A). If X is removed at an appropriate time in development, the projection of Y may subsequently extend beyond its normal zone of distribution into region x, now deprived of an

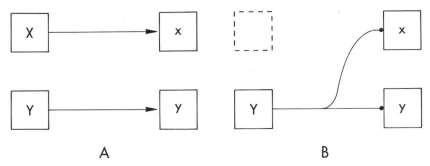

A                                    B

**Figure IV-I**. Plan showing one pattern of abnormal axonal distribution which may result from removal of a region, X. See text for description.

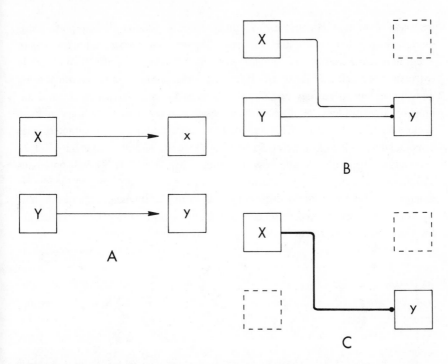

**Figure IV-2.** Plan to show abnormal axon connections which may result from removal of a target nucleus, y with or without simultaneous removal of input, X.

input (Fig. IV-1B). This result could be explained in two ways. First, axons from Y may initially distribute to both x and y, and subsequently retract from x in normal circumstances because of an inhibitory influence exerted by axons from X or because the Y axons are at a competitive disadvantage in making appropriate synaptic connections. This retraction would not occur if the axons from X were absent. Alternatively, the axons from Y normally may distribute only to y but then grow into x if it is deprived of its normal input, possibly because that input is no longer there to block growth from Y or because Y axons are attracted by uninnervated synaptic sites. Another common experimental plan, shown in IV-2, is to remove the region to which an axon population is normally distributed (region x). The axons from X may then distribute to y instead, (Fig. IV-2B), this aberrant termination being enhanced if Y has also been removed (Fig. IV-2C).

    In order to understand the basic mechanisms involved in the development of these aberrant patterns, it is clearly important to know how the axons first grow out, the stage of development of the system at the time the initial lesion is made, whether all axons of a population behave similarly, and to what extent the response of an axon changes as it matures. Another problem in interpreting this kind of study is that usually the interaction between only two axon populations in a region is being tested. Yet there are other axons that may be affected—not only extrinsic ones, but also local ones from cell bodies situated in the terminal area. To what extent these may modify or confound the primary interactions is unknown. The problems relating to the order and disorder of axonal growth will occupy Chapters 9 and 10. Chapter 9 deals with general matters, Chapter 10 with a special case—that of determining on which side of the brain axons may end. In Chapter 11, we will examine how dendrites grow, how their growth processes compare with neurite outgrowth in vitro, and what factors may influence the emergence of the mature dendritic form. Of particular interest is the question of whether dendrites are passive receivers of axonal inputs or whether synapses are the result of active growth by both axons and dendrites.

# CELL BIOLOGY OF NEURITE OUTGROWTH

## Characterization of Growing Neural Processes In Vitro

Since the work of Harrison (1910), it has been recognized that a specialized structure—the growth cone—exists at the terminal part of a growing neural process (or neurite). The growth cone (Fig. 8-1) is best observed in vitro since in this situation its dynamic properties can be studied directly (Bray, 1970, 1973; Bray and Bunge, 1973; Bunge, 1973, 1977; Yamada et al., 1970, 1971). It consists of an enlarged area about 10 $\mu$m in diameter from which emerge ruffles of membrane (foliopodia) and small filopodia (or microspikes) of 0.15–0.3 $\mu$m in diameter and up to 20 $\mu$m in length. In culture the filopodia can be seen to wave about in the medium, expand to their full length in 30–60 seconds (covering a 10° arc in 1–4 seconds), and then retract in a matter of 30–60 seconds (Yamada et al., 1971, Ludueña and Wessels, 1973).

When it waves about, the shaft of each filopodium remains rigid, and all the angular movement occurs at the base (Strassman and Wessells, 1973). A force of retraction of $3 \times 10^{-10}$ dynes has been calculated for an individual filopodium (Nakai, 1960). Filopodia apparently can fuse with each other or with their parent growth cone, neural process, or cell body, but not with other branches of the same or other neurons (Spooner et al., 1974). They act as sites of adhesion between the growing neural process and the substrate, and, as such, provide an anchor to allow further extension beyond the region of attachment. In fact, several investigators (e.g., Bray, 1973) have commented that the only sites of adhesion made by a neuron in culture are at the growth cones. Other functions that have been attributed to the filopodia are as tactile or chemosensory agents testing out the environment into which the axon is to grow.

Another feature of the growth cone is its role in membrane addition. Membrane is apparently added continuously at the growth cone, having been trans-

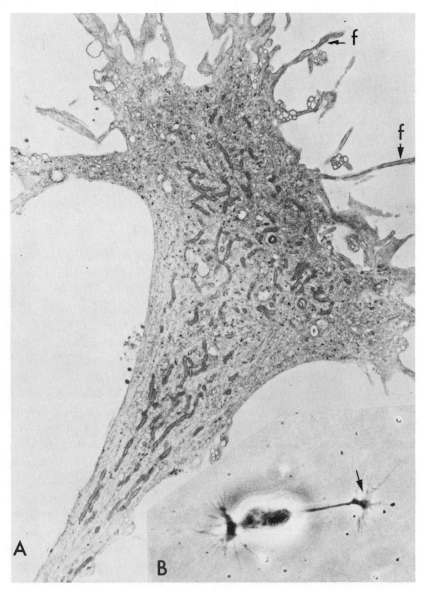

**Figure 8-1.** Superior cervical ganglion cells from one-day-old rats grown in culture. (A) Electron micrograph of neurite growth cone and associated filopodia (f) Note the concentration of organelles in the center of the growth cone and the peripheral zone of microfilaments which extends into the filopodia ×6930 (courtesy M.B. Bunge). (B) Phase contrast light micrograph of similar cell showing growth cone (arrow) ×690. The cell body is the highly refractile area (courtesy D. Bray).

ported from the cell body in association with an intracellular tubular system. Once on the surface it flows back rapidly in the ruffled membrane to occupy a more proximal position on the neurite or to be retracted into the growth cone and recycled. This was demonstrated by putting small particles (about 5 $\mu$m in diameter) of carmine in the culture dish (Bray, 1970). These stuck to neural processes, including growth cones, and a backward transport across the surface of the growth cone at a rate of 70–120 $\mu$m per hour was demonstrated. The same experiment showed that, while there is some indication of a to-and-fro movement of particles on the membrane more proximally in the neural process, there is no net centrifugal movement from the cell body which would suggest membrane addition outside the region of the growth cone. This view is supported by the fact that branches, once formed, do not increase their distance from the cell body. According to these observations it seems that all growth by extension must occur by addition of membrane at the growth cone.

Experiments in which extending neurites were separated from the cell body (Hughes, 1953) have shown that the activity of a growth cone can continue undiminished for a matter of hours. This suggests that beyond the requirement of raw materials from the cell body, the growth cone is relatively autonomous, at least in its mechanical operation, from the central part of the cell. Why this organelle should form in the first place is unclear. It is apparent, however, that almost all new growth cones on a process are formed by division of a single growth cone, thus creating a bifurcation in the axon. Branches are formed only rarely by development of a new growth cone along the course of a process; perhaps in this case a remnant of ruffling membrane is left behind after a division of the growth cone (Bray, 1973). One wonders, however, whether a suitable local microenvironment may be able to induce a new growth cone along the course of a neurite. It is clear that in some circumstances a cut axon will develop a new growth cone at the cut edge, which will behave much as a growing tip of a newly developing process does. The behavior of regenerated growth cones has been demonstrated especially elegantly in a series of experiments by Speidel (summarized by Speidel, 1964) in which he exploited the fact that a tadpole tail is translucent so that the cutaneous nerves can be visualized and their growing processes can be followed after damage. Under these conditions axonal growth is far from rigidly controlled, instead, processes are continuously being emitted and retracted.

These various observations indicate that the growth cone and its associated structures are essential for active nerve growth. What then are its structural components and how do they work to allow directed growth? We can say quite a lot about the elements of neurite extension, largely as a result of observations by

M. Bunge and co-workers (e.g., Bunge, 1973, 1977; Bray and Bunge, 1973; Rees et al., 1976). Wessells and his various colleagues have addressed themselves especially to the subcellular mechanisms of nerve growth, a topic of particular importance when factors that influence nerve growth patterns are considered (summarized in Wessells et al., 1973; see also Letourneau and Wessells 1974; Letourneau, 1975a,b).

Growth cones (identified by phase microscopy while growing in culture and then studied after fixation with the electron microscope) have a set of features and organelles which, although not unique, are unusual in their concentration in such a peripheral part of a neural process (Fig. 8-1A). Apart from microtubules, which are ubiquitous in nerve cells, the central area of the growth cone contains neurofilaments; lysosomes; many coated, dense cored and agranular vesicles; and agranular saccules and reticulum, sometimes continuous with the surface membrane. Toward the periphery there are more tubules and saccules, but more significant is a feltwork of microfilaments that extends into the filopodia. Studies of the surface membrane in freeze fracture preparations (Pfenninger and Bunge, 1974) show that the density of intramembranous granules is particularly low at the growth cone ($85/mm^2$) and progressively increases along the axon to about $660/mm^2$ in the soma. With time the density of particles increases in the presynaptic process to an adult figure of about $700/mm^2$. This suggests that the earliest membrane added at the growing tip is incomplete and matures only by the subsequent addition of intramembranous granules. The low numbers of granules in immature membrane is somewhat surprising in that intramembranous granules generally appear to reflect the presence of glycoproteins. Since growth cones show extensive binding with concanavalin A (a plant lectin that binds especially with carbohydrates, Nicolson, 1974), they would therefore be expected to have abundant membrane glycoproteins and lots of intramembranous granules. Pfenninger and Bunge explain this discrepancy by suggesting that the glycoproteins at the growing tip may be in nongranular form. At the very least the results show differences in the morphological (and therefore biochemical) characteristics of newly formed compared with more mature membrane. Accordingly, the membrane might be expected to behave differently at different points along the length of a neurite and at different stages of maturation. This is of some importance in view of the suggested role of glycoproteins in recognition phenomena. More recent studies (Pfenninger and Maylié-Pfenninger, 1976) have tested the lectin-binding properties of the growth cones of a number of different types of neuron taken from rat embryos. Of the four lectins tested (concanavalin A, wheat germ agglutinin, ricin I, and soybean agglutinin), each exhibited characteristic and different patterns of binding with the growth cones of neurons from superior cervical ganglion, spinal cord, dorsal root ganglion, olfactory bulb, and cerebellum. This implies that the composition of surface carbohydrates on the growth cone of each neural type is different and as such has considerable relevance to the patterns of interaction each type may exhibit with its environment.

## Mechanism of Neurite Elongation

From the preceding discussion, it has become apparent that neurite extension occurs by the addition of new membrane and the protrusion of processes beyond an area of attachment with the substrate. An important part of this process is the formation of filopodia. Wessells and his colleagues have paid particular attention to the role of microfilaments (protein threads of 5–7 nm diameter) in this process (Yamada et al., 1971; Wessells et al., 1971, 1973; Spooner et al., 1973, 1974; Ludueña and Wessels, 1973). In a number of tissues such filaments have been implicated in intracellular contractile mechanisms that alter the overall cell and hence tissue pattern. In one configuration the microfilaments are concentrated below the surface membrane, into which they appear to insert. Such filaments, termed lattice filaments, are abundant in the periphery of the growth cone and are the only organelle consistently found within the filopodia, in which they appear as a polyhedral network with the long axis parallel to the axis of the filopodium. Elsewhere they are packed more symmetrically. The suggestion has been offered that the change in conformation is the result of an active contractile process and leads to the formation of a filopodium. An indication that microfilaments have a role in contractile processes is the finding that in some situations they bind heavy meromyosin and therefore contain an actin-like element (Spooner et al., 1973, 1974). Such binding has yet to be clearly demonstrated for the surface lattice filaments of neurite growth cones. Another indication comes from the use of cytochalasin, a drug which causes retraction of filopodia and cessation of surface undulations of the growth cone. Fine structural examination of material so treated indicates that the microfilamentous feltwork is condensed, as indeed it is in other tissues where tissue or cell-shaping characteristics are disrupted by the drug. But more recent work has shown that, contrary to earlier expectations, cytochalasin may not be entirely specific to microfilaments (e.g., Rhoten, 1973; Croop and Holtzer, 1975). Colchicine, a drug primarily targeted at microtubular protein, has no effect on growth cone dynamics, nor on microfilaments, although it does cause retraction of the neural process connecting the growth cone with the parent cell body. Thus there appears to be some relation between the lattice filaments and the extension mechanisms of the neurite, but it has yet to be proven whether this is a causal relationship.

## Interaction with Substrate

An advancing neurite depends on an interaction with a substrate in order to progress; it cannot grow across a fluid-filled space. Only the growth cone and its associated filopodia are attached to the substrate, however. The neurite itself lies free, bridging the gap between cell body and growth cone. The filopodia at the leading edge of the growing neurite contact the substrate to which they attach. As the growth cone advances, a new wave of filopodia come to attach ahead of the previous set, which gradually retract into the body of the growth cone and lose contact with the surface. Thus, the advancing edge pulls itself along by successive waves of filopodia. The rate of advance varies considerably even for a single cell type in good health. Chick dorsal root ganglion cells, for example, show a twenty-fold difference according to the substrate used (Ludueña, 1973), indicating that variations in the substrate will materially alter growth patterns. A number of different interactions between growing neurites and substrates have been considered.

### a. Mechanical guidance

In tissue culture studies it is a common observation that if there is a scratch across the culture dish, neurites (or migrating cells) will tend to grow along the defect. Weiss (1934) found that the plasma clots in which he grew neural tissue developed stress lines along which the plasma micellae were oriented. Such lines provided a more subtle form of contact guidance than culture dish scratches for the growing neurites. More recent studies (discussed below) indicate that other factors may override the contact guidance phenomenon, at least in plasma clots, but they do not invalidate the fact that neurite growth can be directed by mechanical guides.

### b. Effects of differential substrate adhesiveness

Another factor of some importance is how tightly a growing neurite adheres to the substrate. This has been investigated systematically by Letourneau (1975a,b), who has found a relationship between adhesiveness and growth rate.

Using a variety of substrates—glial cells, highly basic small-chain polypeptides (such as polyornithine and polylysine), collagen, palladium, and the plastic surface of the tissue culture dish—he found that neural processes attach most securely to glial cells and to the basic compounds (even more tenaciously in the presence of high calcium ions), less so to collagen and least to the palladium and plastic. These findings apparently reflect real adhesive differences and not a toxic effect of the less attractive substrates. Letourneau found that the more adhesive substrate increased the rate of growth markedly (50 $\mu$m/hr on polyornithine compared with 7.7 $\mu$m/hr on plastic); and that the degree of branching, as reflected in the number of growth cones per neurite, also increased. If part of the surface of a culture dish is coated with polyornithine and part with palladium, neurites will show a preference in growing over the polyornithine-coated regions.

Thus, the direction of axon growth may be dictated by patterns of differential adhesiveness of the substrates over which an axon grows. The degree of ramification could similarly be determined by the substrate.

### c. Chemotrophic influences

A variety of cell types show real chemotactic responses; for example, slime-mold cells move along a concentration gradient of a specific chemical, acrasin (Barondes, 1975). It has been suggested that comparable chemotrophic action-at-a-distance effects may direct an axon to its target (e.g., Sperry, 1963; Hubel and Wiesel, 1971). Experimental studies of nerve growth have offered no support for the idea and even some evidence against it.

Weiss and Taylor (1944) cut the sciatic nerve of rats and allowed it to regenerate into a Y-shaped artery in which one arm was filled with the products of a degenerating nerve and the other blocked off at its distal end or filled with a tendon. The nerve fibers regenerated nonselectively. It is perhaps unfortunate that one branch was not filled with muscle tissue since the study showed that degenerating nerve tissue does not constitute a special stimulus for guiding regenerating axons; it did not demonstrate whether or not their normal target tissue would provide a chemical attractant.

Nakai and Kawasaki (1959) investigated the effects of various materials scattered in a tissue culture dish through which processes of dorsal root ganglion cells were growing. They watched how axons grew in relation to cell debris and to particles of palladium, nickel, starch, polyethylene, paraffin, and cholesterol. Neurites growing close by would not alter their direction of growth; those growing directly into an obstruction would change their growth direction by retraction of filopodia and development of new filopodia at a different position. In both cases, the neurites extended numbers of filopodia to contact the foreign material. Even after the growing tip had passed, some filopodia would still maintain contact for a while before finally retracting. Nakai and Kawasaki interpret these observations to mean that rather than being subject to any action at a distance, the growth cone filopodia "palpate" objects along the course of growth. The direction of growth may only be altered if the filopodia attach to the object, and if this attachment is strong enough to withstand the force of retraction of the individual filopodia. Under such circumstances the whole growth cone will be directed toward the attachment point. If the attachment is too weak, the filopodia will break contact with the substrate as they retract and the growth cone will remain in the same position. In the light of Letourneau's work it would be interesting to repeat these studies using beads coated in polyornithine, so as to observe what effect a highly adhesive obstruction has on the growth patterns. Rees et al. (1976) have recently followed neurites growing out of a spinal cord explant across a dish of dissociated superior cervical ganglion cells. They found that axons did not obviously home in on their target cells, which suggests, in accord with Nakai and Kawasaki (1959), that axons reach the target cell by chance; but once in contact, they may either form a permanent junction or retract and alter their direction of growth.

The possibility that electrical fields could direct nerve growth, although suggested (see Jacobson, 1970), has never been confirmed.

## Interaction with other cells

This topic is extremely important since neurites in vivo do not grow across collagen-coated plates but across or amongst groups of cells. We need to know what potential interactions are available between an axon and its cellular substrate.

### a. Interaction with nonneuronal cells

Nakai and Kawasaki (1959) showed that filopodia attach strongly to macrophages and that a neurite can be dragged along for some distance by a migrating macrophage. As indicated above, Letourneau (1975a) demonstrated strong adhesiveness to neuroglial cells more so than to collagen; associated with this was an increased rate of growth.

Nerve growth factor (NGF) (see Chapter 4) stimulates attachment of autonomic and dorsal root ganglion cells to substrates and increases the extent of neurite outgrowth as well as the longevity of the neuron in culture. Its function appears to be complemented by material produced by supporting neuroglial cells. The suggestion has been made (Varon, 1975) that there may be neuroglia throughout the central nervous system which produce similar factors. Studies of NGF-like evidence on the effects of gliomas support this idea. But is this effect specific? The active component of NGF is basic and Letourneau's work shows that the effects of NGF can be mimicked (at least with respect to adhesion and neurite outgrowth) by a highly basic medium. Macrophages are also more active as phagocytes when engulfing basic materials. Is then the stimulatory effect of neurite outgrowth by nonneuronal cells and NGF simply due to their provision of a highly basic substrate? Although this is an attractive idea, the real situation is certainly more complicated. Studies of the central nervous system have shown that NGF is effective in promoting axonal growth in only a few systems (see Hendry, 1976). It is clearly important to know the basis for this selectivity of action.

### b. Interaction with other neurites

When the filopodia of one neurite meet those of another, three courses of action are possible (Nakajima, 1965):

1. The filopodia will retract, apparently because of an inhibitory influence.

2. One neurite will follow the other, forming a fiber bundle (the process of fasciculation) if additional neurites contribute.

3. Transitory contact will be made and then the neurite will continue its growth by crossing the other process, there being no permanent connection between the two.

Since all three outcomes can be observed in the same culture dish, it seems that the neurites themselves must exhibit a degree of selectivity. The viscosity of the medium in which the neurite growth occurs is also important, however. In a solid plasma clot, only inhibitory retraction occurs. Dunn (1971) has studied this in detail and considers inhibitory retraction the primary mechanism for radial outgrowth of processes from dorsal root ganglion explants. If two explants are put in a dish close to one another, there is a zone of inhibition of growth between the two (Fig. 8-2A). Such inhibition occurs despite the presence of stress lines running in the clot between the two explants. If, however, the clot becomes liquefied, fasciculation occurs and many fibers are found running from one explant to the other, producing what Weiss (1934) describes as a "two-center effect" (Fig. 8-2B).

The mechanism of fasciculation is clearly an important one since most

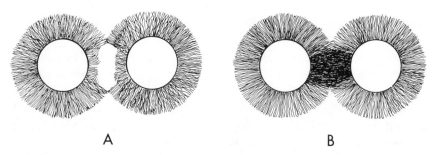

**Figure 8-2.** Outgrowth of neurites from two explants of embryonic neural tissue placed close to one another in tissue culture, showing either (A) mutual inhibition of growth or (B) fasciculation of axons growing between the explants. The two effects appear to depend on the viscosity of the medium.

axons running from one region to another in the brain do so in bundles rather than singly. Weiss (1941) has suggested that a pioneer fiber may grow to the target organ and, once it has made contact, begins to attract later growing fibers. Subsequent work has shown that while there is good evidence for the existence of pioneer fibers in certain systems (Macagno et al., 1973; Goldberg and Coulombre, 1972), fasciculation can occur before the pioneer fibers have made synaptic contact. Using phase microscopy Nakai (1960) described two mechanisms of fasciculation. In the first, one growth cone approaches either another growth cone or a neurite which stimulates that structure to send out filopodia. The filopodia of each process attach and retract, bringing the two neurites into alignment. In the second process, lateral filopodia arise from the shafts of adjacent neurites and attach, retract, and again pull the neurites into alignment. From these findings, it would appear that fasciculation is an active event and that its selectivity is determined by whether filopodia of different neurites adhere strongly enough to one another to withstand the forces of retraction. The viscosity of the medium is presumably important in increasing or decreasing the force of retraction necessary to produce alignment.

Why axons in vivo should sometimes fasciculate and sometimes mutually inhibit one another's growth is uncertain. For a single axon population, it would appear that one mechanism would be operant as it runs in a fiber tract, while the other would take effect in a zone of terminal arborization. Differential adhesiveness of the substrate in different parts of an axon's course could be an important factor in controlling its type of interaction with its neighbors. Changes in the surface proteins along the course of the axon may also be important.

## Summary

Neurites commonly grow in tissue culture by means of a specialized structure, the growth cone. The specializations of this structure are clearly directed toward the requirements outlined at the beginning of Part IV: there is indication of large amounts of internal membrane which can be externalized; the filopodia show particular behaviors depending on the substrate, the relation to other neural processes, and possibly the cell's own intrinsic controls; and extracellular material is internalized in coated vesicles and can be transported back to the parent cell body.

When one is considering the varying behaviors of growing axons in the intact nervous system, it is important to realize that they may depend heavily on the differential adhesiveness shown by filopodia to physical, neuronal, and non-neuronal cellular substrates. When the filopodia retract, the whole neurite will advance if adhesiveness is high or stop (at least temporarily) if it is low. This process is probably also influenced by the microchemistry of each region the growing neurite encounters, including the availability of NGF or related substances.

# 9

# AXON OUTGROWTH IN VIVO

The mechanisms of neurite extension were examined in the last chapter. We saw that in tissue culture the direction and orientation of a neurite's growth and its pattern of branching can be modified by altering the substrate over which the neurite is growing or the medium in which it is growing, as well as by interaction with other neurites. We now have to consider whether these same phenomena are likely to influence nerve growth in the intact brain. First, the question of whether growth processes are similar in the two situations and particularly whether growth cones can be recognized in vivo, must be addressed. Second, the course of axonal growth in a few systems will be described and then a series of questions will be posed concerning the variability and determinants of such growth patterns. The circumstances associated with laterality of axonal growth within the brain are dealt with separately in Chapter 10.

## Identification of Axonal Growth Cones in Vivo

In order to identify growth cones in developing tissues, one would ideally watch the axon growing, fix it, and then study the growing tip, as has been done in tissue culture experiments. This is usually not possible in vivo. The best alternative is to attempt to identify by the light and electron microscope, features resembling those of the growth cones described in vitro. Structures at axon tips that are similar to growth cones have been identified with the light microscope in Golgi-stained preparations of developing brains by a number of investigators (e.g., Ramon y Cajal, 1960, Morest, 1968). The first correlative electron microscopic study of growth cones in vivo was done by Tennyson (1970), who traced the axons of dorsal root ganglion cells into the spinal cord in early fetal rabbits, at a time when the axons are just beginning to penetrate the cord. In small series of adjacent sections, she identified the terminal points of the growing axons. Much as in tissue culture studies, there was an expanded bulbous area 6–13 $\mu$m long and 2–5 $\mu$m wide containing a variety of organelles— neurofilaments, microtubules, ribosomes, agranular reticulum, mitochondria, dense bodies, and clear vesicles 48–100 nm in diameter. Around the periphery could be seen a feltwork of microfilaments extending into outgrowths that appeared very similar to filopodia. Studying serial electron-microscope sections of developing retinal ganglion cell axons of fetal mice, Hinds and Hinds (1972) found similar features at the terminal part of many axons: an enlargement containing a large variety of organelles from which arise finer filopodia and sheets of cytoplasm (or foliopodia), both containing microfilaments and some vesicles. Hinds and Hinds also reported some growing processes that show no major specialization. Similar unspecialized leading processes (although not axons in these cases) have also been identified in migrating neurons of developing monkey cerebral and cerebellar cortices, (Rakic, 1971, 1972a). A related example in the developing visual system of the crustacean *Daphnia* has been reported by Lopresti et al. (1973). In this animal receptor axons grow as a group of eight fibers into a cartridge in the first synaptic layer, which is called the lamina. One of the axons grows faster than the others and has a distinct growth cone, while the others, which apparently use the first axon as a guide, do not have any terminal specialization.

Thus it appears that while some axons in intact tissue do have growth cones at the growing tip, others may extend with no major terminal specialization. The latter group may follow predefined routes, growing along glial channels or pioneer axons and eventually forming the fiber bundles characteristic of the adult brain. This sort of axonal growth is not characteristic of the tissue culture preparations in which growth cones have been studied in detail, and the growing tips of axons in bundles have not been investigated. One further difference between in vivo and in vitro studies is that in culture dishes the substrate is fixed while in the brain it is usually enlarging due to the development of more cells and processes in the region where the particular axon is growing. Such enlargement therefore dictates a need for the addition of membrane along the course of an axon as well as at its terminal growth cone, which is something of little significance in vitro.

### Patterns of axonal growth in vivo

### General comments

An axon may form in two ways: it may extend from a cell body which is fixed in position, or the axon tip may remain stationary and the cell body move away from it. While many axons develop in the first way, it is clear that at least the proximal part of some axons is formed by the second mechanism. Perhaps the best example of these two mechanisms is the granule cell of the cerebellum; the tangential array of the parallel fibers is the result of true axonal extension, while the perpendicular proximal part of the axon is formed by cell migration (Fig. 9-1). An interesting point emerges from cases like this where axon and leading process of the cell body are moving in different directions at the same time. Whatever special affinities may guide the migration, they are apparently different for different parts of the cell surface. A possible example of a comparable growth pattern is found in the extrinsically projecting cells of the cerebral cortex (Stensaas, 1967). Migrating through the intermediate zone on their way to the cortical plate, these cells may send out a tangentially running axon into the intermediate zone. As they continue their migration, the more proximal, perpendicularly oriented part of each axon would represent the trailing process (Fig. 9-2). Altman (1970) has reported that a considerable part of the axon of the granule cells in the dentate gyrus of the hippocampal formation is formed in this manner. He suggests that this growth mechanism is characteristic of microneurons (small interneurons, which contrast with large projection cells). However, the proposed

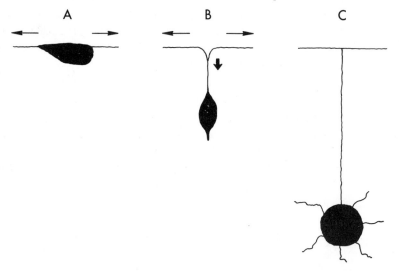

**Figure 9-1.** Diagram to show development of axon of cerebellar granule cell. (A) Axon extends from cell body (thin arrows) adjacent to external granule layer. (B) Cell body migrates (thick arrow) toward Purkinje cells, leaving a trailing process which in C becomes the proximal part of the mature axon.

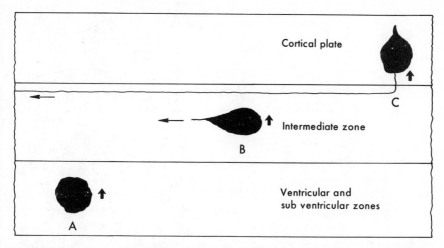

**Figure 9-2.** Sequence of migration of a cortical cell which emits an axon growing tangentially (thin arrow) within the intermediate zone (B), and leaves a trailing process as it migrates (thick arrow) through the cortical plate.

axonal growth patterns for cerebral cortex projection neurons described above, as well as the observations of many interneuron axon plexuses, are not consistent with this generalization.

### Patterns of growth of retinotectal axons

It is important to know how axonal outgrowth is related in timing to the final division of the cell from which it emanates and of the cells with which it synapses, and how soon an axon starts forming synapses once it has arrived in the appropriate region. All may be determinants of the connections eventually made. An interesting comparison is provided by studies of the optic pathway from the retina to the opposite optic tectum (or superior colliculus) in frog, chick, and rat. These are summarized in Figure 9-3.

There are several points of significance. The first is that in frog and bird, ganglion cells are generated by addition of cells around the circumference so that the first formed lie close to the optic disk and the last formed are at the peripheral margin of the mature retina. In the rat this is not the case—cells occupying any region of the retina may appear at any time throughout the generation period.

A second point concerns the timing of innervation of the tectum. In both frog and rat, optic axons form synapses soon after they arrive, while in the chick the axons are delayed in a fiber layer above the synaptic zone for several days

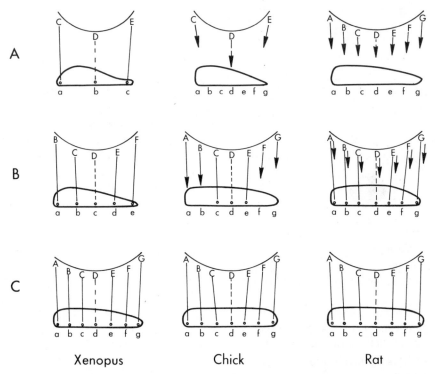

Xenopus                 Chick                    Rat

**Figure 9-3.** Three stages in the maturation of the retinotectal pathway in relation to the pattern of cell genesis in the retina and tectum. The retina is above, temporal to the left; the tectum is below, rostral to the left. The thickness of the tectum signifies its degree of maturation. Optic axons with arrows are growing; those with circles have formed synapses. The dotted line signifies the projection of the central retina. Letters signify stable points on retina and tectum. Data for chick from Kahn (1973) and Crossland et al. (1975). See text for further description.

before finally invading it. This pattern has also been demonstrated in the geniculocortical pathway of rats (Lund and Mustari, 1977); and it is probably the same for monkeys (Rakic, 1977).

A third point concerns the degree of maturation of the optic tectum at the time it becomes innervated. In both chick and rat, most cell division has ceased before the axons arrive. In *Xenopus,* by contrast, axons arrive from the retina long before the more caudally situated tectal cells have completed their final division. A clear-cut gradient of maturation from rostro-lateral to caudo-medial exists. This presents a problem concerning the progressive development of the visual map in this animal, as there is a requirement for a continuously shifting map (since the retina adds cells all around the periphery and the tectum adds them only at its posterior border). Such a shift has been proposed by Gaze and his colleagues (Gaze et al., 1974; Chung et al., 1974). It must be added that a recent study (Jacobson, 1976a) on timing of cell birthdays has questioned the validity of some of the primary observations on the radial symmetry of the addition of cells around the border of the retina. More cells are added ventrally than dorsally, but while this might skew the map of the retina on the tectum, it is difficult to see how it modifies the nasotemporal relationships which are the crucial ones for a shifting map hypothesis.

From these comparisons between species it is clear that the first axons grow out of the eye shortly after the first retinal ganglion cells have become postmitotic. Whether all the axons grow out immediately after cell division (and perhaps migration) is completed is not so obvious. Beyond this, the relation of axonal development to the time of final division of the parent cell, to synaptogenesis, and to the patterns of cell division of the tectum vary between the animals. Since the end result—a map of the retina upon the tectum—is identical in each case, it would appear that correlative timing of axonal growth with the other factors is relatively unimportant. Further support for this view is provided by a study in which the optic nerve was prevented from growing into the tectum of Xenopus until a late stage, at which time a normal pattern of innervation was shown physiologically and the tectum was reported to be of normal histological appearance (Feldman et al., 1971).

### Variability and determinants of axonal growth patterns

These may best be discussed by asking a number of questions.

### a. Is there variability in "normal" patterns?

In the human neuropathological literature there are comments periodically on the appearance of major aberrations of individual pathways, or of unusual tracts, in apparently normal brains. One would like to know whether this variation is normal, showing that programming of pathways is not that precise, whether it reflects genetic variance, or whether it is secondary to pathological disturbances during development.

A number of studies have emphasized the role of genetic variation by showing strain-related differences in the distribution of certain pathways in mouse brains. These include the size and even presence of the corpus callosum, the way in which the fornix distributes relative to the anterior commissure (Wahlsten, 1974), the disposition of fibers in the optic chiasm (to be discussed in the next chapter), and the pattern of axonal distribution of the granule cells of the dentate gyrus (Barber et al., 1974). In the last case, the distribution of axons materially alters the patterning of synapses on the postsynaptic cells. Thus, the granule cells of the dentate gyrus have a major output that can be distributed in three bands in the hippocampus: on apical dendrites of pyramidal cells just above the layer of pyramidal cell bodies, on basal dendrites below it, or on both apical and basal dendrites within it. The relative weighting of these three pathways varies significantly between strains, but not within strains. Such variation may have a correlate in behavior, since there are indications of strain-related behavioral differences (see Wahlsten, 1974; Fuller and Wimer, 1973), which can be selected by breeding. However, the link between a slight change in the synaptic pattern of a group of cells of unknown function and a gross behavioral change is always difficult to make especially when one does not know if other less notable strain-related differences also occur.

In the present context, these findings do indicate that if variation is found in the disposition of a pathway in a genetically mixed population, one cannot assume that this reflects loose developmental control. Exactly how the genetic differences result in the displaced pathways is not known. They could, for example, reflect subtle changes in the timing of cell migrations or of outgrowth of different axonal populations.

### b. What features may constitute a substrate across which axons grow in the developing brain?

Physical Features. Perhaps the best example is the optic pathway from the retina into the brain. This closely follows the course of the optic stalk, as indicated by Kuwabara (1975) and Lund and Bunt (1976). A number of teratogens, if injected into pregnant rodents or rabbits, result in aberrations of optic axon outgrowth. In some cases axons grow into the iris or lens, penetrating the retinal layers at various positions and growing between the receptor layers and pigment epithelium, or leaving the eye at the ora serrata on the peripheral border of the retina (Giroud et al., 1962; Tuchman-Duplessis and Mercier-Parot, 1960; Clavert, 1974). It is thought that this aberration may be due to early closure of the optic fissure which, in effect, deprives the growing axons of their normal tissue plane (Clavert 1974). Whether they innervate cells in the anomalous regions of termination has not been determined.

Coherent Cell Groups or Layers. Axons often change direction abruptly when they reach an interface between two coherent groups of cells. Indeed, the

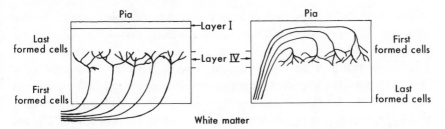

**Figure 9-4.** Diagram of normal (left) and reeler (right) cortex, showing the pattern of distribution of thalamic axons.

delineation of some regions of the brain, such as the thalamic nuclei, depends heavily on the fact that nerve fibers often tend to grow around, rather than through nuclear groups. While this may sometimes result from the different directions of axonal extension and somal migration, certain cell groups might actively inhibit the ingrowth of axons and thus provide a surface along which axon bundles may grow, or, alternatively, they might actually attract axon growth among them. A system in which this can be investigated is the thalamocortical pathway. In early development many of the axons of cells in the thalamus which are destined for the cerebral cortex can be seen growing laterally toward the cortical plate in a stalk which will ultimately become the internal capsule (Lund and Mustari, 1977). Once the axons reach the cortical plate, they spread along its deepest border, mingling with the cells of this region. The axons do not invade the full depth of the cortical plate until somewhat later. Thus, there appears to be something which holds them for a period of development in the deep layers.

A major aberration of the thalamocortical pathway is seen in the reeler mouse (Frost and Caviness, 1974). As indicated in Chapter 7, the cell migration patterns are disturbed so that the first cells to be generated are located most superficially, and the last are deepest. Interestingly, rather than following their usual course in the white matter to the region they will innervate, the thalamocortical axons grow straight through the cortex in fascicles toward the surface (Fig. 9-4). Having reached this layer, they then distribute in an appropriate manner with respect to region, topographic map, and layer (as defined by cell birthday rather than position). There seems to be a special relation between the growing axons and the first-formed cells of the cortical plate, which is prerequisite for orienting the axons' further distribution within the cortex.

**Other Neural Processes.** We have seen from in vitro studies that neurites growing in an appropriate medium will tend to follow other neurites and form a bundle of fibers. Since in the brain axons running from one region to another almost invariably do so in major bundles, one might expect that many axons, rather than following the guidance clues outlined above, would simply follow other axons that have already grown out. Two prerequisites are necessary for such a pattern: (1) a staggered timing in axon outgrowth, and (2) some pioneer axons that have the capability to respond to the region through which they grow in a way that later growing axons cannot. The work of Lopresti et al. (1973) and Bennett and Pettigrew (1974) shows both staggered growth and pioneer fibers in model systems. The first study concerned the growth of retinal axons into the cartridges of cells in the lamina of *Daphnia* and was mentioned earlier in this chapter. Bennett and Pettigrew (1974), studying the development of innervation of the rat's diaphragm, found that a single axon would grow along the length of one myotube, and that further axons would then follow the course of this exploratory axon. While similar mechanisms have not been clearly outlined in the central nervous system, studies by Goldberg and Coulombre (1972) on silver-stained whole mounts show that although the initial growing part of the chick optic nerve appears quite haphazard, the region behind this is organized into bundles, indicative of fasciculation. These studies demonstrated that fasciculation occurs before the first axons have formed synaptic contacts.

Many studies suggest that all the axons of a fiber tract do not grow out simultaneously, but rather sequentially. The patterns of outgrowth of retinal axons have already been described, and these certainly support the idea of a staggered growth. The situation is not so clear in structures such as the cerebral cortex, where cells contributing one pathway are generated at approximately the same time. In these cases, do axons all grow out at the same time, or are some delayed with respect to others? A few studies of corticofugal pathways in rats, indicate that such pathways occupy a smaller area early in development than they do later (e.g. Leong and Lund, 1973), but one is uncertain whether the size difference is due to fewer fibers or smaller fibers in the initial pathways. No study has yet shown in the developing vertebrate brain the existence of pioneer fibers which behave differently from axons appearing later and which lay down the pathways for later axons to follow.

### c. To what extent is axonal growth a passive process that follows extrinsic guides, and to what extent is it determined or directed by the cell body?

Examples may be presented to assert both extremes. Stell and Witkovsky (1973) plotted the course of the axons of a population of ganglion cells in the dogfish retina. The authors found that the axons leave the cell body in any position, grow for a short distance in any direction, and then turn to grow towards the optic disk. The implication is that external influences play a major part in directing the course followed by the growing axon.

A tissue culture study Sobkowicz et al. (1968) showed that different cell types may respond differently to the same environment. They grew slices of embryonic mouse spinal cord in culture and found that the axons of dorsal root

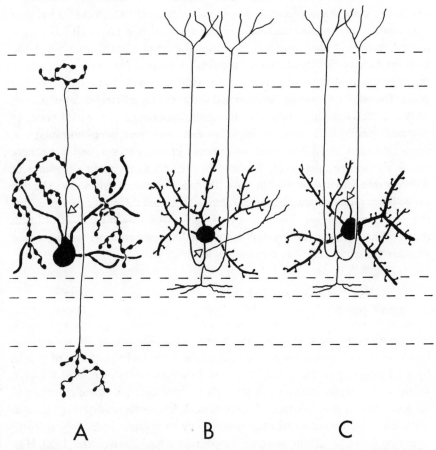

A                    B                    C

**Figure 9-5.** Two principal stellate cell types of a subdivision of layer IV (Layer IV C$\beta$) in the visual cortex of the monkey. (A) shows a cell with smooth dendrites and a beaded axon which arises (arrow) from the superficial side of the cell body and has a heavy local distribution. (B) and (C) show spiny stellate cells with axons (arrows) which distribute mainly in upper cortical layers, arising usually from the deeper aspect of the cell body (B), but occasionally (C) from the superficial side (courtesy J.S. Lund).

ganglion cells tend to grow in whorls around the explant, while ventral horn cells send their axons out radially from the explant. In the course of outgrowth, the axons of the two cell types cross one another, which suggests that they are not responding to different local microenvironments. Such differential axon development might be expected commonly in normal development. For example, in part of layer IV of the monkey's visual cortex there are two classes of stellate cell (Fig. 9-5; and Lund, 1973). They are generated over the same time period (Rakic, 1974) and their axons begin to develop about the same time (Lund et al., 1977). One class characteristically has a descending axon, which curves and then runs superficially to arborize finally in the upper cortical layers. It has only a limited number of local branches. Occasional examples are seen in which the axon leaves superficially but soon "corrects" its course (the predictable site of origin of the axon in these cells contrasts with the situation in retinal ganglion cells). The other cell type has an axon that leaves the superficial border of the cell body and then branches profusely, giving the appearance of a wilted plant, to distribute mainly in its layer of origin. In this case, there is no indication that temporal factors could determine the different growth patterns, and since these two cell types are intermixed, it would be difficult to invoke the presence of different microenvironments. The most likely explanation would seem to be that the cells are programmed early in development to emit different axon types that follow different courses. The recent observation that growth cones of different types of axons have differential lectin binding properties (Pfenninger and Maylie-Pfenninger, 1976) is consistent with this idea.

### d. Is there evidence for "attraction at a distance" in the ordering of axonal growth?

There is no clear evidence for axons being directed by distant cues in culture or in model experiments, and indeed the postulated mechanism of growth by local sampling of the environment and differential adhesiveness of the axon would tend to argue against it. In vivo, there are similarly no good examples to support "attraction at a distance." Several studies have shown that if optic axons are allowed to regenerate into the midbrain by an aberrant path, they will still innervate the optic tectum, forming a topographic map there (Gaze, 1959; Hibbard, 1967). This might reflect a distant attraction by the tectum, but on the other hand, physiological responses in the tectum are weak, which suggests that only a few of the available axons have reached it, and what happens to all the others remains unknown. The outcome could be explained by the axons growing in a random manner, where some would happen to reach the correct terminal area and form connections. The results of recent eye transplantation experiments in frogs are consistent with the view that the tectum does not exert a specific attraction at a distance to growing optic axons (Constantine-Paton and Capranica, 1975, 1976a,b, see below).

### e. Is there evidence to support the concept of tissue gradients as determinants of axon growth patterns?

This question has received considerable attention, particularly with respect to the direction of growth. It is well known that many fiber-tract systems contain axons running in both directions, and one possible explanation is that they are responding differently to a tissue gradient—a graded change in the amount of some material along the axis of growth.

The major interest in this question has come from work on the Mauthner cell of amphibians and fish by Oppenheimer (1942), Piatt (1943, 1944, 1947), Detwiler (1943, 1951), and Hibbard (1965). There is only one Mauthner cell on each side of the brain; because of its large size, a single Mauthner cell can be identified in normal histological preparations and its axon can be followed in silver-stained material for most of its course. Thus, the organization of a single cell in the intact animal can be studied in much the same detail as in tissue culture, without any special histological procedure.

If the segment of brainstem containing the Mauthner cells is transplanted from one embryo to the flank region of another, the axons still grow in an appropriate direction (Piatt, 1944), suggesting that they follow some intrinsic clue in the tissue and do not just follow other axons. In other studies segments of brainstem containing the Mauthner cell bodies have been transplanted from the brain of one embryo to another, in a normal or reversed orientation to a position ahead of the host's Mauthner cells. Transplants placed in normal orientation always have Mauthner axons which run caudally from the cell body and penetrate the host tissue, irrespective of the relative ages of host and donor animals. In the transplants between young embryos, axons will still grow caudally in normal fashion, even if the graft is placed in reverse orientation (Detwiler, 1951). In older embryos, if the graft is placed in a reverse orientation, the axons grow first rostrally to the apparent border of the graft and host and then usually turn to grow caudally (Hibbard, 1965). As summarized in Figure 9-6 these results suggest that a rostrocaudal gradient develops in the brainstem before transplantation, and that axon growth initially follows the gradient within the graft. When the axons meet the host, they come under the influence of the host gradient.

There are a number of problems, however, in drawing this conclusion from

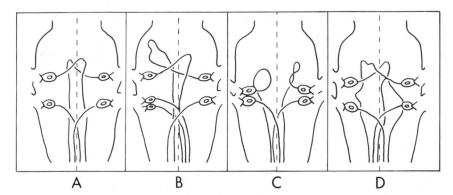

      A                B                C                D

**Figure 9-6.** Dorsal view of brainstem showing four examples of the course of Mauthner cell axons from a segment of brainstem transplanted with rostrocaudal axes reversed immediately rostral to the regular Mauthner cells (redrawn from Hibbard, 1965).

the experiments: (1) No one has defined exactly when the axon first grows out from the Mauthner cell body. The normal silver stains used would not be adequate for showing the very first axons. The apparent reversal of growth seen by Hibbard after rotation in older embryos could be explained if the axons have already grown caudally at the time of transplantation. When the segment of brainstem is removed, the axons are cut and continue a caudal growth, following host axon systems as they regrow back through the transplant. (2) We know nothing about the short-term changes—at what point the Mauthner axons change their direction of growth, and when the host fiber pathways along which they grow penetrate the graft. (3) Hibbard reported 1 of 12 cases of rotated transplants in which the axons did not reverse, but continued growing rostrally. Further study of such exceptions could be very illuminating. (4) It is something of a puzzle, if indeed the graft has a rotated gradient, that host axons or the Mauthner axons originating in the graft grow back through the graft in the rotated condition.

These questions are raised because this series of experiments is often quoted as evidence for the presence of tissue gradients. In fact, the results could be explained largely by the phenomenon of fasciculation—of relating to a host fiber pathway of defined orientation—although the experiment in which the brainstem was grown outside the nervous system would tend to argue against this explanation (presuming again that axons had not grown out at the time of transplantation). The crucial question concerns the time that the axons first grow out relative to the time of transplantation. These experiments are of further interest in relation to the question of laterality, which will be dealt with in the next chapter.

A similar type of study (Constantine-Paton and Capranica, 1975, 1976a,b) involved transplantation of the eye cup together with some forebrain into the brainstem in the position of the primordial ear vesicles of tadpoles of *Rana pipiens* (stage 16–18, Shumway). The animals survived beyond metamorphosis, and the projection of the misplaced eye was tested by injecting it with $^3$H-proline and by recording electrophysiologically. It was found that most axons ran caudally into the spinal cord in a posterolateral position. In one case, they ran rostrally for a short distance toward the optic tectum, but rather than innervating it, curved caudally. Again, the results argue against attraction at a distance, but whether the axons are following a gradient or simply growing along a pre-existing pathway is not clear. We still do not know, of course, what would determine the pre-existing pathway.

## Summary

Axons in vivo, like neurites in vitro, can follow physical cues and guides provided by nonneuronal elements. Their response to their environment, however, is not totally mechanical: different neurons appear to respond to the same environment in different ways. The capacity to form bundles and, subsequently, tracts is an important attribute of axons in intact brains. This capacity would appear to result, in part, from a mechanism of fiber following, much like that indicated by Nakai (1960, see Chap. 8), but it may also derive from an ability of groups of neurons to prevent the ingrowth of axons into the group during certain stages of development. Such behavior would yield the characteristic pattern of tracts skirting round nuclear groups. There is no clear evidence that attraction at a distance directs nerve growth, and, while we have no good explanation as to why axons should grow in one direction rather than the opposite, the concept that gradients of some kind may be involved suffers from a lack of experimental support. It is apparent that quite subtle changes in axonal patterns have a genetic correlate, although it is unclear at present what are the primary targets of the genes in question, and therefore exactly how they modify axonal growth patterns.

# 10

# LATERALITY OF PROJECTIONS

As we have seen, the brain is divided into right and left halves, each half dealing with sensory and motor events occurring in relation to one or the other half of the body. In mammals, the forebrain and upper brainstem of one side are concerned largely with events happening on the opposite side, while each side of the lower brainstem and spinal cord is concerned with events occurring on the same side of the body. (Fig. 10-1). The reason for this curious reversal of representation between upper and lower parts of the central nervous system is unclear, but one important consequence is that many pathways running to and from the forebrain have to cross the midline. The crossing occurs as an interweaving or decussation of the axons arising from similar regions on each side of the brain. Often the decussation is complete and all the axons from one side cross to terminate on the other; the corticospinal (or pyramidal) tract of many mammals is one example of this. Less frequently, the decussation is partial, some axons crossing and others remaining uncrossed.

The problems associated with laterality of termination have attracted considerable interest in recent years because they represent a situation in the intact brain where factors influencing the course of axon growth may be investigated at a site distant from that at which the axon forms synapses. Of particular relevance are the findings that a variety of factors can influence the laterality of projection, either disturbing the logical order at a decussation or causing pathways to cross the midline by anomalous routes.

Apart from providing insight into the determinants of laterality in normal circumstances, these studies also address the question of how axons behave if they are directed to the inappropriate side. Do they end in the usual regions and are they disposed within these regions in a normal fashion? The answer to the first question is yes, to the second, frequently. There are two principles of relevance here. The first is that axons from each side are often segregated into

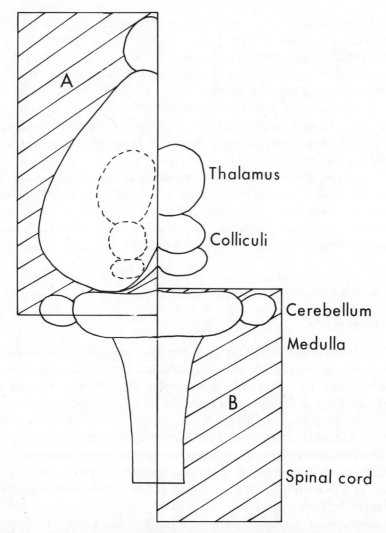

**Figure 10-1.** Dorsal view of rat brain illustrating general plan of laterality of interconnections. Regions within A for the most part relate to the contralateral body and space, while regions within B are more concerned with ipsilateral events. Connections between two different regions within either A or B are usually ipsilateral; connections between a region in A and one in B are usually crossed.

separate layers or laminae in the nucleus of termination. Thus, if axons were directed inappropriately to the ipsilateral side, would they occupy layers usually reserved for ipsilateral axons, or would they still occupy their normal layers although on the wrong side of the brain? The second principle is that many regions have a map of some aspect of the external world, such as of the visual field, impressed upon them. In projections from one region to another the detail of that map is retained. But if a nerve bundle is distributed to the wrong side of the brain, will it map as if on its normal side or will it attempt to compensate for the functional disorganization resulting from its projection to the wrong side? This question assumes particular significance in the visual system, where the logic of the partial decussation relates to the degree of overlap of visual fields seen by each eye and to the normal crossed and uncrossed axons mapping relative to one another with respect to visual field positions.

In this chapter we will examine the question of laterality in some detail. The chapter is divided into two parts; the first deals with normal and abnormal patterns of decussation and the second with abnormal crossing of axons not associated with decussations. Summaries are provided at the end of each part and a general summary is given at the end of the chapter.

## Normal and Abnormal Decussation Patterns

### Mauthner cells

As was indicated in the Chapter 9, Mauthner cells are of interest because there is only one on each side of the brainstem and it emits a large axon which, even in normal material, can be seen crossing the midline to innervate the contralateral side of the spinal cord in a series of branches. Before crossing the midline, the axon traverses a tract of descending axons of various origins, termed the medial longitudinal fasciculus. Once over the midline, it joins the same tract on the other side to continue its caudal course. The pattern of decussation occurs in an apparently normal manner even if the segment of medulla containing the Mauthner cells is extirpated at an early stage before differentiation and is transplanted to the flank region (Piatt, 1944).

Aberration of the decussation pattern was found in a study of embryonic salamanders (*Pleurodeles*) by Hibbard (1965) which was mentioned in Chapter 9. In this study a segment of medulla containing Mauthner cells was transplanted from one animal into a position just rostral to the normal pair of cells in another. If transplanted in a normal orientation, most axons cross normally, but a few either fail to cross or project bilaterally. If transplanted with the rostrocaudal axis reversed, two-thirds of the axons decussate even though this means ultimately that the cell formerly on the left side will now innervate the left rather than the right side of the tail. In other cases, one axon remains uncrossed, one branches and projects bilaterally, or both remain uncrossed (Fig. 10-2). The last pattern commonly occurs if the transplanted cells are situated particularly close to the

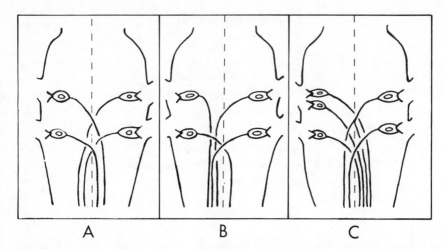

**Figure 10-2.** Decussation patterns of Mauthner axons from brainstem segments transplanted in normal orientation rostral to the regular Mauthner cells. Note in C, a duplication of the Mauthner cell on one side (redrawn from Hibbard, 1965).

host cells, but the significance of this observation is far from clear. In one bizarre case axons run rostrally to decussate in the optic chiasm and posterior commissure, respectively. These results demonstrate a number of points:

1. There is some lability in the pattern of decussation of single axons after transplantation.

2. In these circumstances decussation is not the result of an interaction at the midline between axons from each side (since the axon originating on one side may cross while the other does not).

3. A single axon may branch to supply both sides of the brainstem.

4. Piatt's contention that the property to decussate is intrinsic in the cell seems to be supported by the results.

5. The power to decussate is greater than any intrinsic left-rightness.

It will be of some interest to bear these points in mind as we study the aberrations that can occur in decussations involving whole tracts containing large numbers of axons.

### The pyramidal decussation

The direct pathway from the sensorimotor cortex to the spinal cord, the corticospinal tract, decussates with its partner from the other side in the lower medulla. The decussation may be complete or partial, depending on the animal. Investigation of the gross extent of crossing in humans shows that there is considerable variation, which in some cases may be a secondary effect of a

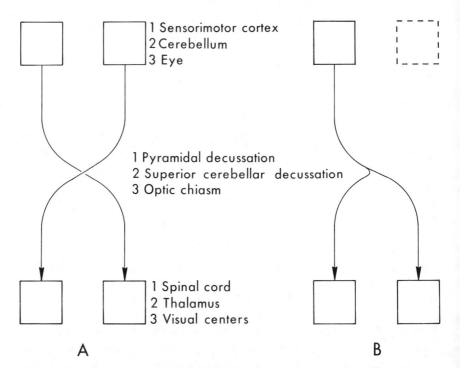

**Figure 10-3.** Effect of removal of the input from one side at an appropriate stage of development in three different systems on the distribution of axons from the other side. (A) normal, (B) unilaterally lesioned.

cerebral lesion (Lassek, 1954). Experimental studies have been undertaken on rats, which normally have a totally crossed pyramidal decussation. After unilateral sensorimotor cortical lesions (Leong and Lund, 1973; Leong, 1976a,b) or forebrain lesions (Hicks and d'Amato, 1970) are made in young rats, the corticospinal tract from the intact cortex projects bilaterally, the abnormal ipsilateral fibers being first identified at the pyramidal decussation (Fig. 10-3). If the sensorimotor cortical lesion is subtotal, the aberrant pathway from the remaining cortex distributes only to the region of the spinal cord to which the lesioned cortex should have projected. In all cases the aberrant ipsilateral pathway distributes into the same laminae as the crossed pathway and forms synapses there.

### Superior cerebellar decussation

The dentate nucleus and nucleus interpositus (two of the deep nuclei of the cerebellum) project to the contralateral midbrain and forebrain. The fibers reach the opposite side by crossing in the superior cerebellar decussation to innervate, among other nuclei, the red nucleus and nucleus ventralis lateralis of the thalamus. If the cerebellar nuclei of one side are removed at birth, those of the remaining side, when tested six weeks later, project axons bilaterally to the nuclei mentioned above (Fig. 10-3) (Lim and Leong, 1975). A sparse ascending projection from the deep cerebellar nuclei is present at birth and it is presumed that more fibers grow through it postnatally, and that these growing fibers may be the ones which project aberrantly.

### Optic chiasm

This decussation is particularly interesting because it is frequently partial, the pattern of distribution of crossed and uncrossed axons relating to the degree of overlap of the visual fields seen by each eye. In addition, variations from this pattern are found in normal animals, in genetic variants, and after lesions made early in development. For these reasons it deserves close study.

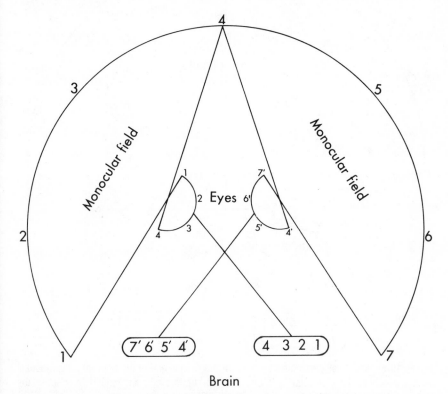

**Figure 10-4.** Visual field, its representation on the retinas and in the brain of an animal in which each eye surveys a separate part of the field.

### a. The logic of decussation patterns of optic axons

In order to understand the variations we must first examine the logic of the idealized optic decussation. The underlying requirement is that one side of the brain should receive information from the opposite visual field, ranging from points directly in front of the animal (the vertical midline) to the far contralateral periphery. If the eyes are placed on the side of the head and each eye sees only the contralateral field, the visual world is represented in the brain as shown in Figure 10-4. Since the lens inverts the visual image, the vertical midline is represented in the most temporal (or posterior) retina, while the periphery is represented nasally (or anteriorly) on the retina. The whole optic pathway crosses

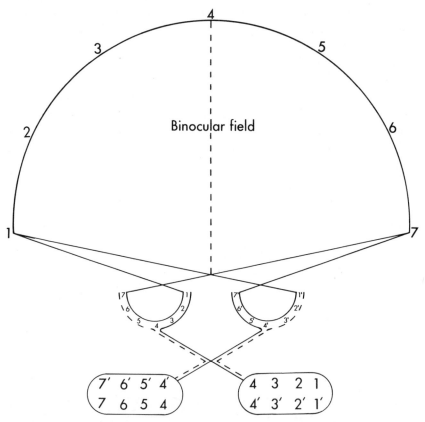

**Figure 10-5.** Visual field, its representation on the retinas and in the brain of a hypothetical animal in which there is total binocular overlap. In this and subsequent diagrams, the vertical midline is indicated by a dotted line.

at the optic chiasm so that the whole retina is represented on a contralateral brain region. In this case the contralateral *visual* field and contralateral *retinal* field are confluent. Such a pattern is found in fish.

The opposite situation is illustrated in Figure 10-5. Here both eyes are frontally placed and see exactly the same visual field. Since all visual information to the right of the vertical midline projects to the left side of the brain and vice versa, the visual afferents are divided so that the input received by the nasal retina (temporal field) projects contralaterally while that received by the temporal retina (nasal field) projects ipsilaterally. This means again that the vertical midline is represented at one edge of the central brain region, but here this line receives input from the midpoint of retinas rather than from the temporal border of the contralateral retina. Further, both eyes contribute to the central visual map on each side of the brain, and their axons are distributed relative to one another in a coherent fashion with respect to points on the visual field. This situation never occurs in real life since in no animal do both eyes survey exactly overlapping areas of space. Instead the overlap of the visual fields (the binocular segment) is partial, leaving a nonoverlapping monocular segment (see Figure 10-6) in which the most temporal part of the visual field on each side is seen only by the most nasal part of each retina, and is relayed only contralaterally in the brain. The relative size of binocular and monocular segments depends on the degree of overlap of the visual fields, the binocular field being large in primates and small in rodents, for example. One point of interest in this system of overlap of central projections is that while the logic of its organization seems related to functional considerations, i.e., the "fusion" of the images relayed by the two eyes, it nevertheless develops before the eyes ever receive a visual input. Thus, in the rat (Lund et al., 1976), cat (Richards and Kalil, 1974), and monkey (Rakic, 1976) a significant uncrossed optic pathway develops before the eyes open and is not disturbed if the eyelids are kept closed. In contrast, it may be added that this does not seem to be the case in the frog. In tadpoles, the eyes are laterally placed in the head (Beazley et al., 1972) and while crossed optic connections to the thalamus are found, no uncrossed fibers have been demonstrated (Currie and Cowan, 1974). By metamorphosis the eyes have changed their position on the head to provide an extensive binocular field. At this time a clearcut uncrossed optic pathway develops. This would seem to reflect an extreme specialization however, rather than a primitive condition.

Throughout this discussion two themes will keep recurring—the mechanisms involved in defining a decussation point; and the "conflict" between the strategy involved in mapping the whole retina on the contralateral half of the brain and the strategy involved in mapping the contralateral visual field.

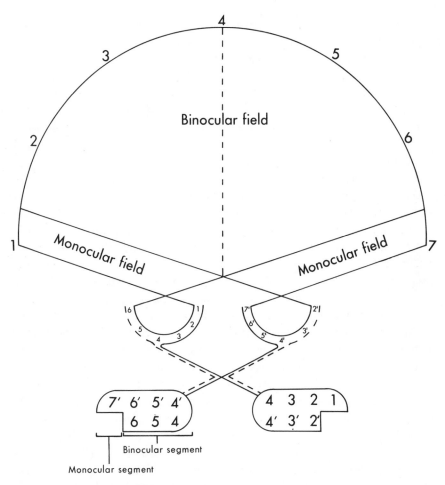

**Figure 10-6.** Visual field, its representation on the retinas and in the brain of an animal with partial binocular overlap.

### b. Retinal cell types in the cat

In a series of physiological studies classifying retinal ganglion cells in the cat, three distinct classes have been recognized. These are termed W, X, and Y (see Stone and Fukuda, 1974) or sluggish, brisk sustained, and brisk transient (Cleland and Levick, 1974a,b). There also appear to be morphologically distinct classes, which correlate with the physiologically identified categories (i.e., $\gamma$ for W cells, $\beta$ for X cells, and $\alpha$ for Y cells—Boycott and Wässle, 1974). The axons

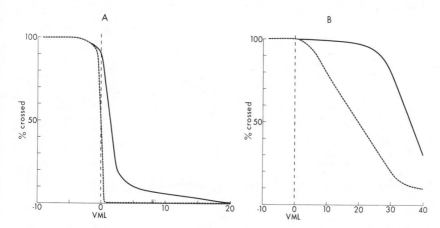

**Figure 10-7.** Laterality of projection of X (brisk sustained) and Y (brisk transient) ganglion cells of the cat retina situated from − 10° in the nasal retina to + 40° in the temporal retina (VML - vertical midline) X-dotted line, Y - continous line. (A) Normal cat, (B) Siamese cat. (Redrawn from Kirk et al., 1976a, Kirk, 1976.)

of the W, X, and Y cells show progressively higher conduction velocities and, therefore, presumably have progressively larger diameters. The Y cells are remarkable in that they have large cell bodies and occur in relatively small numbers across the retina. The W cell population (with the smallest cell bodies) has been divided by Stone and co-authors into two classes (phasic and tonic) although Cleland and Levick (1974b) have added more groups.

Of particular interest is the pattern of central projection for each type of retinal ganglion cell. Studies by Stone and Fukuda (1974) and more recently by Kirk et al. (1976a,b) show that the decussation point varies according to the class. In an overlap zone of about 1° across the vertical midline, X cells project to one or other side of the brain; those in the temporal retina project ipsilaterally while those in the nasal retina project contralaterally (Fig. 10-7A). Y cells, by contrast, show a much broader overlap zone of about 16.5° (Fig. 10-7A); some cells situated temporal to the representation of the vertical midline project contralaterally rather than, as might be expected, ipsilaterally. Of the various categories of sluggish cells (W cells), some are found which project contralaterally up to 10° (sustained) or more (other classes) into the temporal retina.

Thus, while cells in the nasal retina always project contralaterally as expected, those in the temporal retina may not project ipsilaterally and the pattern may vary from one cell type to another.

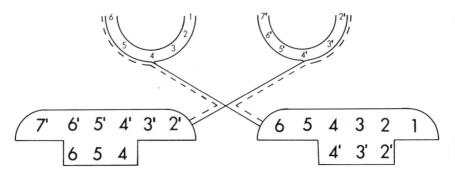

**Figure 10-8.** Pattern of mapping in which the temporal retina projects bilaterally to the brain.

## c. Patterns of central projections

Having found that different ganglion cells (in the cat at least) show different patterns of laterality of projection, it is obviously of interest to know how their axons are distributed in the brain and how the retina is mapped on central regions. There are three possible strategies of mapping:

1. The brain region has a map of the whole contralateral retina, as in Figure 10-4. In this case, there are no uncrossed axons. This may occur even though the visual fields overlap.

2. The brain region has a map of the contralateral visual field, as in Figure 10-5. In this case it receives axons from the contralateral eye nasal to the vertical midline and from the ipsilateral eye temporal to the vertical midline.

3. The brain region has a map of a large part or all of the contralateral retina. In addition, however, it receives uncrossed axons from the temporal part of the ipsilateral eye. This is illustrated in Figure 10-8. In a sense it represents a half-way position between (1) and (2), mapping both retinal fields and visual fields in the same area.

The particular strategy involved varies between animals and from one region to another in the same animal. Table 10-1 provides a summary indicating which of the three strategies is used in particular circumstances. These different patterns seem related to the distribution of different retinal cell types, since in cats X cells project to the main laminae of the lateral geniculate nucleus and W cells project to the superior colliculus; while Y cells project to both (Fukuda and Stone, 1974). It is uncertain, however, whether the differential mapping strategy is determined by the properties of the retina, by the brain region concerned, or by patterns of interaction at the optic chiasm.

**Table 10-1.**   Pattern of retinal map on subcortical visual centers in a selection of animals. See Lund (1975) for source references; squirrel geniculate described by Kaas et al. (1972).

| | LATERAL GENICULATE NUCLEUS OR HOMOLOGUE | | SUPERIOR COLLICULUS OR OPTIC TECTUM |
|---|---|---|---|
| Birds | 1 | | 1 |
| Goldfish | 2 * | | 1 |
| Frog | 2 | (or ? 3) | 1 |
| Rat | 3 | | 3 |
| Squirrel | 2 | | 3 |
| Cat (regular) | 2 | (laminae A, A1) | 3 (partial) |
| | 3 | (MIN lamina) | |
| Cat (Siamese) | 3 | (laminae A, A1) | 3 (more complete) |
| Monkey | 2 | | 2 |

*For explanation of 1, 2, and 3, see text p. 148.

## d. Albinism

Abnormalities associated with albinism are relevant to understanding how the visual input is mapped on the brain. In albino animals and other strains having reduced eye pigmentation, the ipsilateral pathway is reduced in both size and distribution. This reduction was first found in rat (Lund, 1965) and has since been demonstrated in mouse, rabbit, guinea pig, hamster, cat, and mink (see Lund, 1975 for separate references). EEG recordings in albino humans show an asymmetry in the visual evoked response that could be the result of a reduced ipsilateral pathway (Creel et al., 1974), and studies on the anatomical organization of the lateral geniculate nucleus of albino humans indicate highly anomalous cell patterning, a feature that has been correlated with abnormal decussation of axons at the optic chiasm in other animals (Guillery et al., 1975). Detailed investigation of some of these animals shows that the absolute reduction in size of the uncrossed pathway is not the result of a general reduction in the total size of the central optic pathway, both crossed and uncrossed, but rather is the result of aberrant distribution at the optic chiasm of axons originating from cells in the temporal retina.

This distribution has been studied particularly well in the Siamese cat (Kirk, 1976) where, at least for the X and Y categories, there is a substantial increase in the proportion of cells in the temporal retina which project contralaterally (Fig. 10-7B). In the region 0° to 10° from the vertical midline, about 95% of cells project contralaterally; this proportion drops continuously and gradually over more than 40° of representation. What happens when the extra crossed (and reduced uncrossed) temporal axons distribute in the central visual nuclei, in

The Development of Neural Processes

particular the lateral geniculate nucleus? To explore this question, it is necessary first to review the normal organization of the nucleus.

In an ordinary cat the outer laminae of the lateral geniculate nucleus, A and A1, receive crossed and uncrossed optic inputs, respectively (see Fig. 10-9). Lamina A receives an input from the retina nasal to the region of the vertical midline representation on the contralateral eye; A1 receives an input from the temporal retina of the ipsilateral eye. The deeper laminae of the geniculate have

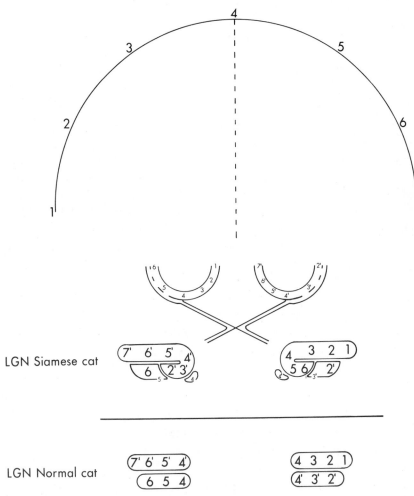

LGN Siamese cat

LGN Normal cat

**Figure 10-9.** The map of the visual field on the lateral geniculate nucleus in Siamese and normal cats.

similar patterns but are less well differentiated and will not be discussed further here. The region of the vertical midline (position 4 in Fig. 10-9) is represented medially on the geniculate and the peripheral visual field is represented laterally. Laminae A and A1 are lined up such that if an electrode is advanced perpendicularly through them, it will be responsive to the same position in the visual field, although as it passes from A to A1 the eye transmitting the image will change. The lateral part of lamina A has no matching A1 and corresponds to the projection of the most nasal part of the retina—the monocular segment.

Guillery (Guillery, 1969, 1974; Guillery and Kaas, 1971) found that in Siamese cats there is a patchiness in the projection to lamina A1, and he suggested that this is due to a strip of temporal retina providing an aberrant contralateral projection (in contrast to strips on either side of it which project ipsilaterally as usual). Nevertheless, the aberrant strip projects to the lamina appropriate for the temporal retina. This interpretation is not in accord with the observations of Kirk (1976) and has in fact been questioned in a recent review (Lund, 1975). The alternative possibility is raised by physiological studies that show continuity of the visual field map from lamina A to an abnormal segment (interpolated between lamina A and the remnants of lamina A1) of the projection in the region normally occupied by lamina A1. The map crosses the vertical midline at the medial border of the nucleus and continues laterally in the abnormal segment as more temporal points on the retina are stimulated (Fig. 10-9). Thus, the abnormal crossed temporal axons show two properties: (1) the axons maintain a continuous map with the regular crossed pathway and (2) they terminate on cells to which they would normally project, albeit on the wrong side of the brain. The 5% or so of the temporal cells that project from the more central retina to the ipsilateral side of the brain presumably project to the medial normal segment of lamina A1, while the increasing proportion of ipsilaterally projecting axons more laterally in the temporal retina occupy the lateral normal segment. The separation of the two segments of lamina A1 receiving ipsilateral axons may in fact be secondary to the increased crossed projection rather than indicative of a discontinuity in the uncrossed pathway as has been implied. Too few recordings have been made from this layer to resolve the question.

Superficially the optic projections of albino rats (Lund et al., 1974), and rabbits (Sanderson, 1975) appear to behave very much like those of Siamese cats, having a discontinuity in uncrossed projection to the homologue of the A1 lamina. More detailed study suggests that they differ in two ways.

    1. There is a reduction in the overall size of the block of uncrossed pathway and little indication of any discontinuity within it.

    2. There is duplication of part of the uncrossed pathway (possibly of

the total visual map), which is seen as a patch situated laterally in the visual map within the monocular segment. Single retinal lesions in the rat can produce degeneration both in the normal part and in the lateral patch. This applies to both crossed and uncrossed axons. Again, the optic fibers seem to be reflecting a conflict as to whether they are mapping with respect to the whole contralateral retina or to the visual field; in the confusion they do both.

This confusion of two strategies of mapping could underlie the anomalies of central mapping in albino animals, but what the primary control may be still eludes us. Clearly, it is related to the pigmentation pattern of the pigment epithelium. Thus the dark eyes of heterochromic rats (rats with one albino-like and one pigmented eye) still show an albino pattern in the uncrossed optic pathway, and examination of the retina shows that while the choroid has melanosomes, the pigment epithelium is nonpigmented (Wise and Lund, 1976). In this strain as well as in some strains of mink (Sanderson et al., 1974), the defect results from the effect of gene combinations exclusive of the albino gene which produce albino-like conditions. This indicates, therefore, that the anomaly is not due to a gene closely linked to the albino locus, but rather is associated with the tyrosinase deficiency attributable to the albino gene and to the other gene combinations. There are two possible relations between pigmentation and patterns of laterality of central projections. One is that the pigmentation defect of the retina may reflect an aberration in the metabolism of the pigment epithelial cells. These cells might have a controlling influence on the adjacent dividing neural retina, and if they are abnormal the pattern of central distribution of the retinal ganglion cells might be disturbed. Alternatively, the pigmentation defect may reflect a more general aberration involving a wider variety of tissues. The close relation of the pigment epithelium to the neural retina may be irrelevant to the defect, and some other cell population (even the aberrantly projecting neurons themselves) may be the source of the chiasmatic variation.

A further aspect of the albino problem is the question of what happens to second or third-order connections—lateral geniculate nucleus to visual cortex or visual callosal connections—when a primary deficit has occurred in the retino-geniculate pathway. Do these pathways show aberrations? Is the geniculate a passive transmitter of anomalous information or does it compensate for the changes? Does the cortex reorganize itself with respect to the geniculocortical map? These questions have been explored only in experiments on the Siamese cat. The results are intriguing, but since they are more directly relevant to the question of mapping than of laterality, they will be described in Chapter 14.

## e. Effects of unilateral eye damage

Unilateral eye removal at an appropriate stage in development results in an enlarged ipsilateral pathway from the remaining eye to the subcortical visual centers (Figs. 10-3, 10-10). Experimental analysis of this phenomenon will be described first for the rat and then briefly for the other animals that have been studied. Since the optic decussation is rarely complete in mammals, the enlarged ipsilateral projection could be due either to expansion of the normal ipsilateral pathway (each axon branching to supply more terminals) or to additional axons distributed ipsilaterally at the optic chiasm. In nuclei like the suprachiasmatic

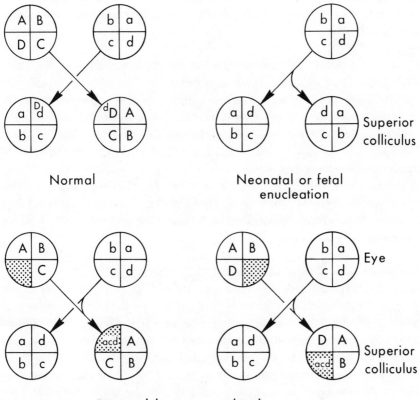

Retinal lesions at birth

**Figure 10-10.** The projection of the retinas on the superior colliculi in normal rats and in rats in which one eye was removed or lesioned (dotted area) early in development.

nucleus (Stanfield and Cowan, 1976) where there is complete overlap in the distribution of crossed and uncrossed projections, separation of one or another mechanism is difficult. This same problem is faced to some extent in the lateral geniculate nucleus, since although the crossed and uncrossed axons may be segregated in their terminal distribution, sprouting of uncrossed axons into laminae normally occupied by crossed axons might still occur. The superior colliculus is possibly the most satisfactory region in which to study this matter because the normal uncrossed pathway is small, especially in albino animals; it distributes sparsely only to the more anterior part of one layer, the stratum opticum, in contrast to both the regular crossed pathway and the enlarged ipsilateral pathway, which end predominantly in the more superficial layers. In addition, the retinal map in the colliculus is much easier to study than that in any other visual center. The results may best be summarized as a series of questions and answers.

### At What Time Are Lesions Effective in Producing an Aberrant Pathway?
Studies by Land et al. (1976) show that in rats if one eye cup is prevented from developing by injecting trypan blue at eight days of gestation and as a result an optic nerve fails to develop on that side, a pronounced aberrant ipsilateral pathway results. If eye removals are performed between five days before birth and birth, the aberrant ipsilateral pathway is still prominent, but of reduced density compared with that shown after total failure of one optic nerve to develop (Lund and Lund, 1971a; Lund and Miller 1975). After birth, the size and extent of the pathway become progressively diminished up to the tenth postnatal day, when enucleation has no major effect on the size of the uncrossed pathway of the remaining eye (Lund et al., 1973).

### What Is the State of Development of the Pathway at the Time of the Lesion?
The first axons from the retina reach the optic chiasm on fetal day 16 and grow across the superior colliculus by fetal day 17, five days before birth in rats. Since retinal ganglion cells are generated between fetal days 12–17 in rats and optic axons grow at about 80–100 $\mu$m/hour, it may be expected that the last optic axons reach the superior colliculus sometime between five and ten days after birth. Thus, the aberrant pathway can be induced by lesions made at any time from before the optic pathway has developed until the last retinal axons have reached their destinations in the brain.

### What Is the Likely Mechanism for Stimulating Aberrant Connections?
It has been shown recently (Cunningham, 1976) that in rats the normal ipsilateral optic

pathway arises as branches of contralaterally projecting axons. The aberrant pathway after enucleation arises similarly. The developmental studies show that at the time of the initial lesion, either at birth or before, the retinotectal pathway is predominantly a crossed pathway (Lund and Bunt, 1976; Lund et al., 1976). As such, the possibility that the aberrant pathway is due to failure of retraction of a system that originally projected bilaterally does not seem to be tenable. The study by Land et al. (1976) shows that the aberrant pathway can be induced without degeneration at the optic chiasm. Thus, it would appear to result from a failure of interaction between axons at the optic chiasm rather than from growing axons following the products of degeneration.

**Where in the Retina Does the Aberrant Pathway Originate?** After removing one eye, the aberrant ipsilateral pathway from the remaining eye comes not just from the area of retina lying temporal to the vertical midline representation, although the heaviest projection arises from this region. Axons also arise from nasal retina (Fig. 10-10), the sparsest projection coming from the upper nasal retina (Lund et al., 1973; Lund and Lund, 1976).

**Can Abnormal Ipsilateral Axonal Growth Be Produced by Partial Retinal Lesions Made in One Eye at Birth?** Small retinal lesions made at birth (Lund and Lund, 1973, 1976) or before (Lund and Miller, 1975) result in a gap in the optic projection to the contralateral superior colliculus in a region corresponding approximately to the area to which the lesioned patch of retina should have projected. This gap is filled exactly by an aberrant ipsilateral projection from the unlesioned eye (Figs. 10-10, 10-11) (Correspondence of the two projections can be tested by injecting one eye with $^3$H-amino acid and lesioning or removing the other eye.) The density of the projection is much higher per unit area than would have been the case had the whole eye been removed at birth. The projection arises from a large part of the unlesioned eye (as indicated in Fig. 10-10) and not just from the area homotypic to the original lesion. Whether it derives from branches of axons that project contralaterally is uncertain, although this might be expected. As such, it would present an interesting situation in which a single axon obeys a different set of cues in mapping its projection on each side of the brain.

The evidence of increased relative density of projection after small compared with total retinal lesions leads to the possiblity that the same number of retinal cells in the intact eye project ipsilaterally after small lesions as project after total eye removal, and that in both cases the axons distribute into the denervated region—in the case of small lesions, a small area of the colliculus.

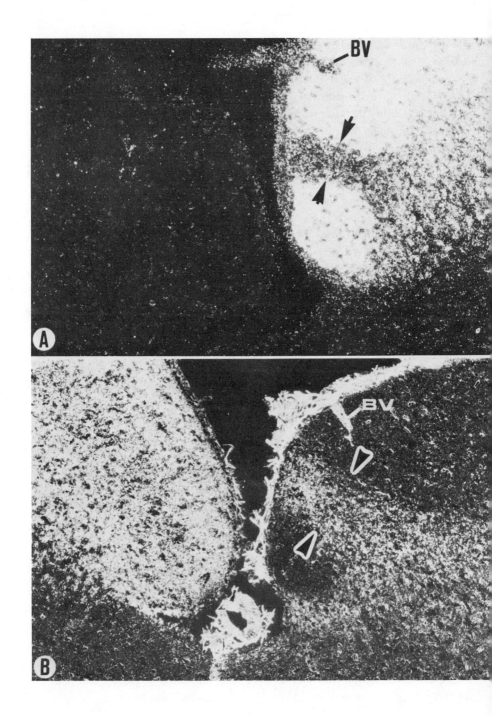

The implications of these results for theories of mapping between regions will be discussed further in Chapter 14.

**Can These Patterns Be Shown in Other Animals?** The first study in which an aberrant ipsilateral optic pathway was suggested was performed by Ferreira-Berutti (1951) in the chick. He prevented one optic nerve from growing to the optic chiasm at three days of incubation when the optic chiasm is just forming and found four to five days later that the remaining optic nerve showed an ipsilateral branch at the optic chiasm. Since birds normally show no ipsilateral optic pathway at all, this was quite remarkable. Raffin and Repérant (1975) have confirmed this finding by enucleating chicks at the 11-somite stage and testing the projection of the remaining eye five days after hatching. They showed, in addition, that the ipsilateral axons end in the same layers of the optic tectum as do the crossed axons, and this is also true for the other subcortical visual centers. The ipsilateral optic pathway in hamsters also expands after enucleation at birth but does not appear to occupy the caudal half of the tectum (Finlay et al., 1977). Enucleation of rabbits at birth produces a very sparse additional ipsilateral pathway from the remaining eye; when tested three months later, physiological responses to visual stimulation cannot be elicited in the ipsilateral colliculus (Chow et al., 1973). The anatomical results resemble those in rats enucleated at five days after birth, and this is not altogether surprising since rabbits are more mature at birth than rats. One would predict that the results in rabbits enucleated before birth would be similar to those in rats enucleated at birth.

### Summary: Decussation patterns

Decussations are of particular interest because they seem to be places where interactions between axons occur far from the regions in which those axons form

---

**Figure 10-11.** Adjacent transverse sections close to the midline of left and right superior colliculus of a rat. The left eye has been lesioned at birth and injected with proline one day prior to fixation; the right eye was removed four days before fixation. (A) dark-field view of autoradiogram showing the projection of the left eye with a gap corresponding to the projection region of the lesion (between arrows) ×150. (B) dark-field view of degeneration showing the projection of the right eye ×150. Note the anomalous projection to the right colliculus (between arrows) filling the gap in (A). A blood vessel (BV) is signified for reference. This corresponds to the experimental plan of Fig. 10-10, lower left (from Lund and Lund, 1976, with permission).

synapses. This allows dissociation of interactions occurring between axons from those occurring between axon terminals and their target cells. Another important feature of decussations is that the course taken by an axon through a decussation can vary from the predicted pattern as a result of a variety of factors, including the particular cell type involved, genetic variations, and lesions made early in development. Further, an axon has only three possible choices at a decussation—to cross, not to cross, or to do both by branching. Though simple systems can rarely be extricated from the complex developing brain, decussations may be used to ask a number of questions about the potential for interaction between axon populations, the factors that guide axons along their correct course and the sequelae of abnormal behaviors at one stage in development. Concerted attention has been given to the optic chiasm because it is a partial decussation and a large volume of control data is available to define the normal patterns of distribution of optic axons. In addition, this is the one place where a variety of factors have been identified which disturb decussation patterns. More limited studies of other systems suggest that the results obtained from the visual system are generally applicable.

One point which emerges from these studies is that although axons may end on the "wrong" side of the brain in particular circumstances, they tend to distribute as if no error had been made at the optic chiasm. Thus, after unilateral enucleation the aberrant ipsilateral retinotectal pathway ends in the layers to which the crossed retinotectal axons distribute and forms a mirror topographic map. Similarly, the crossed temporal axons of ordinary cats distribute in a logical fashion in the contralateral superior colliculus and the aberrant crossed temporal axons of Siamese cats form a logical order in the lateral geniculate nucleus. Some adjustment may occur centrally, however, possibly as a secondary event in response to the altered size of the aberrant pathways. This is particularly noticeable in the albino rat retinogeniculate pathway, in the retinotectal pathway of rats with small retinal lesions made at birth, and in the secondary visual pathways of Siamese cats. The last two will be discussed in more detail later. Nevertheless, the main conclusion is that left-rightness is quite unimportant to an axon. This conclusion has also been reached in several studies of nonmammalian vertebrates in which optic nerves have been uncrossed at the chiasm or eyes transplanted from one side to the other (Sperry, 1945; Straznicky et al., 1971; Stone and Zaur, 1940; Beazley, 1975). In each case of the retinal quadrants project to their tectal quadrants, irrespective of laterality, and this leads to aberrant visuomotor responses which are never corrected.

While we still know nothing about why decussation patterns develop, we can, on the basis of the experiments described here, make some general comments regarding factors that influence such patterns.

These are outlined mainly in relation to the visual system:

1. Although relevant to visual function, the pattern of decussation is established before the animal ever sees a visual image.

2. The position a cell occupies in the retina is not the only factor that determines whether its axon will or will not cross in the optic chiasm. Also of importance is the cell type involved and where it projects in the brain.

3. The decussation pattern is not immutably fixed but can be altered by total or partial lesions of one eye. This aberration can be demonstrated in other decussating systems too, and implies that interactions between axons from each side at the decussation are instrumental in defining the degree of decussation. The studies of Mauthner axons do suggest, however, that other factors may also be involved since the axon of one side can cross without ever meeting its partner from the other side.

4. Genetic factors manifested as an absence of pigment in the pigment epithelium can modify the traditional pattern at the optic chiasm. The exact relationship between the genetic aberration and the chiasmatic disorder is uncertain.

5. To what extent the timing of axonal growth is relevant has not been determined. It is possible, for example, that the decussation pattern could be determined by interaction with axons of each side and that the interaction might only be possible for earlier growing axons. It is also possible that the pattern is set up by the first few axons and is followed by those growing later.

In order to determine the relative importance in the establishment of the normal optic chiasm of the various factors defined above—cell type, axon interactions, genetic variables, timing—it is important to know in some detail how the chiasm develops in the first place. This we do not know, and until we do, discussion must remain speculative.

## Anomalous Crossing Not Associated with Decussations

Several studies have shown that axons which normally innervate a region on one side of the brain will under special circumstances cross the midline by a completely aberrant course and innervate the same region on the opposite side. Such an anomalous crossing may occur, (1) if the terminal region on one side has been removed, and (2) if the parallel input to the contralateral region has been removed. These situations will be discussed separately.

### Removal of terminal region

All the relevant work has been done on the superior colliculus and its nonmammalian homologue, the optic tectum, originating in a series of studies by Schneider and his colleagues on hamsters (Schneider, 1970, 1973; Schneider and Jhaveri, 1974). This group found that if one superior colliculus is extensively damaged at birth, the optic axons which would have innervated that colliculus grow across the midline to innervate the medial part of the opposite colliculus (Fig 10-12C). The axons crossing the midline often do so in a coherent bundle rather than as a sheet along the whole level of the colliculus. The border between the crossed innervation and the normal innervation on the other side is extremely sharp, suggesting that the normal and aberrant pathways cannot mix, but occupy separate territory on the surface of the tectum. If the eye which normally innervates the remaining colliculus is removed at birth or up to six days after birth, the crossed pathway from the other eye spreads across the whole colliculus as shown in Figure 10-12D (So and Schneider, 1976). This becomes progressively restricted the later the eye removal up to 14 days after birth when no spread is detected beyond the medial crossed zone. To what extent the uncrossed axons reaching the colliculus by way of the optic tract interfere with the spreading of the recrossed pathway is not altogether clear. The map of the recrossed projection on the remaining colliculus is roughly mirror topographic, although local discrepancies do occur (Finlay et al., 1977).

Behavioral tests of animals with a recrossed projection (Schneider, 1973) show that their orienting response to sunflower seeds presented in the visual field tends to occur in the wrong direction. Such behavior might be predicted from the physiological studies, as with Sperry's earlier work involving uncrossing of the optic chiasm. In contrast to Sperry's frogs, the misdirected orientation was not totally mirror symmetric. Objects placed in the far temporal field did not elicit a turning to the comparable position on the other side. Thus, the hamsters seem to be able to compensate in part for the misdirected response, possibly by feedback during the eye movement (Schneider et al., 1975). The aberrant looking is abolished by lesions which destroy the recrossed bundle (Schneider, personal communication), suggesting that this indeed is the pathway involved in the abnormal behavior pattern.

Comparable studies have been conducted on rats in which the superior colliculus was removed at various times from fetal day 17 to birth and the projection of the contralateral eye (which normally should have innervated that tectum) was tested 1–2 months later (Miller and Lund, 1975). Differences occur in the recrossed pathway depending on the time at which the lesion was made.

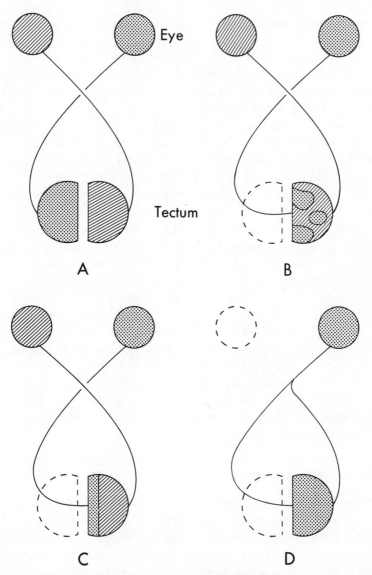

**Figure 10-12.** Retinotectal projections (A) in normal frogs and rodents (omitting the small ipsilateral pathway in rodents); (B) after a tectal lesion in frogs and early fetal rats; (C) after a tectal lesion in neonate hamsters and late fetal rats; (D) after tectal and retinal lesions in neonate hamsters.

Lesions made at the earliest times (before optic axons have reached the superior colliculus) result in a substantial recrossed projection which extends in patches across the intact colliculus (Fig. 10-12B). Although these patches are discontinuous, there is a semblance of topography in that the temporal retina projects anteriorly and the nasal retina posteriorly. In two animals the collicular lesion was partial, leaving a small area of the most posterior colliculus intact. In both, the recrossed projection crosses back yet again to innervate the remaining fragment. The possibility that the regular projection to the intact colliculus may be excluded from the areas innervated by the recrossed projection was not tested. Unilateral tectal lesions made at fetal day 20 (when the optic pathway is already beginning to innervate the superior colliculus) result in a recrossed pathway restricted to the medial border (Fig. 10-12C), much as in Schneider's studies. Innervation by the other eye is excluded from this region. A recrossed optic pathway has not been found after unilateral tectal lesions made in rats at birth.

Several studies of adult fish (Sharma, 1973; Levine and Jacobson, 1975) and frogs (Kicliter et al., 1974) have shown that after unilateral tectal lesions, the optic axons cross the midline and innervate the opposite tectum. This recrossing occurs even after partial tectal lesions. Most anatomical and physiological studies suggest an even distribution of the recrossed projection, implying that there is an intermixing of recrossed and normal optic afferents. One autoradiographic study (Levine and Jacobson, 1975), however, suggests instead a partitioning as in Figure 10-12B. Whether this separation develops with time is not clear since no continuous study has yet been attempted in one laboratory. The work does show quite convincingly that normal terminals can be displaced from their synaptic sites by an additional input.

### Removal of parallel input

There is a large projection from various areas of the cerebral cortex to particular layers of the superior colliculus. Most investigators have found that this pathway is strictly ipsilateral in its distribution but that crossing of corticotectal axons can be induced by a unilateral cortical lesion made in young rats (Fig. 10-13A). After such a lesion axons from the corresponding region of intact cortex, besides innervating the ipsilateral colliculus as usual, also cross the midline to innervate the appropriate layers of the opposite colliculus where they form synapses. This was first shown for the pathway from the somatosensory cortex (Leong and Lund, 1973; Leong, 1976a and b) and has since been shown for the pathway from the visual cortex (Mustari and Lund, 1976). The latter is of particular interest because its topographic order is more clearly identifiable, its

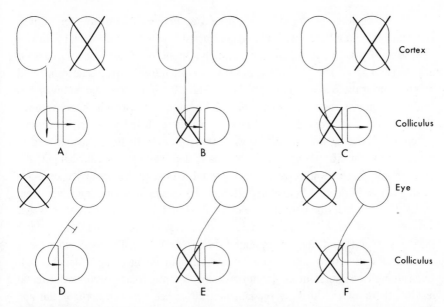

**Figure 10-13.** Effects of comparable lesions (crosses) made during development on the intertectal crossing of retinotectal and visual corticotectal pathways.

developmental history has been studied in more detail, and there are some interesting contrasts with the retinotectal pathway with which it interacts.

The pathway ends largely in the deeper half of the stratum griseum superficiale of the ipsilateral superior colliculus in rats, partially overlapping the optic projection. The cells of origin of the pathway in layer V of the visual cortex become postmitotic on fetal day 17 and the first corticotectal axons to reach the colliculus arrive just before birth on day 22 of gestation. At this early stage the pathway is strictly unilateral. Removal of the visual cortex of one side at any time from fetal day 16 to postnatal day 15 results in axons from the intact cortex crossing the midline to distribute to the deafferented colliculus within the layers in which this corticotectal pathway usually ramifies. Removal of one cortex 20 days after birth results in a small crossed pathway extending just over the midline. The crossed pathway is heavier and more broadly distributed after a cortical lesion at birth if one or both eyes are removed, but eye removal alone is not sufficient to induce the crossing. The crossed pathway honors a mirror topography such that axons projecting medially on the ipsilateral side will project medially on the contralateral side, and so on. This arrangement is least precise for laterally projecting axons, but it is perhaps surprising that they should ever manage to cross the midline at all.

Further studies of the aberrant crossed corticotectal pathway (Jen, Lund, and Mustari, unpublished work) have contrasted it with the retinotectal pathway (See Fig. 10-13). Removal of one colliculus alone does not result in crossed corticotectal axons (Fig. 10-13B), but if that colliculus removal is accompanied by ablation of the opposite cortex at birth (Fig. 10-13C), a crossed pathway can again be demonstrated. This contrasts with the optic pathway which, as we have already seen, can cross the midline after unilateral tectal ablation. A further difference between the two pathways is that while corticotectal axons readily cross the midline if the opposite colliculus is deprived of its cortical input, the same is not generally true for the retinotectal axons. In rats in which one colliculus is totally deprived of retinal afferents by cutting the optic tract, very few axons cross the midline between the colliculi (Lund and Lund, 1976). In frogs, too, removal of one eye at late tadpole or adult stages with survival times greater than one year (Lund and Gaze, unpublished results) results in no intertectal crossing of optic axons (nor indeed, aberrant ipsilateral sprouting at the chiasm). Even though a significant intertectal crossing of optic axons occurs after tectal lesions, there appears to be a fundamental difference between corticotectal and retinotectal pathways in similar experimental situation. These findings are summarized in Figure 10-13 D, E, F.

### Summary: patterns not associated with decussations

We have seen that axons may cross the midline either if deprived of their target nucleus on one side or if the same region of the other side is deprived of the particular afferent population. Curiously, under apparently similar conditions, the two pathways studied show one or another phenomenon but not both. Why this should be so is not altogether clear and deserves further investigation. A number of general remarks pertaining to both conditions may be made:

1. Both appear to result from active sprouting rather than failure of retraction of the crossed pathway.

2. There seems to be a degree of competition between corresponding afferents to the two sides of the brain; which may normally prevent anomalous crossings.

3. All crossed axons end in the appropriate laminae and do not spread to aberrant nuclei; furthermore, they attempt to map in a mirror topographic manner.

4. There is a limited time during which the lesion is effective in inducing the aberration. This appears to be earlier and of shorter duration for the retinotectal than for the corticotectal pathway.

## Overall Summary

From the three sets of experiments described in this chapter—those involving aberrations at decussation, those involving crossing the midline after removal of a terminal region, and those involving crossing the midline after removal of a parallel pathway—it is clear that the course an axon takes is far from being totally determined at the time it leaves the cell body. Studies in which lesions were made prior to the development of particular pathways emphasize that degeneration products from the damaged pathway are not an essential stimulus for sprouting of other axons. Instead, the pattern of distribution of an axon appears to be controlled to some degree by interaction with other axons that normally limit its options. If imbalances are introduced in development by lesions, the axon may distribute more broadly. Its distribution is also subject to genetic controls and relates not only to where the cell body is situated in an array but also to what sort of cell it is.

Even though an axon is directed to the "wrong" side of the brain, it still ends in the same regions and layers as it would on the normal side and maps in a mirror topographic manner, unaware that it is on the wrong side. Any compensation for the error which may occur appears to be secondary in nature.

In each case where variability can be induced by experimental intervention in mammals, there is a limited time course during which that intervention is effective. It varies according to the pathway concerned and appears to relate to its maturational state. What determines the cut-off point is unknown. It could be the cessation of natural growth of the axon or the inception of synapse formation or myelination; but such explanations are speculative. We shall want to know more about this matter when we discuss the failure of regeneration in adult brains in Chapter 17.

# 11

# DENDRITIC DEVELOPMENT

Dendrites provide by their general disposition, orientation, branching patterns, and degree of spininess, a major characterizing feature of neurons. It is apparent too from a number of studies that the pattern of dendritic distribution of a cell may play a part in the pattern of synaptic contacts the cell makes. Several relationships between axons and dendrites can be envisaged.

The first depends on the fact that, given a geometrically ordered field of like axons, a cell placed within that field will, by virtue of its dendritic orientation, have the potential for making contacts in a particular fashion with the axon population. Thus in Figure 11-1A, dendrites oriented perpendicularly to the afferent axons are likely to receive a minimum number of contacts from each of a maximum number of axons. Practical demonstration of this principle is provided by the Purkinje cell of the cerebellum whose dendrites are oriented like fans perpendicular to the parallel fibers (Fig. 2-13). However, a dendritic field placed parallel to the axons could potentially receive a maximum number of synapses from a small number of axons (Fig. 11-1B). The climbing fibers proposed for the cerebral cortex would be of this kind. If the axon set has a spatial order, the receiving cell might be able to bias its input by having an oval or elongated dendritic field. Such a mechanism has been proposed to explain orientation specificity in the visual cortex (Colonnier, 1964). This sort of mechanism places emphasis on dendritic orientation as the determining factor in connectivity rather than on any specific interaction between axons and dendrites in the area.

A second relationship, in contrast, requires that a dendrite ramify specifically in certain layers where particular afferents are also distributed. One example is provided by the retina of ground squirrels (West, 1976) where the inner plexiform layer, in which ganglion-cell dendrites ramify, is divided into nine zones. The two cell types which have terminals ending on ganglion-cell dendrites, the bipolar and amacrine cells, can be divided into a series of subclasses

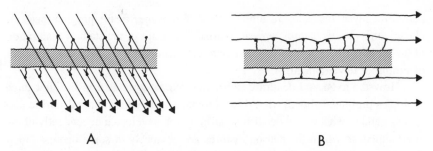

**Figure 11-1.** Two situations showing the significance of relative disposition of axons and a spiny dendrite (shaded area). (A) Many axons running across the dendrite, each making a single contact. (B) Few axons running along a dendrite, each making many contacts.

which ramify in only certain of the nine zones. Similarly, there are some 15 classes of ganglion cells, each of which shows a particular laminar distribution of dendrites. This pattern might ensure that certain ganglion cells receive most of their input from particular classes of amacrine and bipolar cell. It is clear from correlative electron microscopy, however, that the branching pattern does not determine the relative proportion of input from the two cell types, since different classes of ganglion cells having dendrites ramifying in one layer may have nearly all their input from amacrine cells, or, alternatively, half from amacrines and half from bipolars. In the latter case, the specificity of dendritic distribution determines the classes of cells with which a ganglion cell synapses, but other factors determine the detailed patterning of inputs.

In the visual cortex of monkeys, a somewhat similar situation seems to occur (Lund and Boothe, 1975; Lund et al., 1977). Each lamina has a characteristic axon plexus which distinguishes it from other laminae. The apical dendrites of pyramidal cells with somas deep in the cortex have branches which ramify in only certain laminae; therefore such cells are likely to receive a major input from the axons ramifying in those laminae. Furthermore, as the apical dendrite passes through some layers, it may show a marked reduction in spininess. This is particularly noticeable for the large pyramids of layer V as their apical dendrite passes through the layers in which geniculate axons ramify. Thus, again, the dendrites show a specific patterning which appears to reflect local distribution of certain afferents.

A further example of a relationship between cells and afferents is provided by Rakic (1972b) for the stellate cells which lie within the molecular layer of the cerebellar cortex. Those generated earliest, having cell bodies lying deeper in the molecular layer, have dendritic fields extending throughout the layer; the more superficially located ones have local dendritic fields restricted to the superficial part of the layer. Instead of resulting from different specifically programmed cell classes, the various patterns appear to arise as a result of the state of maturation of the parallel fibers at the time of stellate-cell maturation and are thus a product of timing of development.

How the axon and dendritic plexuses become correlated in the first three examples presented above, what role afferents play in determining dendritic patterning, and to what extent dendritic configuration develops independently of the environment in which the neuron matures are obviously of some interest. Such questions may be approached in two ways: by studying the patterns of normal development, and by studying aberrations in development resulting from genetic disorders or from specific deafferentation during development.

## Mechanism of Dendritic Extension

First we will consider the question of whether the mechanisms of dendritic extension in vivo compare with those shown for neurite outgrowth in vitro and for axonal growth. Many authors have commented on features of immature dendrites (see Morest, 1969b for a comprehensive list). Recent studies on dendrite growth stem largely from the work of Morest, who used the Golgi Rapid technique. In a number of regions he has shown that there are large expansions along the course of immature dendrites and at their terminations from which filopodia arise. Since these compare so closely with the images seen in tissue culture at the growing tips of neurites, he considered them to be growth cones. Dendritic growth cones have also been identified in a number of recent electron microscopic studies (Vaughn et al., 1974; Hinds and Hinds, 1972; Skoff and Hamburger, 1974). Those authors found swollen bulbs at the ends of growing dendrites with filopodia arising from them. Both bulbs and filopodia contain fine filaments, but, in general, the large numbers of membrane-bound vesicles found in other growth cones are absent. This is perhaps surprising in light of the suggestion that intracellular membranes are the precursor of new surface membrane.

Vaughn et al. (1974) were particularly interested in the relation of the growth cones to synapses. They found that at earlier times (embryonic day 13) 80% of the synapses in the developing mouse spinal cord were associated with growth cones or their associated filopodia and the remaining 20% were on dendritic trunks. Over the next few days there was a progressive reduction in the proportion of synapses on growth cones (to 30% at embryonic day 16) and a concomitant increase in synapses on dendritic shafts. From these findings they developed a model in which synapses are formed first between axons and terminal filopodia and then as growth continues, these synapses eventually become situated along the dendritic shaft.

A second type of fine structural profile identified by a number of workers (e.g., del Cerro and Snider, 1968) has also been described as a growth cone. It usually consists of an evagination along the course of a dendrite in which is found an aggregation of irregular vesicles. Similar structures identified along the course of growing neurites in culture (Bray, 1973) have been considered to be zones of ruffling membrane left behind by the growing tip and, as such, are capable of initiating new branches. Similar vesicular arrays are also seen in growing axons, both in vivo and in vitro (see Rees et al., 1976; Bunge, 1977).

It appears from this evidence that developing dendrites in vivo contain the usual substructural components associated with growth in vitro, and the possibility that growth occurs both at tips and by addition of new lateral branches must be entertained.

## Sequence of Maturation of Dendrites (Fig. 11-2).

It has been shown in a number of studies (Morest 1969a,b; Marin-Padilla, 1972, 1974; Berry and Bradley, 1976; Lund et al., 1977) that the dendritic tree of a neuron goes through an ordered sequence of development and that this sequence appears to be similar for a wide variety of neurons of quite different adult morphology. Although the sequence of maturation is continuous, four stages can be defined, each based on the appearance of dendritic form and appendages.

*Stage 1:* At the earliest times, the dendrites are short, varicose structures of uneven diameter. Fine hairs often protrude from the expanded regions, giving a ''barbed-wire'' appearance; at the terminals of each dendrite are typical growth cones. Frequently at this stage, the dendrites do not arise from the cell body in the same pattern as in the adult. This is particularly obvious for the Purkinje cell (e.g., Berry and Bradley, 1976) where at early times dendrites are seen growing out in all directions, while later the dendrites all arise from one apical trunk. However, in the visual cortex of monkeys, the dendrites, even in their initial growth, never extend into layers uncharacteristic of the adult form of that particular cell type.

*Stage 2:* The more proximal parts of the dendrites assume a more even diameter and become invested with a number of long, hair-like processes, which may also be seen on the cell body. The more distal dendrites may still have a barbed-wire appearance with growth cones at the tips.

*Stage 3:* Regardless of whether the particular cell type is ultimately invested with spines or not, it goes through a stage in which the dendrites

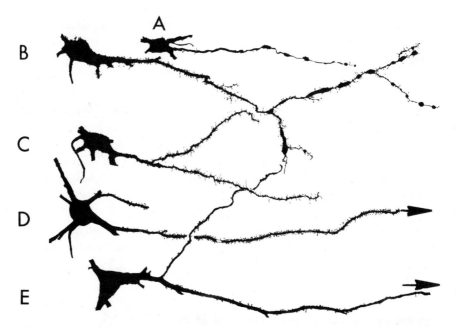

**Figure 11-2.** Maturational sequence of basal dendrite of large pyramidal cell of layer VI (Meynert cell) of monkey visual cortex ×235. (A) Stage 1, (B) stage 2, (C) transition from stage 2 to 3, (D) stage 3, (E) stage 4. Arrows indicate that dendrite extends further than drawn (from Lund et al., 1977, with permission).

develop a substantial population of dendritic spines, shorter than the dendritic appendages seen in the previous stage and frequently having dilated tips. In the cerebral cortex, those cells which ultimately become spine-free in adults also have spines over their cell bodies.

*Stage 4:* There is a gradual reduction in spine numbers, some cells attaining a relatively stable spine population, others becoming spine-free. In the mouse visual cortex (Ruiz-Marcos and Valverde, 1969), a peak in spine numbers is reached at 19 days postnatal, with a substantial decline over the next 30 days and a slow decline thereafter (Fig. 11-3A). Feldman and Dowd (1975) have found in rats that the reduction in spine number continues through adulthood, and there also appears to be concomitant loss of dendrites with age (Vaughan, 1977). In the visual cortex of monkeys (Fig. 11-3B) the highly spiny stage reaches its peak at between five and eight weeks postnatal, there being considerable variation between individual neurons of the same class. The subsequent decrease, initially rapid, occurs up to at least nine months postnatal (Boothe and Lund, 1976). Whether there is a continued slow decrease beyond this time is not known.

A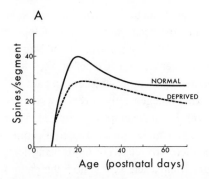

B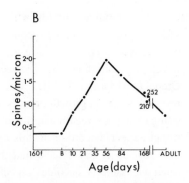

Age (postnatal days)                         Age (days)

**Figure 11-3.** Changes with development in density of dendritic spines on pyramidal cells of the visual cortex. (A) In mice, in normal development (solid line) and with visual deprivation (dotted line); data from dendritic segment 425 $\mu$m from cell body; abscissa, postnatal days (from Ruiz-Marcos and Valverde, 1969). (B) In monkeys, normal development; data from dendritic segment 100 $\mu$m from cell body; abscissa on logarithmic scale (f = fetal day, B = birth) (from results of Boothe and Lund, 1976).

The significance of these events has not been established, although a number of suggestions may be made. The early appearance of the dendrite is presumably related to its growth processes. The acquisition of long, hair-like processes might be related to an early attempt to make contacts, i.e., to ''search out'' growing axons entering the domain of the dendrite. Those which make contact might then retract into a shorter, more characteristic dendritic spine or even retract to the dendritic trunk. What is the significance of dendritic spines? In the adult cerebral cortex (Colonnier, 1968) they are contacted mainly by terminals containing round vesicles and contacts are asymmetric. The dendritic trunk of spinous cells makes only symmetric contacts, with terminals containing flattened vesicles while the dendritic trunk of nonspinous cells has both kinds of afferents (LeVay, 1973). Thus the presence of spines reflects the presence of a particular sort of synapse on certain cells. Why spines should occur only on particular cell types and, indeed, what their functional significance may be are far from clear. During development, it is possible that at least all the synapses made by terminals containing round vesicles are onto spines, and that in some cells these spines persist while in others they retract onto the dendritic trunk.

One final point is the increase and then reduction of dendritic spine numbers with maturity. Does this loss imply a progressive loss of certain sorts of synapses with maturity? Electron microscopic studies on the density of synapses per neuron in kitten cortex (Cragg, 1975a) shows that synaptic numbers go through a similar increase and reduction in the early postnatal period, so the possibility must be entertained that there may be an overproduction of synapses and subsequent pruning. Such a sequence has already been shown for developing neuromuscular connections (e.g., Brown et al., 1976) and appears to be true for developing climbing fiber connections in the cerebellum (Crepel et al., 1976; Crepel and Mariani, 1976). It would be interesting to know whether there are two processes—one a growth process and the other a degenerative event—which are going on at the same time during development. From the graph of Figure 11-3 one might expect, according to this hypothesis, that more spines are being formed than lost up to day 19 and that thereafter there are relatively more being lost. Such a possibility becomes especially significant when we see how the pattern of development is changed during conditions of sensory deprivation.

## Effects of Manipulation During Development

In general, removal of a major input to a cell during development has been reported to cause a truncation of the dendritic field and in some cases a reduction in numbers of spines. In all cases, however, the cells are still recognizable as a particular type. Cell type also appears to be recognizable in the extreme case of deafferentation—removal of a tissue explant and growing it in vitro (Privat, 1975). The effects of genetic disorders, irradiation, and drugs have also been investigated. These may cause abnormal cell body position, defects in disposition of radial glial cell processes, and deafferentation. Sorting out what is the primary cause of the observed aberrations is often difficult. Most research has been directed at two regions, the cerebellum and the cerebral cortex, and for this reason the discussion that follows will be restricted to these two.

### Cerebellum

Interest in dendritic shape in the cerebellum has been centered largely on the Purkinje cell for two reasons. First, the dendritic field is specially oriented with respect to the parallel fibers which terminate on the dendritic spines, and second, the cells of origin of the parallel fibers (the granule cells) can be removed quite selectively in a number of ways. As a result the role played by this pathway in determining dendritic characteristics can be assessed. Depletion of the granule-cell population occurs in certain genetic mutants (weaver, reeler, and staggerer mice), after certain drug treatments (such as with cycasin—Hirano et al., 1972), after X-irradiation (e.g., Altman and Anderson, 1972, 1973), and after treatment with small DNA viruses (in particular ferrets infected with feline panleukopenia virus—Herndon et al., 1971; Llinás et al., 1973, and hamsters infected with Kilham rat virus PRE-308—Oster-Granite and Herndon, 1976). Unfortunately after most of these treatments there are still a few remaining granule cells (albeit in abnormal positions) and the influence the few axons of these cells have on the dendritic development of the Purkinje cell is difficult to assess.

The selective effect of all but the genetic aberrations on granule-cell development seems to depend on the late division of the granule cells compared with most other types in the cerebellum and the susceptibility of dividing cells to the particular treatments. In the case of both cycasin treatment and irradiation, it is possible that disruptive effects may also be experienced by postmitotic cells. For the mutants, the primary influence of the gene involved has not been identified. However, with one exception (the staggerer mutant, to be discussed in more detail later in this chapter), those mutants in which granule-cell loss occurs show the same changes in Purkinje-cell morphology as the animals whose granule-cell loss was produced in other ways. The main effects are a loss of laminar alignment of Purkinje-cell bodies and a haphazard distribution of their dendritic trees. Dendritic spines, on the other hand, appear unaffected by the deafferentation. Loss of laminar alignment is generally assumed to be secondary to a failure of the cerebellar cortex to increase its surface area by the usual proportion, rather than resulting from the absence of a containing influence provided by the parallel fibers.

The disordered dendrites may be considered both with respect to their general distribution and to the planar pattern of orientation perpendicular to the parallel fibers. In conditions of severe granule-cell loss, the general orientation of the Purkinje-cell dendrites is frequently quite abnormal and the higher-order branches (third and fourth order) are frequently missing. Dendrites can originate from any part of the cell body; sometimes more than one dendrite trunk arises from the soma, perhaps from opposite sides. In those cells having dendrites on the apical surface as usual, it is not uncommon to see them cascading down again once they reach the surface—the so-called "weeping willow" appearance noted by Altman and Anderson (1972, 1973). In cells having basally directed dendrites which run into the subjacent fiber layers, the dendrites have a rather different character from those more superficially placed, being longer and thinner.

As regards planar pattern, Eccles (1970) suggested that it was determined by the parallel fibers and, indeed, in the weaver mutant (Rakic, 1975b) and the virus-infected hamster (Oster-Granite and Herndon, 1976) where there are no parallel fibers, the Purkinje cells do not have a planar dendritic field. Furthermore, in the irradiated cerebellum in which some parallel fibers persist, but run in an abnormal direction (Altman, 1973b), the Purkinje-cell dendrites are nevertheless oriented perpendicular to these misdirected fibers. While these findings suggest a correlation in alignment of Purkinje-cell dendrites and parallel fibers, they do not, of course, explain how it happens, although several interesting ideas have been offered (see Herndon and Oster-Granite, 1975).

It is somewhat surprising that despite the absence of their regular input, the spines on the dendrites still develop as usual. Electron microscopic studies of the spines show that they still have patches of subsurface density on their membrane which appear identical to a typical postsynaptic density. (Hirano and Dembitzer, 1973; Hanna et al., 1976). However, in most cases no presynaptic process abuts the spine; instead it is enveloped by a glial process. In a few cases anomalous terminals attributed to climbing fibers or mossy fibers are identified on the spines. It should be added that "naked" spines have also been identified on Purkinje cells grown in tissue culture in the total absence of granule cells (Privat, 1975). Sotelo (1975) has noted that some of the unoccupied spines have regions of "postsynaptic" specialization much more extensive than normal, and has suggested that the postsynaptic cell continues making the synaptic specialization material until the process is blocked by the acquisition of a terminal.

As mentioned earlier, the staggerer mutant mouse provides an exception to these observations. It lacks granule cells as do the others, but in this animal, the Purkinje-cell dendrites are smooth, lack the characteristic spiny branchlets on which parallel fibers usually end, and are closely packed together. Electron microscopy of the cerebellar cortex of this animal shows that there are no (Sidman, 1972) or very few (Sotelo and Changeux 1974a) synapses from granule-cell axons to Purkinje-cell dendrites, but that all other synaptic inputs to the dendrites, including those from climbing fibers, are normal. Intramembranous particles which might have represented unfilled postsynaptic sites appear to be absent in this mutant (Landis and Reese, 1977) in contrast to the weaver mutant already mentioned. Studies of various developmental stages (Sotelo and Changeux 1974a) show that granule cells migrate normally, form a characteristic array of parallel fibers, and even receive their regular input. While the parallel fibers show adhesion sites with Purkinje-cell dendrites they rarely show synaptic specializations, although significantly, synaptic contacts are seen between parallel fibers and stellate cells, as should occur normally. Subsequently, large numbers of granule cells die, even ones which already have synapses onto them. The interpretation of these changes is that the primary defect resides in the Purkinje cell, which is unable to make synaptic contacts, presumably due to a highly specific deficit of the cell surface, with parallel fibers. As a result of failing to form their usual large numbers of contacts with the Purkinje cells, the granule cells undergo a process of retrograde degeneration. Whatever the primary mechanism, it is clearly very precise since the Purkinje cells receive their other afferents as usual and the granule cells make contacts with other cell types.

In this case, therefore, the failure of the normal dendritic configuration is due to an intrinsic defect of the Purkinje cell itself rather than to a primary loss of afferents, and these results in no way invalidate the conclusions of the other studies regarding the autonomous development of dendritic spines. Given the literature suggesting a lability in spine development in other regions, the results for Purkinje cells seem oddly at variance; spines develop in the almost total absence of afferents at any time in development. It is possible, of course, that the spines on particular cell types may behave differently. It must be added, however, that the various studies of the effects of deafferentation of the cortex to be discussed below never show complete spine loss, only a relatively subtle statistical change of the kind that would be hard to replicate for Purkinje cells. The apparently autonomous development of the postsynaptic specialization in the Purkinje cell is particularly surprising and it will be discussed further in Chapter 12 in relation to the process of synapse formation.

### Cerebral cortex

Several topics have been studied in the cortex: the role of cell position in dendritic development, the role of afferents on dendritic disposition, the effects of deafferentation on dendritic spines, and the abnormal dendritic patterns described in certain cases of mental retardation. The role of function on dendritic development, although of direct relevance here, will be discussed in Chapter 15.

### Cell position

In the mouse reeler mutant (also discussed in Chapter 7), there is a tendency for cells to distribute more broadly than normal in the cortex with respect to their "birthday" and to show an overall upside-down arrangement compared with normal, such that the first generated lie most superficial and the last deepest. Golgi studies of the cerebral cortex (Fig. 11-4) show large numbers of pyramidal cells with apical dendrites oriented in all sorts of abnormal planes. Some are laterally disposed; others are inverted (it should be added that inverted pyramids do occur normally but are encountered quite infrequently, see Van der Loos, 1965; Globus and Scheibel, 1967a). Nevertheless, despite abnormal orientation, their basic pyramidal character is unchanged.

A second example concerns a human case of lissencephaly (Williams et al., 1975) in which the cortex was not folded into the usual gyral patterns. Many cells

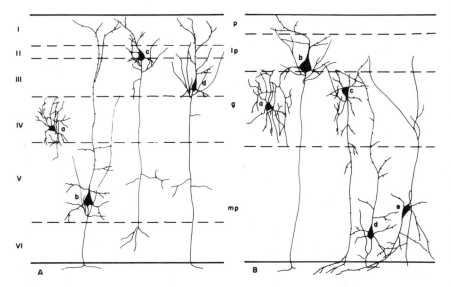

**Figure 11-4.** Drawings from Golgi preparations of corresponding cell types in (A) normal and (B) reeler mutant cortex (from Caviness, 1977, with permission).

destined for the cortical plate failed to migrate through the intermediate zone (Fig. 7-5). In Golgi studies of this material (Fig. 7-6), the cells deep to the intermediate zone show a diversity of dendritic arbors (pyramidal, nonpyramidal, spiny, nonspiny) comparable to that of normally placed cells. What axons may have innervated these cells was not determined.

### Relation of dendritic disposition to afferents

One study has addressed this question directly. It involves the spiny stellate cells of layer IV of the visual cortex of the mouse. The projection from the lateral geniculate nucleus ends within this layer and the dendrites of spiny stellate cells ramify only within the region of geniculate axon distribution. Valverde (1968) found that if he removed one eye at birth and studied the dendritic patterns of these cells in the opposite visual cortex six weeks later, the dendrites appeared to avoid layer IV and instead distributed in the laminae immediately above and below. He interpreted these results to indicate that in the absence of an input, the cells searched elsewhere for afferents. There are, however, some major problems with this interpretation: (1) there is a substantial uncrossed visual pathway which should not have been limited by the lesion; (2) studies on congenital anophthalmic mice (Kaiserman-Abramof et al., 1975), indicate that there is still a geniculocortical pathway even in the total absence throughout development of a retinal input to the lateral geniculate nucleus; (3) unilateral eye removal at birth causes an extra ipsilateral optic pathway from the normal eye to the opposite lateral geniculate nucleus, which would presumably reduce the efficacy of the deafferentation, (Lund and Lund, 1971a for rats; unpublished results confirm this for mice). Assuming that the results are not due to difficulties associated with interpreting Golgi-stained preparations, one is led to suppose that it is not the absence of a geniculocortical pathway which has stimulated the disorder, but more the absence of a normally functioning crossed visual input. Such a view might be supported by the finding that in dark-reared rats stellate cells of layer IV showed a modified dendritic distribution compared with normal (Borges and Berry, 1976).

### Dendritic spine development and deafferentation

As was indicated in Chapter 3, removal of an input to a group of neurons in an adult brain may lead to a loss of dendritic spines. Reduction in the normal number of spines also seems to occur after specific deafferentation early in development in some cortical cells. Globus and Scheibel (1966) found after lesioning the lateral geniculate nucleus or removing one eye in newborn rabbits, that there were significantly fewer spines than normal along the length of the apical dendrites of pyramidal cells with somas in layer V. The spines on oblique branches were unaffected by these procedures but were affected by lesions of the corpus callosum made in young animals (Globus and Scheibel, 1976b). Two points deserve further comment. The first is that these authors use the method of spine change as an indicator of connections. These connections clearly need not be monosynaptic, as is evidenced by the effects of enucleation, but how many inter-mediary synapses are involved is unclear. The fact that evidence is lacking in rodents and cats for a direct input to apical dendrites of pyramidal cells from the geniculate might suggest that an intermediate intracortical neuron is involved. The second point is that these authors refer to the spine changes as a spine loss; the alternative possibility of a failure of maturation was not considered.

Similar reductions in spine numbers of layer V pyramidal cells (in particular those with somas in the deeper part of layer V) have been observed in the visual cortex of mice (Valverde, 1968) and rats (Ryugo et al., 1975a) after unilateral or bilateral eye removal at birth and in the somatosensory cortex after removal of vibrissae at birth (Ryugo et al., 1975b).

In summary, the spine-change literature suffers from a number of weaknesses—not accounting for plasticity of connections after lesions, not knowing exactly how many interneurons there are between the lesion and the cell being studied, not knowing whether there is true loss of spines or simply failure of maturation, and not knowing whether spine reduction means a reduction in number of certain synapses. Without knowing these, the significance of the data is hard to assess.

## Dendritic Development in Mental Retardation

In these studies, Golgi staining of the cerebral cortex in various mental retardation conditions in humans has shown marked changes in dendritic morphology. Each study has concentrated on pyramidal cells, particularly on the degree of spininess of the apical dendrite (Fig. 11-5). In Patau syndrome (D, 13-15 trisomy) (Marin-Padilla, 1972, 1974) one case (a newborn) showed thinner apical dendrites than normal and fewer spines as the dendrites passed through the upper cortical layers. Whereas in a normal newborn human cortex, the spines tend to be longer than in adults, they are even longer in this case, suggesting that the embryonic stage involving elongation of spines has persisted. In addition, the cortex contains fewer cells than normal and a significant number of dead and dying cells. In a mongoloid brain (18 months old) by contrast, the spines are very small with short stalks and segments of dendrite are found with no spines at all (Marin-Padilla, 1972). In a study on retarded children with normal karyotype, Purpura (1974) found that the principal deficit was reduction of spine numbers (sometimes total absence). Some cases showed long, thin spines, suggesting a persistence of an embryonic condition. The magnitude of the effects varied considerably from cell to cell and in different parts of the same cell. Reduced spininess was also noted by Huttenlocher (1974) in a further set of brains of retarded children.

In all these, it is difficult to say what the primary cause of the dendritic changes is. Presumably a reduction in the normal afferents could occur. Most intriguing is the apparent retention of the embryonic condition; it must be hoped that this matter will receive further attention with time.

## Summary

It is apparent that, as for axonal growth, the major part of dendritic extension occurs at growth cones at the tips of the dendrites. The possibility that there is also growth of proximal dendrites and new branch formation along the course of

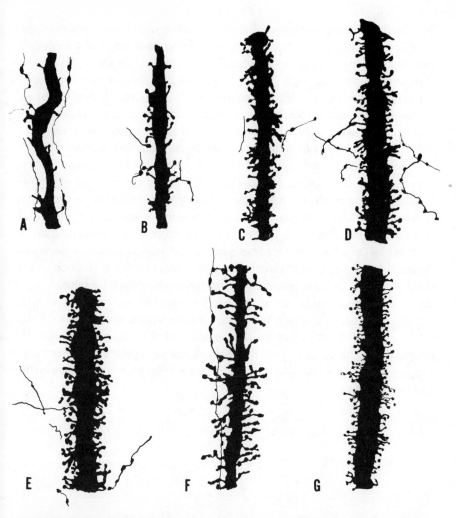

**Figure 11-5.** Drawings from Golgi preparations taken at a comparable position along the apical dendrites of large pyramidal cells with somas in layer V of human cortex. (A–E) Developmental series from normal humans (5-month fetal, 7-month fetal, newborn, 2-month postnatal, 8-month postnatal, respectively); (F) newborn Patau's syndrome; (G) 18-month-old mongoloid (from Marin-Padilla, 1972, with permission).

a dendritic trunk must also be considered, although this manner of growth is apparently of smaller proportion and perhaps of less significance in determining the shape of the dendritic field. The common sequence of dendritic maturation (regardless of cell type) is particularly interesting, especially the stage in which the dendrites develop numerous filiform appendages. Are these processes involved in searching for afferents and do they retract once they make a synaptic contact or in some cases persist as dendritic spines? We do not know. This matter is important in two respects. First it implies that in some neurons, at least, synapses may be made along the course of a dendrite rather than just at its growing tip. Second, the persistence of long dendritic appendages in some mental retardation states might suggest either an absence of appropriate input, or an inability to change to a more advanced step in the maturational process.

One important conclusion from the studies considered here is that the primary character of dendritic organization (whether the cell be pyramidal, Purkinje, or stellate; or whether it has spiny or nonspiny dendrites) is not dependent on whether the cell body attains its normal position or normal complement of afferents. On the other hand, the disposition of the dendrites can apparently be modified by the normal afferent fiber patterns and perhaps (since these are also disturbed in several mutant varieties described here) by the disposition of the radial glial cells. Of some concern is the interpretation of dendritic spine changes after injury. These changes are at best likely to occur through several synapses, and the fundamental question of whether a reduction in spine number really means a loss of synapses still remains unanswered. The autonomous spine development of Purkinje cells, even in the absence of the regular afferent pathway, suggests that spine development and maintenance is not always dependent on the presence of afferents.

# THE INTERACTION BETWEEN AXONS AND TARGET CELLS

The large number of detailed connectivity studies, particularly those made over the last 20 years, have shown that any one region of the brain sends axons only to a limited number of other regions. Within any region of termination it is not uncommon for axons from a single source to distribute only in a particular part of that region. Such local distribution is most often found when the terminal region is segregated into layers or laminae. Each lamina may have a characteristic set of afferents which distinguish it from all other laminae, and within any layer individual cell types may receive inputs not shared by other cell types. In addition, there is frequently a spatial order to a region and this spatial order is preserved in its projection to other regions.

How do these specific connections arise and are they invariant or subject to modification? Five principal possibilities can be entertained with regard to the formation of connections:

1. Axons are directed to a particular region by the various mechanisms suggested in Part IV. Once in that region, they may terminate on cells which have available contact sites. Availability of sites could be controlled by a timed development of sites and coincident timed ingrowth of axons. As such, major emphasis would be given to the relative timing of developmental events as the cause of specificity.

2. Once they have reached a region, axons form connections only with certain cells which they recognize. The final pattern may occur as a result of axons ramifying broadly and then retracting those branches which fail to find the right cell. Alternatively, axons could be attracted specifically to the appropriate synaptic sites and make no further branches. In either case the mechanism would require a recognition process between the growing axon and its target cell.

3. Neurons could connect quite broadly and those making sufficient appropriate connections would survive, while those making too few or none at all would die. This process, too, would require some form of recognition,

but unlike (2) above, inappropriately connected neurons would be eliminated, not just particular axon branches of certain neurons.

4. Within any region an axon has the capacity to project quite broadly,
but this potential is normally limited by competitive interactions exerted by
other axons and terminals in the area. This possibility would emphasize the
role of interaction between axons in forming restricted connections. Such a
mechanism may play a part in (1) and (2) above, and associated with it is the
final possible mechanism, the role of fiber ordering.

5. The pattern of distribution of an axon in a nucleus might reflect its
relative position in the afferent tract in which it runs; the partitioning of a
nucleus according to innervation from different sources might reflect the
separations imposed by tracts growing in from different regions. Such a
mechanism implies that the geometry and relative positioning of fiber tracts
is the primary determinant of specificity.

In order to assess the relative merit of each of these proposals, we will
consider first the basic cell biology of neural interaction in development, then the
specificity of interaction with respect to regions and laminae, and finally, the
problems associated with matching maps in connecting regions.

# 12

# TRANSNEURONAL INFLUENCES IN DEVELOPMENT

At some point in its growth process, an axon is likely to contact another neuron with which it may form synapses. Associated with this is a train of secondary events, which serve to consolidate the connection if appropriate or eliminate it if inappropriate. In this chapter we will consider first the actual process of synaptic formation and then the various events associated with it. We will consider possible mechanisms of intercellular recognition which trigger synapse formation, the chemical changes that occur in the partners of a newly formed synapse, and finally the control of cell numbers associated with formation of synapses.

## Synapse Formation

The order of events associated with synapse formation is difficult to define, largely because the structures involved are only resolvable by electron microscopy, which is essentially a static technique. It does not allow one to define exactly how and when particular organelles contribute to the sequence of development, especially in a brain region where individual synapses are at different stages of maturation. The most successful attempt so far in studying the course of synaptogenesis is a tissue culture study of spinal cord afferents innervating isolated superior cervical ganglion cells (Fig. 12-1) (Rees et al., 1976). Growing axons were followed by phase contrast microscopy and fixed at different stages prior to contact, shortly after contact, and at longer times thereafter.

Immediately prior to contact no specializations were recognized in either the axonal growth cone or the postsynaptic cell (Fig. 12-1A), and this would argue against any growth toward a morphologically identifiable site, resolvable with regular transmission electron microscopic methods. When one or more filopodia make contact, there is cessation of all filopodial activity for up to 30 minutes. In

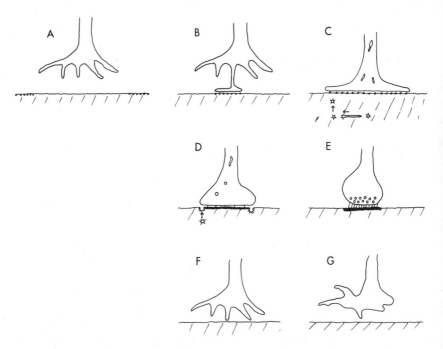

**Figure 12-1.** Drawings of a growing axon approaching a cell, making contact, and either forming a synapse (B–E) or failing to establish a contact and moving away. The dots on the membrane (A–C) signify distribution of receptor molecules. The proposed contribution of the Golgi membranes to the postsynaptic density is indicated in C and D.

some cases the contact area broadens, there being punctate zones in which the intercellular gap is as small as 7–10 nm (Fig. 12-1B). Neither gap junctions nor membrane-associated changes in either cell are found. Frequently filopodial activity will resume and the growth cone will continue on, leaving the contact intact. Sometimes appositions are transitory: the contacting filopodia retract and growth continues elsewhere (Fig. 12-1 F,G). Of those cells making contact, a postsynaptic specialization begins to appear within six hours and gets progressively larger with time. Rees et al. suggest that this may derive from coated vesicles associated with the Golgi apparatus which move to the surface to apposition points, there to exocytose: the coat of the vesicle may then provide the synaptic specialization (Fig. 12-1C,D). Other studies show a preponderance of coated vesicles in vivo open to the surface at times when first synapses are being formed (e.g., Altman, 1971; Lund and Bunt, 1976). The synaptic cleft widens to the normal adult distance of 18 nm within 48 hours.

In the growth cone, presynaptic dense projections develop along with the postsynaptic specialization. Synaptic vesicles are first seen 18 hours after first contact, and these become more numerous with time. By 36 hours, the organelles traditionally associated with the growth cone are no longer found in the presynaptic terminal area.

Studies on the development of the cerebellum in vivo suggest, in contrast to these findings, that postsynaptic specializations may develop in Purkinje cells before contacts are made from parallel fibers (Altman, 1971). Consistent with this view are the various studies in which animals are reared with almost total absence of granule cells (see Chap. 7 for further details). In these cases the Purkinje cell spines develop perfectly normal postsynaptic specializations despite the apparent absence of an input onto them at any stage of development (see Chap. 11). Sotelo (1975) noted in addition that some of these naked postsynaptic sites are much longer than normal and suggested that the cells continue to make the material of the postsynaptic specialization until the axons arrive to stop production. Of course, this view is the complete opposite of that indicated by the experiments of Rees and co-workers, and may indicate that different synaptic associations may follow different sequences of development, in particular with respect to synapses on spines as opposed to those on dendritic shafts.

As was mentioned in Chapter 3, receptor molecules for acetylcholine (and perhaps for other transmitter substances) appear to reside in the postsynaptic surface membrane rather than in the postsynaptic density. Study of the postsynaptic density may therefore give little indication of when the receptors are present, where they are generated, and how they behave during the process of innervation, although it would show where they finally end up. Most of the work on receptor sites has been done on cholinergic neuromuscular junctions, because the receptors bind irreversibly with a derivative of a snake venom, $\alpha$-bungarotoxin. This substance can be conjugated with fluorescent compounds or tritiated, and so the site of binding can be visualized. Devreotes and Fambrough (1976) have found that muscle receptor molecules may be produced within the muscle cell and transferred to the surface. Recent studies have used fluorescent conjugates of $\alpha$-bungarotoxin to study the distribution of receptor molecules on the membrane of innervated and noninnervated muscle fibers taken from *Xenopus* embryos and grown in culture with or without neural-tube cells (Anderson and Cohen, 1977; Anderson et al., 1977). The technique allowed the authors to study the distribution of the receptor sites several times in the same cell. In noninnervated muscle cells, there are several patches of membrane with a high density of receptor sites which move about relative to one another. Such lateral movement of membrane-associated proteins has a precedent in studies of other cell types (see Singer and Nicolson, 1972; Edidin and Fambrough, 1973).

Once innervated, however, the high-density area of receptor sites is found only along the region of innervation, apparently as a result of migration of molecules from other places on the surface. Thus the receptors appear in the surface membrane prior to innervation. Innervation apparently causes a movement of receptor proteins to the area of innervation (Fig. 12-1 A–C) and, at least in some cells, a migration of material from intracellular sources to associate with the same area of membrane.

An interesting observation is that neurons make contact with muscles even though the receptor molecules are blocked by $\alpha$-bungarotoxin (Cohen, 1972). Other studies, using a variety of transmitter blocking agents, have shown that synapse formation per se does not require functional activity. For example, Model et al. (1971) showed that if cerebellar tissue is grown in vitro with xylocaine in the culture medium, morphological synapse formation appears to continue normally even though no electrical activity is recorded. On removal of the xylocaine, normal electrical activity follows. Therefore, the events associated with synapse formation appear to bear little or no relation to the functional activity with which that synapse will be subsequently involved.

To summarize, studies on in vitro axon growth and synapse formation provide no support for the existence of specific chemical attractants that direct an axon from a distance to its appropriate cell, but rather suggest that contacts are made by chance. However, once a contact is made, a specific mechanism involving some sort of membrane recognition occurs and either the development of a formal synapse results or retraction of the filopodia and breaking of the contact occur. The actual mechanism involved in recognition is not clear at present, but some views as to its possible nature will be presented in the next section. The studies on muscle, particularly those of Anderson and coworkers, support the view that receptor sites are indeed present early and become organized by the afferents. However, the associated but not necessarily directly related postsynaptic density appears to vary in its time of appearance according to the cell type involved or the particular circumstances of the experiment.

## Cell Recognition

Recognition between nerve cells appears to be important in determining the disposition and connectivity of neurons. We have already considered in Chapter 7 the role of intercellular recognition in cell sorting and migration; it may play a part in the axonal interactions proposed in Chapter 10; and in Chapter 14 we will discuss its role in establishing topographically ordered projections between different brain regions. In this chapter we are more concerned with the recognition associated with the formation of synaptic connections. While it is apparent that the specific recognition pattern is different for each of the events outlined above (cell sorting, for example, requires affinity between like cells in a region, while synapse formation requires affinities between cells of different regions), the basic mechanisms involved are likely to be similar.

Although no clear evidence exists at present for a molecular mechanism of recognition, a number of interesting theories have been generated. These all point to a major role being played by certain glycoproteins associated with the external coat of the cell membrane. The glycoproteins have a polypeptide backbone which is embedded in the lipid bilayer of the membrane and a series of oligosaccharide side chains made up of a variety of sugars (see Fig. 3-4). The sugars are added by enzymes which are specific not only for the sugar to be added but also for the previous sugar in the chain. Therefore, the heterogeneity of the glycoproteins provides a diversity of organisation which may form a substrate for specific recognition. As indicated earlier, it appears from studies of binding with specific lectins that the axon tips of different classes of developing neurons have different arrays of glycoproteins (Pfenninger and Maylié-Pfenninger, 1976). This is an essential prerequisite if glycoproteins are to be implicated in synaptic recognition; but what mechanisms are responsible for the recognition process?

One of the earliest suggestions was that apposing surfaces contain complementary antigen-like and antibody-like molecules which interact to form complexes and result in adhesion between cells (Fig. 12-2A) (Weiss, 1947). There are abundant antigens on the cell surface, although antibodies have yet to be shown. A second possibility concerns an enzyme–substrate interaction the substrate being present on the surface of one cell and the enzyme on the surface of the other (Fig. 12-2B) (Roseman, 1974). Special attention has been given to the glycosyl transferases responsible for adding new carbohydrates to the side chains of glycoprotein molecules. One requirement is that the enzyme must itself be situated on the surface membrane, and this does not appear to have been proven unequivocally at present. Another proposal is that intercellular adhesion might result from hydrogen bonding between chains of carbohydrates of the two cells concerned (Fig. 12-2D). For such a mechanism to be specific, much depends on the three-dimensional configuration of the oligosaccharides and little is known of this.

Another possible mechanism for specific binding is that the binding molecules are relatively nonspecific but are usually blocked from indiscriminate binding by masking molecules. One such blocking molecule which has been

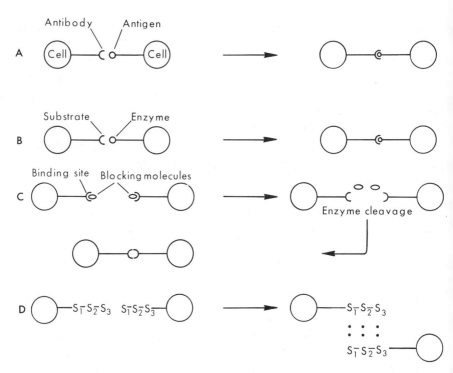

**Figure 12-2.** Diagram illustrating mechanisms hypothesized as subserving intercellular recognition. (A) Antigen–antibody reaction, (B) substrate–enzyme interaction, (C) enzyme cleavage to remove molecules which block binding proteins, (D) hydrogen bonding (after Roseman, 1974).

proposed is sialic acid (see Lloyd, 1975). Recent work on the retina and brain of developing chicks has isolated a surface molecule which can be cleaved to provide a smaller molecule which promotes cell binding (Rutishauser et al., 1976). Therefore, the specificity of the cleavage mechanism would determine the specificity of binding (Fig. 12-2C). Three possible relationships in binding exist: (1) the binding molecules on the two cells interact with one another directly, (2) they both bind to an intermediate molecule, or (3) the binding molecules on one cell bind to a receptor on the other.

The problems associated with applying these theories directly to the specificity associated with synapse formation are several. It is technically difficult to isolate pure populations of cells which are going to interconnect; further, it is possible that only certain parts of the surface membrane (e.g., axon tip and dendritic surfaces) may show the special affinity and these affinities may change with time, as indeed certain experiments suggest (Gottlieb et al., 1974; Barbera,

1975). Beyond this, although an affinity is shown between two cell groups, it is never quite certain what relation it bears to specific synapse formation. In none of the studies so far undertaken has a rigorous attempt been made to test binding properties between interconnecting and noninterconnecting cell groups. This matter will be taken up again in discussing local retinotectal specificities.

One final point of considerable importance here is that cell binding is not the final process in synapse formation. There follows a series of secondary events: synaptic vesicle production, transmitter synthesis, and mechanisms associated with cell maintenance. How then does surface binding lead to these intracellular events? Again there is no clear answer, but various experiments on a variety of cell types favor the concept of some form of "membrane messenger." In one form, this messenger resides on the cytoplasmic side of the membrane where it is released into the cytoplasm and serves a regulatory role in protein synthesis if the external glycoproteins become bound to other cells. Roseman (1974) has reviewed the various options in a comprehensive manner and the reader is referred to his review for further discussion.

## Transneuronal Influences

As mentioned above, a series of secondary events follow the formation of appropriate connections. One of these is a dramatic increase of amounts of enzymes involved in transmitter synthesis in both the pre- and postsynaptic cell. Such effects are frequently selective and involve only certain enzymes. An associated point is that a neuron may survive only if it makes appropriate connections, both afferent and efferent, during a critical stage of development. This suggests that some measure of cell maintenance is provided by such connections. In this section we will examine the various studies pertaining to these matters involving the superior cervical ganglion, the ciliary ganglion, and neuromuscular innervation. Finally in vitro studies of transmitter specificity and cell death in the central nervous system will be reviewed.

### Superior cervical ganglion

Some consideration has already been given to this ganglion, along with other adrenergic systems, in Chapter 5. Most of the cells of the ganglion which innervate a variety of structures, among them the iris and salivary glands, have norepinephrine as their transmitter. This is synthesized from tyrosine by the rate-limiting enzymes, TOH (tyrosine hydroxylase) and dopamine-$\beta$-

hydroxylase. In many studies the levels of tyrosine hydroxylase have been used as an indicator of the level of transmitter synthesis. Assays for the synthetic enzyme for acetylcholine, CAT (choline acetyltransferase) give a good indication of enzyme levels in the terminals of preganglionic axons, since there are only a small number of cholinergic neurons in the ganglion itself. Using these two assays (for CAT and TOH) the effects of experimental manipulations on enzyme synthesis in both preganglionic and ganglionic neurons can be determined. One difficulty in interpreting the results is that if a change in enzyme level occurs it may be due to a number of events. It may arise from a change in the number of synapses (each of which requires a fixed amount of transmitter), from a change in the amount of transmitter per synapse with no change in the number of synapses, from a change in cell metabolism which is not directly correlated with synaptic requirements, or from a change in cell numbers. It is not always obvious which of these are specifically involved.

In normal development the levels of TOH and CAT run at low levels up to postnatal day 2 in the mouse at which time they both show an increase up to day 14 (Fig. 12-3A). This correlates well in timing with the course of formation of synapses (Black et al., 1971). Immunotitration studies indicate that the increase in TOH is due to a real increase in the number of molecules produced and not to an unmasking of enzyme already present. Since choline acetyltransferase is produced in the cell body of the preganglionic neurons, it might be expected that the events occurring at the terminals are in some way being relayed back to the cell body to control the rate of enzyme synthesis. This could be achieved either by maintaining an equilibrium in the amount of circulatory enzyme which would be constantly modified by the growth of the neuron, or it could result from retrograde transport from the synapse or the postsynaptic cell of a material which influences the rate of enzyme synthesis. These options have not been investigated. However, destruction of the postsynaptic cells by 6-hydroxydopamine, antiserum to NGF (nerve growth factor), or by axotomy early in development, all prevent the development of normal choline acetyltransferase levels in the preganglion terminal (Fig. 12-3 F,G). It is not clear, however, whether this is due to a special trophic effect of the postganglionic cell on the presynaptic terminals; whether it is the result of degeneration of the terminals along with the ganglion cells; or whether it is due to failure of transmission and hence lowered requirements for synthesized enzyme without any change in numbers of terminals.

Besides presynaptic neurons being affected by events in the postsynaptic cell, it is clear that the postsynaptic cell is affected by presynaptic events. Thus if the preganglionic nerve is cut, the TOH level in the ganglion cells does not show the usual increase (Fig. 12-3D). This effect can also be mimicked (as in adults)

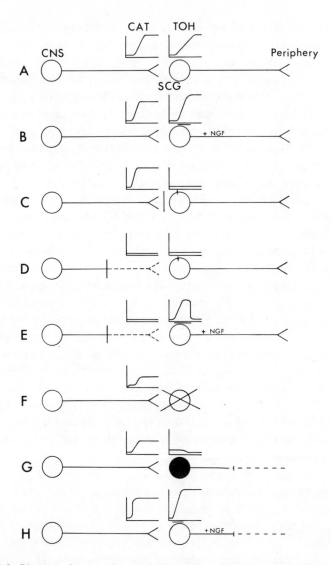

**Figure 12-3.** Diagram of a superior cervical ganglion showing a preganglionic nerve, a ganglionic cell, and its axon to the periphery. The development of levels of enzymes choline acetyltransferase (CAT) and tyrosine hydroxylase (TOH) are shown for various conditions: (A) normal, (B) normal + NGF, (C) with ganglion transmission blocker, (D) cut preganglionic nerve, (E) cut preganglionic nerve and NGF administered during period indicated by bar, (F) eliminate ganglionic neurons, (G) cut axons of ganglionic neurons, (H) cut axons of ganglionic neurons and administer NGF. Arrows or lines under graphs indicate period during which treatment is given.

by nicotinic ganglion blocking agents (Fig. 12-3C; Hendry, 1973), suggesting that the trophic effect is activity-related in some way. Preganglionic nerve section causes only a modest reduction in cell numbers (Black et al., 1972). Thus the failure of enzyme changes in this case is not the result of a massive loss of postsynaptic cells, but is largely due to a specific failure of surviving cells. NGF, which as we will see is important for ensuring the survival of the ganglionic neurons, has only transient effects on ganglia in which the preganglionic nerve is cut (Fig. 12-3E; Hendry, 1973). There is no indication that it can compensate for the absence of afferents, indicating that the mechanisms whereby NGF and transynaptic transmission affect enzyme levels are independent.

In the normal superior cervical ganglion of rats there is a decrease in numbers of neurons between postnatal days 6 and 30 (Hendry and Campbell, 1976). This reduction can be prevented by administration of NGF between postnatal days 6 and 21. If the postganglionic axons are cut, there is extensive neuron cell death and levels of TOH are correspondingly low (Fig. 12-3G; Hendry, 1975). This effect is rather more dramatic than that produced by axotomy in adults (see Chap. 4) where synaptic disconnection in the ganglion results, followed by regenerative recovery. The cell loss in young animals appears to occur during the time that the axons of ganglionic neurons are making synaptic connections in the periphery. Administration of NGF subcutaneously through this period has two effects: it prevents normal cell loss, apparently by preventing cell death rather than by stimulating further neuronal division; and it counters the effect of axotomy. These two effects are shown in Figure 12-4. Further studies in which the submaxillary gland was removed from four-day-old mice (depriving ganglion cells of part of their target tissue) showed that levels of TOH in the ganglia were higher if a cellulose pellet containing NGF was placed in the region of the gland (Fig. 12-3H). This effect was local, involving only the ganglion on the side on which the pellet had been placed (Hendry and Iverson, 1973). Since NGF can be transported in a retrograde direction (Hendry et al., 1974), it is thought that in making connections (or perhaps even on growing into the appropriate area), nerve growth factor is taken up by terminals and transported to the cell body where it has a direct effect on control of enzyme synthesis. The possibility of direct effect on the cell body is supported by a number of studies where the response to axotomy is prevented by local application of NGF to the ganglion. If administration of NGF is stopped after day 21 in rats, then dramatic cell death is not seen (Hendry and Campbell, 1976).

These results suggest that at the time a neuron should be beginning to make connections, the cell body becomes particularly sensitive to NGF transported from the target organ. If the axons do not reach the target tissue in time or if they

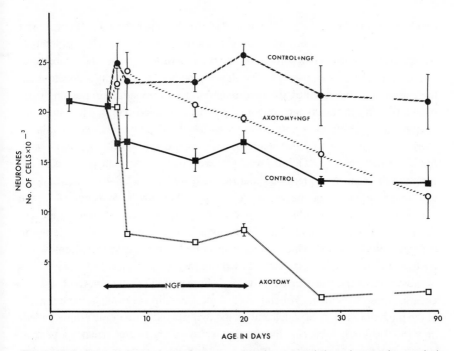

**Figure 12-4.** Changes in numbers of neurons at early postnatal times in superior cervical ganglion of normal rats and after axotomy of ganglionic cells and for treatment with NGF during the period indicated (from Hendry and Campbell, 1976; with permission).

do not transport enough NGF back to the soma (perhaps because of too few branches), the neuron dies. Once this sensitive period is over, the neuron is somewhat less dependent on NGF for its survival, although as we saw in Chapter 4, NGF is still needed for the neuron's normal functioning. It should be added that, apart from these effects, NGF also promotes peripheral sprouting of growing axon tips. It is possible that the increased peripheral field which might be expected in normal animals given NGF might have a secondary effect in sustaining more neurons in the ganglion by retrograde transport. A more extensive review of this literature has been provided by Hendry (1976), and the reader is referred to his review for further detail.

### Ciliary ganglion

The ciliary ganglion of the chick is another place in which the effect on the development of ganglion cells of both afferents and connections with the periphery has been investigated (Landmesser and Pilar, 1974 a,b; 1976; Pilar and

Landmesser, 1976; Chiappinelli et al., 1976). The ganglion lies in the orbit adjacent to the optic nerve. It receives preganglionic nerves from the oculomotor nucleus. These end on two sorts of cells, which in turn innervate the iris muscles and choroid muscles in the eye. The development of functional connections in the ganglion is measured by the percentage of cells responding to preganglionic stimulation; these results are supported by electron microscopy. The initial development of contacts occurs from stage 26–31 (Hamburger and Hamilton, 1951). At stages 34 to 39 as many as 50% of the cells in the ganglion show failure of transmission. This is correlated with two events: half the cells in the ganglion die (and it is presumed that this half undergoes transmission failure); and the postganglionic nerves develop their functional connections in the periphery. Associated with the formation of peripheral connections is the development of acetylcholinesterase in the muscle and choline acetyltransferase in the ganglion cell somas which is then transported to their terminals. Removal of the eye before it becomes innervated by the ciliary ganglion results in a rapid loss of most cells of the ganglion over the same time period in which normal cell loss occurs. Along with the loss of cells there is degeneration of some of the terminals of the preganglion nerves which end on the dying cells. Further quantitative work on normal animals indicates that the number of ciliary nerves reaches a peak of six per ganglion cell just prior to the period of cell death; there is a reduction in numbers correlated with the period of cell death; and a final drop occurs after the period of cell death. The authors propose that initially the ganglion cells make too many axons for the terminal field they are to innervate. Some cells die because the axons are unable to make a sufficient number of contacts, and this in turn may deprive some preganglionic cell terminals of postsynaptic sites and these too may die. Later, possibly those postganglionic collateral branches which themselves are unable to make synapses may also degenerate. Such overproduction and pruning of axons has a parallel in the development of neuromuscular innervation (Bennett and Pettigrew, 1976; Brown et al., 1976) where excess innervation of muscle fibers becomes progressively reduced with age.

Of particular interest is the similar pattern of results in the ciliary ganglion and superior cervical ganglion, especially with respect to the importance of the periphery for the survival of the neurons in each ganglion. Two significant differences are that in the ciliary ganglion the transmitter for the postganglionic neuron is acetylcholine and that these neurons are insensitive to the effect of NGF. Is it possible that a maintenance substance other than NGF is being supplied by the ciliary muscle and transported back to the cell somata of the ciliary nerves? Given the similarities in the two sets of experiments, this possibility would not seem unreasonable, but such a substance has yet to be identified.

## Neuromuscular connections

It has been known for some time that in many mammals limb muscles can be divided into two classes: those giving a fast twitch and those giving a slow twitch. Associated with these physiological differences are differences in histology of the muscle fibers, ATPase activity, $Ca^{2+}$ uptake by the sarcoplasm, and components of the muscle myosin molecule (see Close, 1972, for comprehensive review). In addition it was found (Eccles et al., 1958) that slow-twitch fibers of adult cats are innervated by tonic neurons discharging at a rate of 10–20/sec, while the fast-twitch fibers are innervated by phasic neurons firing usually at a rate of 30–60/sec. This discovery led to the possibility that the properties of the muscle fiber response charactristics are to some extent dictated by the functional innervation patterns. This proposal was tested by cutting the nerves to fast and slow muscles in the leg and either allowing them to regenerate to the muscle they usually innervate or cross-suturing them so that the "fast" nerve would grow back into the "slow" muscle and the "slow" nerve would innervate the "fast" muscle. After such cross-innervation it was found that the muscle changed its physiology to correlate with the type of nerve innervating it (Buller et al., 1960). Associated with this are a variety of biochemical changes (see Buller et al., 1969; Close, 1972, for extensive review).

Two ways were suggested in which the afferent nerve could affect the properties of the muscle: either the stimulation rate might alter the properties of the muscle or the nerve releases some trophic material which alters the muscle behavior. Support for the latter comes from some experiments in which fast-nerve implants placed in the soleus muscle apparently altered the muscle behavior even though they failed to make functional synaptic contact with it (Fex and Sonesson, 1970). These results have been criticized for not providing unequivocal evidence of physiological change (Close, 1972).

While the results described so far appear to parallel the changes seen in the superior cervical ganglion after afferent transmission is modified, further work indicates that the innervation not only affects the degree of muscle responsiveness, it also appears to dictate the kind of proteins made. A biochemical study by Samaha et al. (1970) showed that after cross-innervation, the proteins in myosin of the fast and slow muscle appear to change their character from those characteristic of one muscle type to those characteristic of the other. In an extension of this work, Sréter et al. (1975), investigated a certain part of the myosin molecule, the light-chain component, after cross-innervation of the nerve to soleus (a slow muscle) and the nerve to extensor digitorum longus (a fast muscle) and after cutting the nerves and allowing them to regenerate back to their regular muscles.

This work was done in rabbits where the physiological effects of cross-innervation are unambiguous; unlike Samaha et al., Sréter et al. studied the physiological effects along with the biochemical changes. In normal muscle three light-chain components were identified after gel electrophoresis of extracts of soleus muscle. These have characteristic molecular weights which differ from the three light-chain components extracted from extensor digitorum longus. If the nerves to each muscle are cut and allowed to regenerate back, the light-chain component patterns remain essentially normal. A soleus muscle innervated by the nerve to extensor digitorum longus shows not only its own three light-chain components but also the three normally associated with extensor digitorum longus muscle. The same is true of the other cross combination. The result is a little surprising because although each muscle retains its old myosin component, its contractile properties are those of the newly acquired myosin. One further observation made by these workers was that, in accord with earlier studies, if a fast muscle is stimulated chronically over a period of 15 weeks with a slow-twitch pattern, it becomes transformed into a slow muscle, physiologically and biochemically.

The important point to be gained from these results is that a muscle appears to make a new protein if the pattern of activity of the afferent nerve is altered; therefore, stimulation patterns may control the kinds of proteins being synthesized. One minor caution must be expressed. It is possible that both muscle types make all six light-chain polypeptides, and the three uncharacteristic of the particular type may be made in sufficiently small quantities as to be undetectable. Thus, rather than making a new protein, the muscle may selectively change the rate of synthesis of an existing protein.

## Dependency of Transmitter Type on Environment of Cells

An interesting counterpart to the cross-innervation experiments is found in the studies on the role played by the environment in the type of transmitter neural crest derivatives synthesize. For example, Le Douarin et al. (1975) found that if presumptive adrenergic neuroblasts taken from the thoracic neural crest are transplanted to the brainstem (where neural crest derivatives eventually synthesize acetylcholine), they too would synthesize acetylcholine rather than norepinephrine. These results could be interpreted in two ways. There may be two populations of cells, one destined to synthesize norepinephrine and the other acetylcholine, and depending on the local conditions one degenerates and the other survives or vice versa. Alternatively, single cells might have the potential to make both transmitters and local conditions stimulate production of one at the expense of the other.

Transmitter types have been studied further by growing sympathetic neurons dissociated from the superior cervical ganglion of newborn rats in various culture conditions (see Patterson et al., 1976; O'Lague et al., 1976). When grown without other cell types, the cells develop the properties of sympathetic neurons, synthesizing and accumulating adrenergic transmitters. When grown in the absence of nonneuronal cells (e.g., cardiac muscle, neuroglia), there is a massive (up to 1000-fold) increase in the amount of acetylcholine synthesized, and cells in the culture make cholinergic synapses (nicotinic and muscarinic) both with themselves and with striated muscle fibers present in the culture dish. This effect can also be produced by growing the neurons alone in a conditioned medium from a dish in which nonneuronal cells have been grown. Further study on the physiology and pharmacology of single cells (Furshpan et al., 1976; Reichardt et al., 1976) indicates that individual sympathetic neurons have the capacity to synthesize both norepinephrine and acetylcholine at any one time. When grown with nonneuronal cells, they tend to function as cholinergic neurons; when grown alone, they function as adrenergic neurons.

Such plasticity is particularly interesting because it occurs so late in development (after most of the neurons in the ganglion have become postmitotic) and because it suggests that at least some neurons may have the potential to use a number of transmitters but that this potential is limited by the local environment in which the neuron finds itself. Clearly the culture conditions are artificial; if the experiments are directly referable to in vivo development, one might logically expect there to be no adrenergic neurons anywhere in the nervous system since development of neural crest cells always occurs in an environment of nonneuronal cells. Le Douarin et al. (1975) suggested that the tissues through which the neural crest cells migrate may be important in dictating transmitter characteristics and it is possible that by chance the tissue culture studies have selected nonneural cells which are more characteristic of parasympathetic migrating channels. Nevertheless, the results demonstrate an interesting principle and suggest that the options a neuron expresses with respect to its particular transmitter depend in part on the environment in which the neuron differentiates.

## Cell Death in the Central Nervous System

We have already seen that cell death occurs during the course of development of two peripheral ganglia, it occurs at the time connections are normally made, it can be increased by removing the peripheral target organ, and in one case it can be prevented by NGF. The implication is that the neurons must innervate their target region before a certain stage in order to obtain a "maintenance substance" which prevents their death. Such a mechanism would be a useful one in eliminating axons which have connected inappropriately and it is obviously of interest to know to what extent these findings from peripheral structures relate to development of the central nervous system.

The first definitive investigation of these problems concerned the dorsal root ganglion cells of the chick spinal cord. It was known that there were more cells in the ganglia of segments serving limbs (i.e., cervical and lumbosacral) than in the intervening thoracic segments. Hamburger and Levi-Montalcini (1949) found that early in development, the cell numbers were similar at all cord levels. However, a substantial cell degeneration peaking at the fifth and sixth day of incubation was found in the thoracic region; this would also occur during the same period in the limb-bearing segments if the limb bud had been removed. The cell loss in the thoracic region could be prevented by transplanting a supernumery limb to this region. A similar pattern was also found for the motor neurons of the cord, there being a small cell loss in the limb segments and a larger cell loss elsewhere. Removal of a limb bud caused motor neuron degeneration over the same period as the time of normal cell death.

From these results emerged the concept that neurons are produced in excessive numbers and that those having a terminal field with which to connect, survive, while the others degenerate. Since the wave of degeneration occurs at the time when axons are known to be ramifying in their peripheral regions (Oppenheim and Heaton, 1975), the expectation would be that some kind of direct interaction with the periphery determines the neuron's survival. A similar series of events has been defined for motor neurons of tadpoles (Prestige, 1970). Here there is substantial cell death normally, which is exaggerated by extirpation of a limb bud. Again, the normal and lesion-related cell death follow the same time course. One point of note here is that in a second series of experiments the numbers of axons in the ventral root were counted. These followed the same reduction in numbers with time as did the cells, suggesting strongly that some cells which ultimately degenerate have, in fact, already sent their axons out to the periphery.

This matter has been investigated further in the centrifugal visual pathway of chicks. An irregularly folded sheet of cells on each side of the midbrain, the isthmo-optic nucleus, sends axons to the opposite eye where they innervate amacrine and displaced ganglion cells. When the nucleus first appears (tenth day of incubation), it contains more than 20,000 cells, but less than half of these survive beyond the seventeenth day (Figs. 12-5 and 12-6A,B; Clarke et al.,

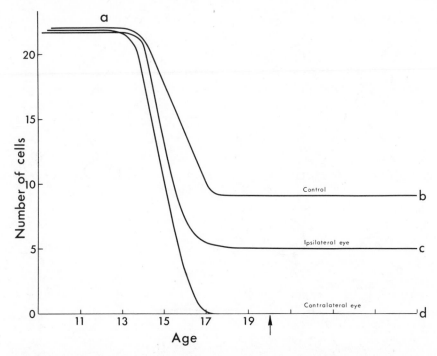

**Figure 12-5.** Changes in neuron numbers with age in isthmo-optic nucleus of control chicks, and ones with an eye removed early in development: a–d relate to examples given in Fig. 12-6. Arrow on abscissa indicates time of hatching (after Crossland et al., 1975).

1976). After injecting horseradish peroxidase into one eye, Clarke and Cowan (1975) found most cells in the contralateral isthmo-optic nucleus to be labeled, confirming, in another system, that cells destined to die had indeed sent axons to the terminal area, although whether they had formed synapses remains to be seen. Three additional interesting results emerged from their study.

First, at early stages, cells which still projected to the eye were lying as much as 800 $\mu$m from the confines of the nucleus. These ectopic cells were all eliminated at later times.

Second, some cells at early stages project ipsilaterally and this ipsilateral projection is lost as development progresses (Fig. 12-6A,B). This transitory ipsilateral pathway may be due to some cells projecting only ipsilaterally or to cells projecting bilaterally. In either case, the axons projecting in this anomalous manner are eliminated, possibly together with the cell body of origin. This result suggests that the pathway has some measure of lateral specificity, and is in marked contrast to the various experiments described in Chapter 10.

A third point of interest is that if one eye is removed early in development,

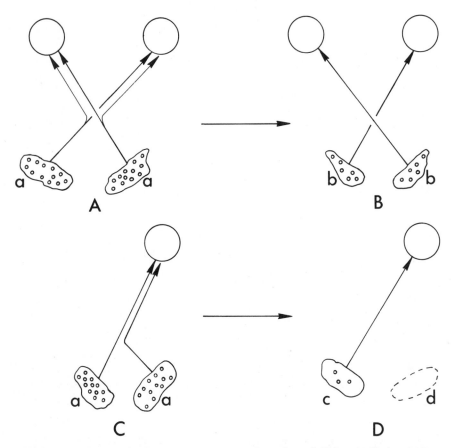

**Figure 12-6.** Diagram of cell numbers in isthmo-optic nucleus of chick and retinal projection of nucleus. (A) Normal early development, (B) normal adult, (C) early development with one eye removed, (D) later development with one eye removed. Cell numbers at a–d correlate with designated positions of Fig. 12-5.

many more cells in the isthmo-optic nucleus contralateral to the enucleation send ipsilaterally directed axons into the intact eye (Fig. 12-6C). These appear to compete with the regular contralateral projection to that eye, since in these cases developmental cell loss in the contralateral isthmo-optic nucleus is greater than normal. Later the ipsilateral pathway disappears, the cells in the nucleus die, and the nucleus of the other side remains reduced in size (Figs. 12-5, 12-6D; Clarke et al., 1976).

Whether there are trophic agents in the central nervous system which function like NGF remains to be seen, but the close parallel between these results and those on the superior cervical ganglion suggests that they probably exist.

Other work on neuronal death concerns the role of afferents in maintaining cells. In studies on the retinotectal pathway of chicks (Kelly and Cowan, 1972) and mice (DeLong and Sidman, 1962), it has been found that removal of an eye early in development causes a reduction in the number of tectal neurons. Detailed study of the course of this transneuronal degeneration in chick (Kelly and Cowan, 1972) shows that if an eye is removed prior to the time that its axons reach the tectum, development of tectal cells continues normally until the stage is reached at which they should become innervated, at which point they die. Such cell death is never as extensive as the retrograde cell death, but possibly other afferents may serve to maintain some cells. The fact that cell death occurs at the time of normal innervation strongly suggests that the cells of the tectum are programmed to die if they do not receive an appropriate input by a certain stage of development.

## Summary

From these results, a number of points may be made: (1) There is good evidence that in some regions more neurons are produced than are finally needed. The proportion of redundant neurons varies according to the region. (2) A substantial amount of cell death occurs during development, at approximately the time the cells in question are innervating their target region or are themselves being innervated. Such cell death can be enhanced by removing either the target nucleus or the specific afferents, respectively. (3) This natural process appears to eliminate inappropriately connected cells. (4) An increased innervation of a region does not lead to neuronal hyperplasia. (5) The degree of normal cell death varies considerably from one region of the brain to another. (6) Given the similarity in the behavior of neurons in the brain compared with those in the superior cervical ganglion, the question arises as to whether there are trophic agents analogous to NGF which ensure the survival of cells in certain interconnected brain regions. (7) It would appear that the specificity of connections in the mature nervous system is achieved not only by cells actively making highly specific connections, but also by elimination of those cells which, for one reason or another, have become inappropriately connected. The selectivity of cell or axonal death is clearly of importance in this context and will be considered further in Chapter 13.

# 13

# SPECIFICITY OF CONNECTION PATTERNS

When an axon population courses through the brain, it does not innervate everything along its path. Instead, it is very selective. It ramifies and forms synapses only in certain regions, and within any single region, it may project only to part of it. In addition to this, the axon may synapse only on particular cells or on particular parts of the dendritic field of those cells. The various mechanisms by which a restricted projection may arise have already been outlined in the introduction to Part V and include specific recognition between axons and target cells, interaction between afferents, and the geometry and timing of afferent nerve growth. We will now see how these are relevant to particular experimental situations.

## Regional Specificity

There are a number of examples of an axon innervating its usual target tissue in a reasonably normal manner even if it reaches it by a highly anomalous route. Thus, if an optic nerve regenerates through the oculomotor nerve root, it still innervates the optic tectum, giving a weak but normally organized visuotopic map (Gaze, 1959). Further examples include those in which axons cross between tecta, forming an ordered projection within the usual area. These projections were discussed in detail in Chapter 10.

The number of cases in which axons have been shown to terminate in uncharacteristic regions is small. In the experiments already mentioned (Constantine-Paton and Capranica, 1975, 1976 a,b) in which an eye cup was transplanted into the position of the ear vesicle in tadpoles, axons grow caudally in the spinal cord and appear to ramify within the grey matter. Interestingly, they do not innervate the vestibular nuclei adjacent to the region of the implant, although another ear vesicle placed in this region does innervate them. In other words, innervation is not totally random and it does not necessarily occur in any region which has been deprived of its normal input by the transplantation procedure. Nevertheless, it can occur to some extent in quite anomalous positions.

The main studies of aberrant innervation to anomalous regions in mammals have concerned the distribution of retinal afferents after tectal lesions made early in development. In hamsters, removal of the superior colliculus at birth results in an expanded optic projection to several regions which normally receive optic afferents and a major projection to two regions which receive few or no optic afferents—nucleus lateralis posterior (which receives afferents from the superior colliculus) and the medial geniculate nucleus (a thalamic nucleus processing auditory inputs) (Schneider, 1973; Schneider and Jhaveri, 1974; Kalil and Schneider, 1975). The latter projection only occurs if the auditory afferents to the medial geniculate nucleus have also been removed. Similar results have been obtained after comparable lesions made in fetal rats (Baisinger et al., 1977).

These results are interesting because they show that despite a major restriction of the terminal area available to optic axons, the axons do not project in a random fashion throughout the brainstem, but instead innervate some nuclei along the course of the optic pathway, even if these nuclei are totally unrelated to visual function. As to the underlying mechanisms, there are two things which appear necessary for the sprouting: an input deprived of its terminal field and a nucleus deprived of its normal afferents.

Particular attention has been paid to nucleus lateralis posterior. The optic projection to this nucleus is heavier if the tectal lesion is made earlier, including times when only a few optic axons have reached the colliculus. The projection also appears to be heavier if the optic projection which crosses the midline to the remaining intact superior colliculus (see Chapter 10) is restricted (Fig. 13-1). This last point has led Schneider to suggest that axons tend to occupy a finite volume of neuropil ("terminal space"): if that "space" is reduced in one position, it is likely to be expanded elsewhere (see Schneider and Jhaveri, 1974; Devor and Schneider, 1975). Extended to its natural limit, it might be suggested that a neuron has an intrinsic limitation as to the amount of axoplasm it can make or to the number of synapses it may form. This is an interesting point obviously worth intensive investigation.

Unfortunately, the basic experiments from which the theory derives are concerned with whole populations of neurons, probably at different stages of maturation, and nothing is known of the numbers of synapses involved nor indeed the number of axonal branches made by each neuron. Furthermore, there are some experiments which contradict the basic idea. For example, we saw in

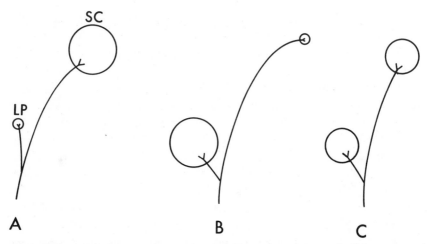

**Figure 13-1.** Diagram illustrating Schneider's theory of conservation of terminal space. (A) Normal animal with a large optic projection to the superior colliculus (SC) and a small (if any) projection to nucleus lateralis posterior (LP) of the thalamus, (B) result of unilateral tectal ablation—a small recrossed tectal projection and a large LP projection, (C) result of eye + tectal lesion (Fig. 10-12D)—a large recrossed tectal projection and a smaller LP projection.

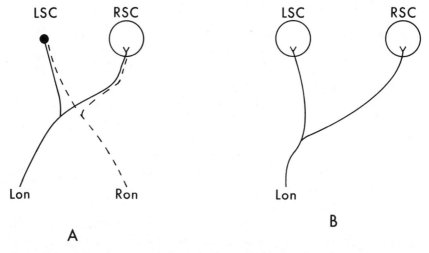

**Figure 13-2.** (A) Diagram of projection of left and right optic nerves (Lon, Ron) to the left and right superior colliculi (LSC, RSC), (B) removal of right eye doubles terminal space occupied by left retinotectal afferents.

Chapter 10 that if one eye is removed from an immature rat, the axons of the remaining eye will project bilaterally in the brain, effectively doubling their "terminal space," i.e., the volume of tissue in which they ramify (Fig. 13-2). Of course, this says nothing about the total volume of the individual axons or of the total number of synapses they may make, but this question is not addressed in Schneider's theory. In addition, the sprouting experiments described later in this chapter all tend to argue against the theory.

However, some support has been provided by experiments in which the area where olfactory afferents distribute is foreshortened by a transverse lesion across the base of the brain (Devor, 1975, 1976). In such cases, the olfactory afferents spread into adjacent cortical regions to which olfactory afferents do not normally project, as well as innervating the olfactory brain regions proximal to the cut. It is not clear whether these regions are deprived of their regular inputs by the lesion. If they were, the results might be thought to demonstrate a nonspecific sprouting into deafferented areas, rather than to provide direct support for a theory of conservation of terminal space.

## Specific Local Axon Distribution within a Region

Local areas can often be defined within a region in which particular axon populations show a restricted distribution. A number of regions have been studied in detail to see what factors may define such local partitioning of projections and to what degree the boundaries between two inputs may be modified after experimental intervention. All the cases to be described here concern the separation of projections in various regions into laminae, in particular in the optic tectum, superior colliculus, lateral geniculate nucleus, neocortex, hippocampal formation, and prepyriform cortex.

### Optic tectum

Maturana et al. (1959, 1960) identified five functional classes in the optic nerve of frogs (*Rana*): (1) sustained edge detectors, (2) convex edge detectors, (3) changing contrast detectors, (4) dimming detectors, and (5) dark detectors. Each class ramifies at a different depth in the tectum within a circumscribed lamina except for the changing contrast and dark detectors, which overlap in their terminal field. Since the dendrites of many cells in the frog tectum span the whole set of afferent laminae, it is possible, although unproven, that inputs from different functional groups may end at different levels on the same cells rather than necessarily ending on different cells. These authors (1959) note that if the optic nerve is cut and allowed to regenerate, then the laminar distribution of different response types is restored. Keating and Gaze (1970) reinvestigated this question in *Rana* and defined three response types—sustained edge detectors (1 and 2 of Maturana et al.), changing contrast detectors and dimming detectors— and found these to be tightly ordered in layers from surface to deep. On regeneration, the same relative order was maintained, although somewhat less tightly, even though the tectal layers may have shrunk.

Hunt (1975b) found that the laminar separation is maintained even if the visual map is totally disrupted by a "scrambling" procedure. By contrast, a study (Chung et al., 1973) in which frogs (*Xenopus*) were exposed to stroboscopic illumination for several weeks showed the opposite result: the visual map was normal but laminar separation of functional inputs was no longer evident. Whether this is due to actual changes in the depth distribution of optic axons in the tectum or to major changes in physiological properties of cells in the retina is difficult to determine; the results tend to suggest the former.

### Superior colliculus

The superior colliculus is a complex laminated structure (Fig. 13-3). The upper layers receive inputs from visual structures, the intermediate from other sensory systems, and the deeper layers from motor systems. (There are, in addition, brainstem afferents which terminate in several layers, but these will not be considered further here.) In the stratum griseum superficiale of rodents the visual cortical afferents end mainly deeper and the retinal projection is heavier more superficially. If the eye is removed at birth in rats, a dense, visual cortical projection occupies the whole depth of the stratum griseum superficiale (Lund and Lund, 1971a, Mustari and Lund, 1976). This is due to cortical afferents ramifying more heavily in the zone of dense retinal termination rather than just to shrinkage of the denervated superficial layers. However, it is not certain whether the additional terminals are the result of active sprouting or a failure of retraction of a corticotectal projection which at the time of lesioning is much more extensive than normal. Schneider and Jhaveri (1974) have indicated that if the visual cortex is removed from newborn hamsters, optic afferents end in higher density in the deeper part of the layer.

Evidence for or against sprouting between the visual and nonvisual layers is poor. It has been indicated that if the superficial layers of the superior colliculus of hamsters are removed at birth, optic axons distribute instead into the deeper layers. This could be explained by proposing that optic axons exhibit a modified

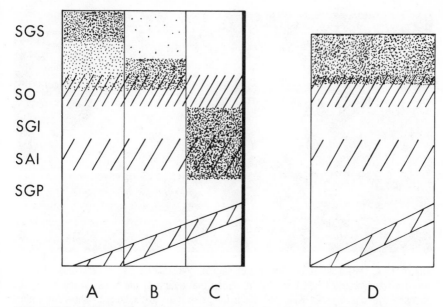

**Figure 13-3.** Section through superior colliculus showing principal layers: strata griseum superficiale (SGS), opticum (SO), griseum intermedium (SGI), album intermedium (SAI), griseum profundum (SGP). Layers with a preponderance of myelinated axons are indicated by stripes. The distribution of certain afferents in rodents is indicated. (A) From retina, (B) from visual cortex, (C) somatosensory afferent, (D) visual cortical pathway after contralateral eye removal at birth.

laminar specificity, or by the possibility that the lesion leaves some superficial cells intact—either because it was not quite deep enough or because at the time of the lesion, the cells had not yet migrated to the surface layers. Studies in which all the extrinsic afferents to a set of layers are removed and the projection to adjacent layers is tested have not been undertaken. For example, are the somatosensory inputs still confined to the intermediate layers if the retinal and visual cortical afferents are removed from the upper layers, or do the deeper-level afferents now sprout more superficially?

In a further study in which slits were made across the tectum of fetal rats, it was noticed that there were pockets of cells in the deeper layers of the superior colliculus, which in Nissl preparations resembled the cells of the superficial layers; in addition such cells received optic projections (Fig. 13-4; Miller and Lund, 1975). It was suggested that they were in fact superficial cells which still attracted optic afferents but had either been pushed down or prevented from migrating by the lesion. To test this possibility, a further attempt was made to

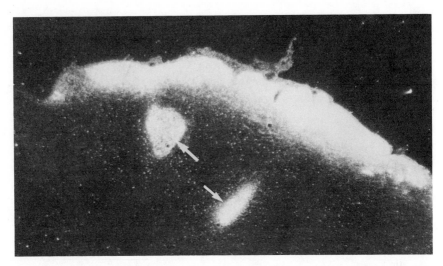

**Figure 13-4.** Dark-field autoradiogram of sagittal section of superior colliculus of rat in which a transverse slit had been made across the tectum of the fetus. When adult, the opposite eye was injected with ³H-proline (one day prior to fixation) to show the retinal projection (light areas) to the midbrain. Besides the regular projection at the surface there are two anomalous regions of optic innervation (arrows).

transplant tissue between fetal animals. Although technical difficulties have so far limited progress, a system of transplantation was developed which can be used to test many questions concerning special affinities which may exist between groups of neurons.

Successful transplantation of developing mammalian nervous system has been accomplished by a number of authors. Among the earlier studies, those of Dunn (1917) and Clark (1940) strongly suggest successful transplantation with good differentiation of the transplant. For success, both the transplanted tissue and the host must be immature, and for this reason rodents such as rats are ideal animals since at birth many systems have yet to complete development. One of the difficulties with the earlier studies was that there was no good way of labeling the transplant and the question of whether host tissue displaced in the implantation procedure was being misidentified as transplant must always be raised.

The use of ³H-thymidine to label transplants has helped a lot and the studies of Das (1974, 1975) attest to the fact that tissue can be transplanted from fetal or newborn rats into the cerebellum of newborn rats, a structure which is still embryonic at birth, and in which the transplanted tissue can be clearly identified, developing much as it would in tissue culture.

Lund and Hauschka (1976) have used this approach to investigate whether tissue placed in a region where axons normally arborise would automatically be innervated by those axons. Pieces of optic tectum, including all layers (visual and nonvisual), were transplanted from fetal rats to the region of the superior colliculus of newborn rats. Even though embedded in the visual layers of the superior colliculus, these transplants are not automatically filled with optic innervation. Instead, one or more discrete regions of the transplant receive the optic projection (see Fig. 3-8). These regions are usually histologically distinct from the rest of the transplant. Optic terminals forming synapses are readily identified with the electron microscope. Thus there is selective innervation of the transplant by host afferents, possibly to regions which normally would have been innervated by the specific pathway. This last point clearly needs further support but is rather important. If true, it would imply that even when taken out of the context of their regular position, tectal cells may still express selective affinity for optic axons.

In addition to receiving connections from the host, the transplants are very active in sending axons into the host nervous system. It would be interesting to know to what extent these efferent projections reflect the region of origin of the transplant, or the region in which it has been placed. Transplants of tectum placed slightly distant from the optic pathway, either deep in the superior colliculus or on the inferior colliculus, do not generally become innervated by optic axons, suggesting that (1) some sort of selective recognition process is occurring and (2) it is important to have the appropriate terminal region along the course of the growing axons.

One interesting extrapolation from these studies is that tissue taken from fetal animals may be maintained in tissue culture for as long as two weeks and then implanted into a newborn animal where it forms connections with the host brain (Lund and Hauschka, 1976, unpublished observations). Such an approach should allow one to monitor the affects of particular in vitro manipulations by testing the capacity of the tissue to make appropriate connections in the intact brain. In none of these transplantation studies have neuronal somata been identified which migrate from the transplant to the host. However, in those preparations grown in culture in a medium containing tritiated thymidine, there is a pronounced division of cells which do migrate for considerable distances from the transplant into the host brain. These appear to be nonneuronal cells, probably astrocytes. In one additional report (Das and Altman, 1971) migration of neurons from transplants was reported, but these too could well have been glia or other nonneuronal cells.

## Separation of crossed or uncrossed inputs
## to the lateral geniculate nucleus

In a series of studies the question of laminar separation of inputs to the lateral geniculate nucleus of the cat has been investigated (Guillery, 1972b; Kalil, 1972; Hickey, 1975). As was indicated in Chapter 10, in the normal nucleus there is a series of laminae separated by a relatively cell-free fiber-rich area. Lamina A receives input from the contralateral eye, A1, and from the ipsilateral eye; the deeper laminae (C, C1, and C2) also receive alternating crossed and uncrossed inputs. If an eye is removed in an adult cat there is transneuronal atrophy of the cells located in the laminae to which that eye projects. If the eye is removed in a cat in the first postnatal week, while trans-neuronal changes are severe in the denervated laminae, certain cells along the border adjacent to the innervated layer appear quite normal. Are they perhaps being "sustained" by an input crossing the laminar border from innervated to denervated layers?

Anatomical studies using autoradiographic and degenerative methods show this to be likely (Fig. 13-5). The sprouting occurs largely into the more lateral parts of the binocular segment of the denervated layer; it extends only a short distance into the denervated layer and its extent coincides well with the region of surviving cells in that layer. There is a small amount of sprouting from lamina A1 into the monocular segment of A, which is presumably to inappropriate parts of the visual map. In addition, there is sprouting from the optic tract into the underlying monocular segment of lamina A. If one eye is removed at 14 days postnatal or later, there is no significant spread of projection from normal to deafferented laminae.

These results suggest a partial invasion of the deafferented laminae by the axons distributed in adjacent laminae and by the optic tract. This assumes, although it does not appear to have been tested, that at the time of initial eye removal, the projections are clearly segregated. Why the extended projection should spread only a short distance into the deafferented lamina is not altogether clear. Hickey (1975) suggested that since transneuronal cell changes are rapid in young animals, sprouting occurs as long as there are healthy cells to be inner-vated and that cells beyond a certain state of atrophy are no longer capable of attracting axons. Since this phenomenon only occurs in young animals, it is also conceivable that it represents the last stage of the growth phase of optic axons.

A related study (Rakic, 1976) has shown that in monkeys at fetal day 78 the optic projection from each eye appears mixed and is not clearly segregated into the six distinct laminae characteristic of the adult, although such segregation has

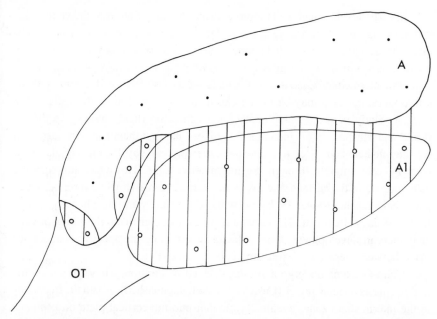

**Figure 13-5.** Cross section of laminae A and A1 of the lateral geniculate nucleus of an adult cat in which the contralateral eye was removed shortly after birth. There is pronounced cell shrinkage in lamina A (dots) except in areas into which the ipsilateral optic projection from the remaining eye (vertical lines) has spread. Normal-size cells indicated by circles, OT = optic tract (after Hickey, 1975).

taken place by day 124. In addition, if one eye is removed before that time, the projection of the other eye, when tested later, occupies the whole geniculate. Such a spread may be due to chiasmatic rerouting or to a failure of sorting, but in that case it would clearly not be due to interlaminar sprouting.

### The cerebral cortex

The cerebral cortex is composed of a series of layers, defined largely by the packing density and size and shape of cell bodies occurring at different depths. Besides this, each lamina has a unique pattern of dendritic branching and axonal terminations. For example, afferent axons from specific thalamic nuclei such as the lateral geniculate nucleus or nucleus ventralis posterior end predominantly in layer IV. Callosal axons from the contralateral cortex end heavily in the upper layers and more lightly deeper, in some cases avoiding layer IV. (See Chap. 2 and Lund et al., 1975, for further discussion of lamination.)

In early development, cells migrate in a radial direction from the ventricular and subventricular zones toward the surface, while axons grow in a tangential direction around the intermediate zone. Since cells destined for individual layers are formed within a limited time span, it is possible that the layer-IV cells might cross the intermediate zone just as the thalamic axons which will innervate them arrive there; contacts may be made; and then the axons follow the cells to the surface. Study of the development of the geniculocortical pathway in the rat (Lund and Mustari, 1977) indicates that indeed the first thalamic axons arrive in the intermediate zone deep to the visual cortex on fetal day 18, the day that most layer-IV cells migrate through the intermediate zone. However, although the cells migrate to the surface, the axons move much more slowly over a period of one week, reaching layer IV by postnatal day 4 and ramifying heavily in this layer 6–8 days after birth. Thus it would appear that coincident migration of cells and fibers in development is not a factor determining the specific connections made between geniculate axons and cells of layer IV.

Callosal axons undergo a similar delayed maturation. They first cross the midline in rats on fetal day 21 (Cusick and Lund, unpublished). Again the fibers stay in the intermediate zone, eventually moving into the cortical plate on postnatal day 4 and innervating their characteristic layers on days 6–8 (Wise and Jones, 1976). Again the cells which they innervate have also reached their appropriate layers some days earlier. Since thalamic axons end in layer IV of the somatosensory cortex and callosal axons tend to avoid this layer (Wise, 1975), Wise (1976) tested the idea that if the corpus callosum is cut early in development (before connections are formed), thalamic connections might spread beyond their usual region of termination. This does not seem to be the case. Conversely callosal axons do not spread into the thalamic projection area after early thalamic lesions.

A further study on the relation of cell migration patterns to connections (Frost and Caviness, 1974; Caviness, 1977) has used the reeler mouse. Here the cortex develops in an inverted fashion with the first-formed cells occupying layer II and the last layer VI; layer IV is formed at the same time as usual. The thalamic axons follow an unusual course as they enter the cortex. Instead of running up into layer IV from the white matter, they course in bundles obliquely through the cortex to the surface, where they curve downward to innervate layer IV (see Fig. 9-4). Thus, despite this major modification in growth pattern, the innervation is quite specific with respect to layer. In this, as in Wise's study, experimental modification of cortical organization, whether by removing inputs or by modifying lamination, does not seem to alter the specificity for terminal sites. A real affinity appears to exist between axons and particular cells.

### Hippocampal formation

The hippocampal formation is a particularly interesting place in which to study specificity of lamination, because here the laminar organization is solely due to laminar separation of different groups of axons on the dendrites of a particular cell type (see Chap. 2 and Fig. 2-14). The majority of studies have concentrated on the dentate gyrus. The granule cells of this region receive inputs from ipsilateral and association axons (from regions CA 3 and 4 of the hippocampus), from the septal nuclei, and from medial and lateral entorhinal areas. The septal afferents stain well with the cholinesterase method but have not been demonstrated very satisfactorily by degeneration techniques. The medial and lateral entorhinal afferents (via the perforant pathway) are mainly from the ipsilateral cortex, but recently a small number of crossed perforant axons have also been found (Goldowitz et al., 1975). Each of the inputs has a distinct laminar distribution (Fig. 13-6).

If the ipsilateral entorhinal cortex is removed completely, the other afferents to the dentate gyrus all show modification (Cotman et al., 1973; Lynch et al., 1972, 1973, 1976; Steward et al., 1973; Zimmer, 1973a,b) (Fig. 13-6). A substantial crossed perforant pathway develops, which is due to terminal sprout-

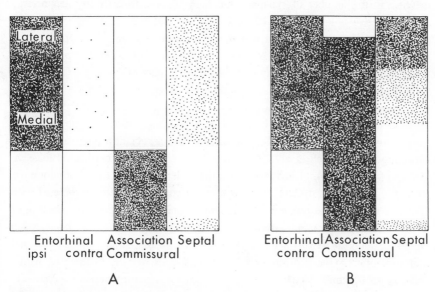

|        | Entorhinal  Association  Septal |                                | Entorhinal Association Septal |
|--------|---------------------------------|--------------------------------|-------------------------------|
|        | ipsi   contra  Commissural      |                                | contra   Commissural          |

A                                                                B

**Figure 13-6.** Laminar distribution of afferents in molecular layer of dentate gyrus of hippocampus (cell bodies of granule cells at bottom of picture). (A) Normal rats, (B) after removal of ipsilateral entorhinal afferents.

ing of the small numbers of crossed axons already in the area. Commissural and association terminals come to end more distally along the dendrites. Since by the time the entorhinal lesions are made, the commissural and entorhinal afferents are already strictly segregated as in adults (Loy et al., 1977), the extended distribution of commissural afferents after such lesions is clearly due to true sprouting of axons that extend their area of terminal distribution. Whether at an earlier stage the two inputs are mixed and then segregate out has not been determined. If the medial perforant pathway is left intact, allowing normal innervation to the intermediate third of the dentate gyrus, commissural and association axons do not grow through the normally innervated region to the denervated lateral perforant pathway termination area in the distal segment of the granule cells. In other words, sprouting does not apparently occur in response to a distant denervation, even along the same dendrite.

Electron microscopy of acetylcholinesterase-positive synapses (from the septal nuclei) shows a thirty-fold increase in such synapses mainly in the outer quarter of the molecular layer after removal of the entorhinal cortex (Cotman et al., 1973). However, light microscopic studies of acetylcholine esterase and choline acetyltransferase, (Nadler et al., 1973) show only a transitory increase in the amounts of each enzyme. The authors interpret this to be caused by an increase in the number of synapses that then become functionally less active. Zimmer (1974b) has offered the alternative suggestion that the increase in synapses may be transitory and that other terminals may displace septal axons. The problem really arises from using a biochemical label to define a pathway. It is impossible to correlate activity level and total number of synapses. Clearly this subject needs further study.

If the commissural axons are removed, no downward sprouting of perforant axons is observed. This could be due to intrinsic restrictions or to sprouting by association and local axons. To test whether the association pathway blocked sprouting, both Lynch et al. (1974) and Zimmer (1974 a) performed massive lesions of CA 3 and 4, effectively isolating the dentate gyrus from both commissural and association afferents while leaving the perforant pathway intact. Lynch et al., using degeneration methods after entorhinal lesions report no sprouting of the perforant pathway input; Zimmer, using Timm's method, suggests that the inner part of the dendritic tree is filled in with axons showing the staining characteristics of perforant fibers. One possible problem (apart from staining differences which should also be considered) is that it is extremely difficult to obtain a complete lesion that removes all association axons; a slight difference in the extent of the lesion in the two studies could have differential axonal effects. A further point is that the lesion damages the axons of the dentate granule cells. This complicates the issue somewhat since there may be retrograde changes which affect the dendritic development or cause proximal axon sprouting of dentate granule cells into the inner part of the molecular layer. Retrograde changes have not apparently been investigated. The possibility of a response by local cells in the dentate gyrus to partial deafferentation of the granule cells must also be entertained, and indeed studies of Nadler et al. (1974) suggest that these cells do increase their activity after entorhinal lesions.

Gottlieb and Cowan (1972) have found that the relative density of terminals of commissural and association axons within their normal region of distribution in the dentate gyrus varies from one part to another. They suggest that the two pathways compete for a finite number of synaptic sites and that the relative timing of ingrowth of each is the important determinant of how many synaptic sites they capture. The relative density of terminals depends on autoradiographic grain counts which, as already mentioned, may not be valid. Furthermore, one pathway is probably demonstrated by slow transport and the other by fast transport of labeled materials. In addition, this interpretation depends on there being a finite number of synaptic sites and a differential ingrowth of the two pathways which bears a closely matched temporal relation to the timing of maturation of the cells. While the matter of differential growth has been addressed (Fricke and Cowan, 1977), nothing is known about numbers of synaptic sites. Thus further work is necessary to support this interesting idea.

### Prepyriform cortex of rats

This region, situated on the base of the brain, is similar to the hippocampus in that it has a deep layer of cells with dendrites extending into a molecular layer. Into the molecular layer ramify two main projections, one from the olfactory bulb, which is distributed predominantly to the distal parts of the dendrites, and the other from adjacent cortical regions which distributes more proximally, there being little overlap between the two in an adult rat. If the olfactory bulb is removed at birth, the cortical afferents do not restrict their distribution to their usual sublamina, but instead ramify throughout the whole molecular layer (Westrum, 1975).

## Cell Specificity

One suggestion arising from the studies described above is that regional or laminar separation of inputs may in some cases be the result of special affinities existing between populations of axons and target cells. That growing axons selectively or preferentially connect with only certain postsynaptic cells may be seen in a number of cases. In the cerebellum of mutant mice, for example, although granule cells are drastically reduced in numbers, mossy fibers nevertheless end predominantly on these cells, even though they are in abnormal positions. Despite the presence of large numbers of postsynaptic specializations on Purkinje cells, for which there is no matching presynaptic component, diversion of mossy fiber terminals to these sites occurs only infrequently.

A particularly good example of cellular specificity of connection has been given for the ciliary ganglion (Landmesser and Pilar, 1970, 1974a). This ganglion has two types of cells: one which innervates the choroid muscles of the eye and the other which innervates the muscles of the iris and ciliary body. The afferent nerves from the central nervous system are of two types, distinguishable by anatomical, physiological, and pharmacological characteristics, and each type innervates only one cell type. If the nerves are cut and allowed to regenerate, each group reinnervates the appropriate cell type, although a few fibers were identified which appeared to make anomalous connections. Thus there appears to be a high affinity between each axon group and its appropriate target cells. However, this special affinity is not absolute.

## Neuromuscular Specificity

In neuromuscular systems, the special affinity between particular axons and their muscles appears somewhat less precise. Nerves can be redirected onto inappropriate muscles, innervate, and even (as we have already seen) change the properties of the muscles.

Why, then, should muscle innervation be so precise? This matter interested Weiss in the 1920s and 30s, and while the detailed concepts he developed are difficult to reconcile with modern neurophysiology, the basic idea that muscle fibers respond selectively to innervation from appropriate rather than inappropriate nerves still appears to hold to some extent (see Weiss, 1936).

The problem is particularly relevant to this chapter if individual muscles are compared with separate nuclei of the brain. There is an obvious experimental advantage to studying neuromuscular connections in that the appearance and performance of single nerve inputs to individual muscle fibers can be recognized both physiologically and anatomically. These connections can perhaps be more readily manipulated than central nervous synapses. Some muscle fibers receive input from single axons, while others have a multiple input, and it is the latter category which, as several authors have suggested, may serve as a realistic model for synaptic events in the central nervous system.

Recent interest in this area has been stimulated by the experiments and provocative conclusions of Mark and his colleagues (see Mark, 1974, 1978).

In one series of studies they investigated the interrelation of two nerves innervating the rotator muscles of the eye of the carp and goldfish. As shown in Figure 13-7, the superior oblique muscle (SO), innervated by nerve IV (trochlear), causes clockwise rotation of the eye while the inferior oblique muscle (IO), innervated by a division of nerve III (oculomotor), causes counterclockwise rotation. If the fish swims downward (Fig. 13-7B), nerve III and the IO muscle maintain the eye in a stable position, but if the fish swims upward (Fig. 13-7C) nerve IV and the SO muscle keep the eye straight. If nerve IV is cut and allowed to regenerate, normal eye orientation is restored. If, on the other hand, nerve IV is cut, the IO muscle is removed, and nerve III is directed to regenerate into the SO muscle, the eye rotates too far (Fig. 13-7) when the fish is tipped down, the nerve having formed functional synapses with the wrong muscle.

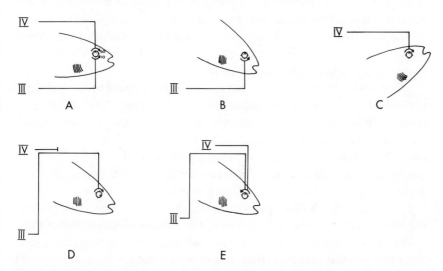

**Figure 13-7.** Role of cranial nerves III and IV in rotation of the goldfish eye. (A) Stimulation of III causes contraction of inferior oblique muscle (IO) and counterclockwise rotation of the eye; stimulation of IV causes contraction of superior oblique muscle (SO) and clockwise eye rotation. (B) Eye position corrected by III and IO. (C) Eye position corrected by IV and SO. (D) If IV cut and SO innervated by III, overrotation of eye when head pointed down. (E) IV allowed to grow back, giving normal eye position, the effect of nerve III being suppressed. The arrow in the eye indicates the dorsal pole.

The crucial point in the experiments results from seeing what happens if nerve IV is allowed to regenerate back into the SO muscle which is anomalously innervated by nerve III. The behavioral studies show that normal orientation patterns soon result (Fig. 13-7F). This has been confirmed by S. Scott (1975). Electron microscopic studies suggest that the synapses of the aberrant IIIrd-nerve innervation persist in this dual innervation. Mark and colleagues suggest that a mechanism of suppression of otherwise normally appearing synapses of nerve III by nerve IV has occurred. They point out that from this data one cannot therefore guarantee that synaptic connections seen with the electron microscope are functional. They go on to hypothesize that "mutual repression of different classes of synapses by characteristic chemical markers released in proportion to impulse activity would provide a mechanism for learning with no change in fixed synaptic connections" (Marotte and Mark, 1970). In contrast, S. Scott (1975), in further investigation of this system, has found that the aberrant IIIrd nerve can still drive the dually innervated muscle and, furthermore, that the effects of stimulating both nerves III and IV are to some extent additive. These results, together with recording from single fibers, indicate that individual muscle fibers are still innervated by functional synapses. Further, the mean quantal content of the excitatory junctional potentials of the foreign synapses does not appear to be different from the normal ones (Scott, 1976). These results would not be anticipated from Mark's model. What, then, is the basis for the behavioral suppression and is the physiological suppression proposed by Mark untenable?

Support for Mark's viewpoint has recently been provided by Yip and Dennis (1976a) in cross-innervation studies on the adult newt. They removed the forelimb flexor muscle and transplanted its nerve into the antagonist muscle, anconeus, which it then innervates if the normal nerve to this muscle is also cut. If the normal nerve is prevented from regenerating, the aberrant innervation was found to drive the muscle with increasing efficiency over time. If the normal nerve is then allowed to regenerate, synaptic efficiency of the aberrant junctions is greatly affected. This is due to a decline in the amount of transmitter released per impulse, there being a reduction in the number of quanta rather than a reduction in the amount of transmitter per quantum. Thus, although the aberrant nerve may still stimulate the muscle, it does so with enough failures to cause it to be relatively ineffective under normal physiological conditions. Unfortunately, anatomical supporting evidence is not yet available to show that the aberrant synapses are indeed still making contact in this case. Certainly, if the normal nerve is again cut, the aberrant pathway restores its efficiency quickly, suggesting at least that the nerve terminals are close by, if not in direct contact with muscle membranes.

A number of more recent studies (see Mark, 1978) provide additional documentation of suppression of inactive neuromuscular synapses. In all of these, the suppressive effect appears to influence transmitter release rather than the receptivity of the postsynaptic membrane, but what is the underlying mechanism? How could the "right" nerve suppress the presynaptic efficacy of the "wrong" one? Mark (1978) suggests that calcium influx into terminals which occurs at the time of transmitter release might be blocked in some way, and this would have the effect of blocking transmitter release. This is one of several possibilities. Another related question is whether suppressed terminals eventually degenerate. This is not known.

It should be added that suppressive mechanisms are not universal for all multiply innervated muscles. For example, Frank and Jansen (1976), working on the perch, cut a fin nerve and directed it into the gills. The nerve innervated these, provided that the normal innervation of the gills was also interrupted. In cases in which the gill nerve was allowed to regenerate back, double innervation of the gills was noted. Sometimes stimulation of gill or fin nerve produced gill muscle contractions of equal size while in other animals one or other nerve was more effective in driving the muscle. Both innervations were found to drive single muscle fibers. The authors concluded that foreign synapses need not be inhibited by the correct nerve.

The reason why suppressive effects occur in some cases and not in others—and indeed why suppressive effects occur at all—might be found in the developmental history of the particular system. We have seen already that individual muscle cells are initially innervated by several axons and the number is reduced over the succeeding weeks (Redfern, 1970; Bennett and Pettigrew, 1974; Brown et al., 1976). It is possible that some of those which are lost are inappropriate afferents: in regeneration, for example, some of the axons which first innervate the denervated muscle are from inappropriate neurons and these are later lost (e.g., Yip and Dennis 1976b). The first step in eliminating inappropriate afferents could be by a suppressive effect. Such an effect would only be necessary between nerves which (1) share an affinity for the muscle and (2) have courses which come into close proximity in normal development.

How may this phenomenon relate to the central nervous system? One obvious difference is that neurons seem to have quite broad affinities for muscles, as is evidenced by the various cross-innervation studies (see Close, 1972), while neurons in the central nervous system apparently have more selectivity of innervation between particular subpopulations, which could in part be achieved by a suppression phenomenon. In addition, in Chapters 15 and 16 we will see that there are connections which, although functionally ineffective in driving cells, can be unmasked by lesions of dominant pathways and these too might be existing in a suppressed state, in the manner envisioned by Mark.


## Discussion

From the results described in this chapter, a number of factors can be identified which may be important in limiting the extent of terminal distribution shown by an axon.

1. Connections cannot be made between neurons unless some special affinity exists between them. Some of the possible mechanisms responsible for special affinities have been indicated in the previous chapter. Here a further issue arises and that concerns the possibility that special affinities do exist between some cells which do not normally interconnect. The reasons that they do not normally interconnect even though they have the potential to are several: (1) there may be a hierarchy of affinities and axons may only connect with cells lower in the hierarchy if higher ones are unavailable, (2) other afferents may have already occupied the available synaptic sites on some cells, (3) there may be interaction between axons which mutually limits their distribution, and (4) the particular axon may not grow close enough to the region containing cells that it has the potential to connect with. At present, individual experiments offer some support for each of these possibilities, and it is likely that all play a part in restricting what might otherwise be a rather broadly programmed pattern of special affinity.

2. There is substantial plasticity of afferent distribution to cells with which the axons have an affinity (even if this affinity is not normally expressed): plasticity of afferent distribution cannot occur into regions containing cells with which they have no affinity. This statement suffers from a logical weakness in that affinity is defined by the potential to form connections, thus introducing a circular argument. Nevertheless, it is worth making because it emphasizes the contrast between rigidity and plasticity of brain organization.

3. There may be intrinsic controls which limit the amount of axoplasm and or the number of synapses a neuron can make. This is a natural extrapolation from Schneider's theory of "conservation of terminal space," and while the theory as stated does not hold up well to careful inspection, it remains possible that it could be applied, with some restatement, to the behavior of single cells. The apparent exceptions in which neurons expand their terminal field could be accounted for by proposing that they still make the same number of synapses or that the full self-limiting potential is not normally expressed because competitive interactions between axons prevent any one axon from making its full complement of synapses.

These statements are obviously speculative; although consistent with current studies, they lack experimental verification. We will see in the next chapter that these same three proposals may well be important not only in defining regional or laminar specificity but also in determining how axons distribute relative to one another in a homogeneous sheet of cells.

# 14

# THE CORRELATION OF MAPS
# IN THE BRAIN

There is a primary requirement for an animal to know not only the specific quality of external stimuli to which it is exposed, but also where those stimuli are located and what their spatial order is. Knowing these qualities, the animal can then direct a response in the right direction. Appropriate to this need, a map of the animal's external world is preserved throughout the pathways in the brain leading from the primary sensory nuclei to the systems involved in directing an oriented response. The most obvious maps are those of the visual world (which we have already mentioned a number of times) and of the body surface. These are not the only ones, however: auditory signals are spatially coded according to tones (as well as for position), and olfactory signals are coded according to particular chemicals. In addition, many regions of the brain preserve their own spatial order in their projections to other regions, even though the logic of such projections is not always immediately obvious.

The level of resolution of a map is often extremely fine. This is certainly the case in the visual system where, in humans, the level of acuity which can be distinguished corresponds to the width of a retinal receptor. Since acuity measurements depend on processes occurring at a cortical level, this detailed map must be preserved through a series of synapses, without any loss of precision. How does this happen? What controls dictate the establishment and maintenance of such a precise pattern?

These questions were first asked in reference to the visual system by Sperry and Stone who exploited the observation of Matthey (1925) that if an optic nerve of adult newts is cut, it regenerates and vision is restored. Sperry (1944, 1951) noted that the axons appear to grow across the region of the cut in a disorganized manner, and yet he found, as had Stone and Zaur (1940), that the final pattern of connections formed was sufficient for the animal to regain its normal visuo-motor orienting responses to a lure as well as normal optokinetic nystagmus responses to a moving drum. In other words, despite the apparently haphazard nerve growth, the optic axons and second-order connections are able to make appropriate patterns of connections such that the animal can respond normally to visual signals. Subsequent neurophysiological studies of Gaze (1959) showed that this order is reestablished by the optic axons themselves, since the regenerated optic terminals distribute over the optic tectum, giving a normal visuotopic map (Figs. 14-1A, 14-2B).

In a further series of experiments, Sperry (1943a,b; 1944) cut one optic nerve of adult newts and a variety of frogs and toads, and rotated the eye through 180° so that the original dorsal retina was now ventrally placed and the anterior (or nasal) retinal posteriorly placed. The advantage of frogs and toads over newts is that they react more precisely to lures, so that the visual behavior can be tested more accurately. The question asked in these experiments was: would the re-

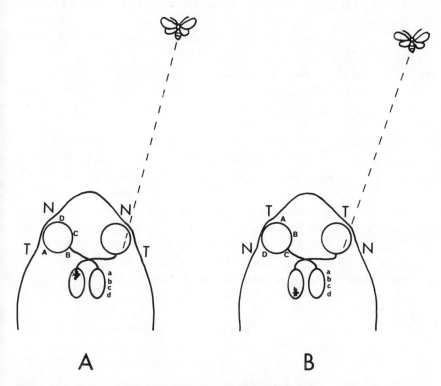

**Figure 14-1.** Eyes and optic tecta (OT), viewed from above, in (A), a normal frog and (B), one in which both eyes have been rotated through 180° and temporal retina (T) is now anterior and nasal retina (N), posterior. The projections of a fly and of four retinal points ABCD are indicated on right and left tecta, respectively.

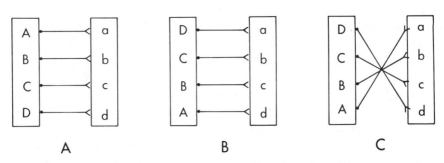

**Figure 14-2.** Schematic of retina and tectum showing four positions across the retina, A–D connecting to four positions across the tectum a–d. (A) Normal, (B) after eye rotation prior to stabilization, and (C) after eye rotation post-stabilization (see Fig. 14-1).

generating axons from the inverted eye make connections that would adapt for the inversion and make functionally appropriate connections, or would points in the retina reconnect with the same points on the tectum as before, even though the eye was turned upside down? Testing the orienting response to a fly on a lure, Sperry found that no adaptation occurred. Thus when the fly was presented above and in front of the frog, the frog would strike below and behind and so on. Gaze (1959) confirmed by physiological methods that the retinotectal map is, in fact, organized as might be predicted from the behavioral experiments (Figs. 14-1B, 14-2C).

In additional experiments (Sperry, 1945; Stone, 1948, 1953) on adult newts and late tadpoles, eyes were transplanted from one side to the other with or without rotation. This meant that the nasotemporal axis was reversed, but the dorsoventral axis was normal, or that the dorsoventral axis was rotated and the nasotemporal was normal. In these cases, as in the case of uncrossing the optic chiasm, the visual responses showed no indication of accommodation to the aberration; this was again borne out by physiological studies mapping the visual field on the tectum (Straznicky et al., 1971; Beazley, 1975).

These results indicated two things to Sperry. First, in view of the disorganization of regrowth, the distribution of axons in the optic nerve would not seem to be a factor determining their pattern of mapping in the tectum, and second, function clearly does not dictate connections, at least with respect to the formation of a map. He therefore proposed (Sperry, 1944) that regrowing optic fibers must possess specific properties of some sort by which they are differentially distinguished in the visual centers according to their respective retinal origins. This statement has been elaborated in a series of extended reviews (e.g., Sperry 1951, 1965) and has been expounded most succinctly in a short paper (Sperry 1963). In this he postulated that neurons must carry some kind of individual identification tags, presumably cytochemical in nature, by which they are distinguished one from another almost, in many regions, to the level of the single neuron; and further, that the growing fibers are extremely particular when it comes to establishing synaptic connections, each axon linking only with certain neurons to which it becomes selectively attached by specific chemical affinities.

With these experiments, Sperry provided two things. First, he showed that studies concerned with the development and restoration of maps were useful in providing insight into the general problems of neural specificity, and second, he offered a hypothesis, the testing of which has provided a focus for much developmental neurobiology. For this reason, we will devote considerable attention to the subsequent experiments on retinotectal mapping, before considering two other systems in which patterns of mapping are of special interest—the geniculocortical pathway and the peripheral cutaneous innervation.

## Retinotectal Maps

A number of events might be expected to take place in the process of establishing a coherent map of the retina upon the tectum. In early development, the cells of the retina and of the tectum develop a spatial identity by which they "recognize" a positional difference with respect to their neighbor. In order to account for the eye reversal experiments already mentioned, it is necessary that the cell's spatial identity become irrevocably fixed at some point: this process is sometimes misleadingly referred to as "specification." Finally, the axons connecting the two regions must be able to relate to the spatial identity cues they have acquired to provide an ordered and coherent map of the retina on the tectum.

In order to investigate these processes further, discussion will be divided into several categories: the nature of topographic specificity, the evidence for special affinity, the development of positional identity (both with respect to polarity and radial position), and the effects of various manipulations performed on adult fish and amphibians and in the later stages of development of birds and mammals.

### Nature to topographic specificity

There appear to be two components to the information a cell uses to establish its positional identity, and these are to some extent dissociable by various experimental manipulations. The first concerns polarity—the definition of dorsoventral and mediolateral axes. Polarity seems to be determined very early in development; and indeed once even four cells are present in a sheet of cells destined to be the retina, a crude polarity could potentially be established. The second component of positional identity is radial specificity, i.e., knowing where a cell lies with respect to the center and the edge of the map. In an amphibian retina where growth is by addition of cells around the periphery, a cell cannot "know" its radial position until cell division has ceased (unless it has some knowledge of the final cell number). As a result, this parameter might be expected to be more susceptible to modification at much later stages of development, and, as we will see, this appears to be so. That these two components may be dissociated has been curiously overlooked in the literature, but a recognition of this separation goes some way toward resolving some of the fundamental contradictions which have arisen in recent years.

The problem of positional identity is not restricted to the nervous system. It has received considerable attention in studies of *Hydra,* insect cuticle, and slime molds. Largely from those studies there has emerged a number of theoretical papers which attempt to provide mechanisms consistent with known data to explain how positional identity may be acquired. Among these are papers by Wolpert (1969, 1971), Goodwin and Cohen (1969), Crick (1970), and McMahon (1973). Of these studies Wolpert (1971), in particular, has made a significant attempt to relate the theoretical plan to the visual system, emphasizing that the mechanisms whereby an ordered change across a sheet of cells (a gradient) is established are likely to be the same, wherever encountered. The most frequently proposed mechanism requires that a substance be produced at one end of a group of cells (source) and be removed at the other end (sink). The substance must be freely diffusible between cells, and a certain proportion of it must be taken in by each cell, such that those closest to the source will absorb a maximum amount and those closest to the sink will absorb least. It is expected that the molecule in some way alters the behavior of the cells in a graded manner so that each cell expresses its positional identity in a unique fashion. One attraction of the proposal is that if the number of cells in the sheet is reduced and the sink is closer to the source, the map will accommodate to the change and form in logical order within the reduced space. Such accommodation could provide a basis for the size disparity experiments to be described later.

The various models based on this mechanism vary in requiring active or passive diffusion, or proposing a metabolic effort by the cell to maintain a constant amount of the substance against a concentration gradient. In order to provide polarity information, it is necessary to have at least two different molecules and some authors have suggested that the actual situation may be even more complicated. The mechanism proposed by Goodwin and Cohen (1969) differs in that it requires a periodic double "signal" to be emitted at some position. The signals travel across the sheet of cells at different rates and the temporal delay between them will increase, the further away the cells are from the source: this phase difference could signal the positional identity. Another approach by McMahon (1973) suggests that rather than substances being freely diffusible between cells, it is the surface contact which is important and the asymmetry is determined by the distribution of receptor molecules on the surface membrane.

One further concept derived from general studies which is applicable to the visual system is the rule of distal transformation (Rose, 1962; Wolpert, 1971). It has been particularly applied to insect limb regeneration and states that after being cut off, a limb will regenerate only segments distal to it. Thus a proximal limb grown in culture will regenerate all distal segments, producing a whole normal limb again, while a distal segment will provide a mirror reduplication of itself.

A requirement of all these theories when applied to the retina is that the diffusible material or whatever is involved must provide a graded and irreversible change in some parameter of the ganglion cells such that they will establish subtly different connections. Several possibilities may be considered. The cells may make different surface proteins, which result in their having different affinities for cells of the tectum, as proposed by Sperry. Alternatively, the different surface proteins may allow cells to recognize their relative position with respect to their neighbors. The latter proposal forms the basis of the "arrow" model of Hope et al. (1976). Further proposals include the possibility that electrical activity in the optic nerve may be used to maintain spatial identity (see Chung, 1974).

All these proposals require that the cell acquire some sort of positional label which it uses to establish its ordered projections. The alternative must also be considered: that each cell occupies a position on the map, as a passive phenomenon, that axons are ordered within the optic nerve relative to the geometry of the cell body, and further that this order is responsible for the ordered map on the tectum. Such a passive positional identity, although not consistent with the eye rotation experiments already described or with translocation studies to be discussed later, cannot be totally overlooked.

### Evidence for special affinity between retinal and tectal cells

Among the various hypotheses of neural interrelations, some attention has been given to testing directly the special affinity model proposed by Sperry. The important point is to study the interrelation independent of the interferences of fiber ordering. This has been achieved in several ways.

In an ingenious study, DeLong and Coulombre (1967) took pieces of developing chick retina and placed them on the surface of the optic tectum. The fragments grew axons into the optic tectum and the authors were interested to see whether axons would grow into any region of the tectum irrespective of the position of the retina from which they were taken or whether the axons from each piece of retina would course across the surface to find the appropriate part of the tectal map and terminate there. The results suggested the latter, and this study has been presented for many years as powerful support for Sperry's hypothesis. Unfortunately, Goldberg (1974), working in Coulombre's laboratory, discovered a different interpretation for the extended fiber passage from pieces of transplanted retina to areas of tectum. In development, the chick tectum rotates through 90° between days 7 and 13, and also becomes displaced from the midline. Goldberg suggested that the retinal transplants become attached to the skin and shortly after transplantation connect with the tectum immediately below them. This area of tectum moves differentially relative to the skin and the axons between the two elongate, giving the impression of a directed tract. Most, if not all, of Delong and Coulombre's observations may be explained in this way.

A different approach to the question of direct tests of the concept of special affinities between retinal and tectal cells has been attempted in a recent set of studies (Barbera et al., 1973; Barbera, 1975). In chick, the dorsal retina projects to the ventral tectum and the ventral retina projects dorsally. Retinas from 7–12 day chick embryos were divided into dorsal and ventral halves. The cells were dissociated, labeled with $^{32}$P, and then put into a rocking dish; a set of dorsal and ventral half tecta were pinned onto the bottom of the dish. After a period of exposure, in which the dissociated cells made many chance contacts with the surface of the half tecta, the tecta were washed off to remove nonattached cells and the radioactivity due to adhering retinal cells was measured. From this an estimate of the number of adhering cells could be given. In general, about 10,000 or less of the million retinal cells plated made contact with both appropriate and inappropriate tectal halves. In the best cases, there were twice as many contacts with the matching pairs (i.e., dorsal retina/ventral tectum or ventral retina/dorsal tectum) compared with the other combinations. No difference in the ratios is found if the age of the tecta is varied from 8–14 days, or if they are innervated or not innervated from the eye prior to removal.

Nonneural tissue shows no special adhesiveness to the tectum, but cerebrum shows a preferential adhesion to the ventral tectum. Some sort of affinity between cerebrum and tectum might be expected since these regions do interconnect; but why it should only be expressed in the ventral tectal half is not altogether clear since the connections involve both halves of the tectum. Besides the neural retina, the pigment epithelium taken from dorsal and ventral halves also shows selectivity in its tectal adhesiveness; therefore, the suggestion has been made that it is the pigment epithelium which carries polarity information and that such information is transmitted to the adjacent neural retina (Keating, 1976).

These affinity studies are both exciting and disappointing. They show a pattern of selectivity which in most cases mimics that found in the intact animal, but a map based on such small differences in adhesiveness would surely be very imprecise and on its own would fail to provide the resolution required of behavioral and physiological studies and predicted by Sperry's hypothesis. Since the retinal sample is composed of all cell types, and the ganglion cells (the only ones connecting with the tectum) are likely to be in a minority, it is possible that they show a better selectivity but are masked by other types showing less or no selectivity. Barbera (1975) discounts this proposal on the grounds that there are proportionately more ganglion cells at earlier times, but the selectivity of adhesion does not change with age. It is possible that the differential adhesiveness is subtly graded and cannot be easily detected with the rather gross procedure used in this study. It is a pity that the studies did not investigate the effect of far-ventral compared with far-dorsal retina, since there is a clear discontinuity between the two and the difference in adhesion between matching and nonmatching parts ought to be greater.

A similar approach has been followed by Balsamo et al. (1976) in two experiments. In one, they collected the supernatant from tectal, retinal, and cerebral fragments grown in a medium containing $^{14}C$ glucosamine. They presumed that the labeled supernatant came mainly from cell-surface molecules, but of course the $^{14}C$ could also be released into the supernatant from cells which died while in culture. The supernatant from each region was put on monolayer cultures of cells dissociated from different retinal quadrants, tectal quadrants, or cerebrum. After washing, the radioactivity was measured, giving an indication of what degree of binding there was to the monolayer. The authors found that tectal factor binds poorly to monolayers of cerebrum or tectum, but that it binds well to retina. Binding is best to appropriate quadrants, although the differential binding between matching and nonmatching quadrants is on the order of 3 to 2 or less.

A second approach employed the phenomenon of cell capping by the plant lectin, concanavalin A (Capping is a phenomenon in which receptors on the membrane which bind to the lectin tend to cluster in groups, forming the appearance of a cap when labeled). Since the retinal binding factor inhibits capping in

80% of the retinal cells which show this phenomenon, the authors tested whether it would show a local effect if taken from isolated quadrants of tectum and put on monolayers of cells from retinal quadrants. They found, as before, a preference for matching quadrants—this time of about 2 to 1 at best.

It is somewhat disconcerting that the preference in affinity in matching compared with nonmatching quadrants is still at such a low level and also that in this case there is only a low level of binding for both cerebrum and tectum, since cerebrum connects to tectum and there are ample intratectal connections. If indeed this assay is a true reflection of connectivity affinities in the brain, one would expect good binding in both these cases. Unless this anomaly can be explained, one has to wonder how relevant these findings are to in vivo development. Along the same lines, one may question whether the cell-body region of a neuron whose dendrites and axon have been wrenched off will show the same affinities as the axon tip. Since, in the intact brain, axons often have a number of branches showing quite separate courses, and these differ from dendritic outgrowth patterns, it might not be unreasonable to postulate that different surfaces of the same cell may express different affinities.

If these studies show a crude reflection of a cellular basis for spatial recognition (based on differential adhesiveness or some type of surface membrane binding), then is spatial specificity itself loosely programmed? Are additional influences needed to provide a detailed map? Or do these results bear little or no relation to the mechanisms involved in the correlation of maps of different brain region? These questions will be borne in mind in the later sections of this chapter where various manipulations of the retina and tectum have been carried out in an attempt to disturb the mapping between them.

### Development and stabilization of positional identity

This matter has been addressed by a number of authors, but the studies of Szekely (1954) and Jacobson (1968a) and in particular the series of papers by Hunt and Jacobson (summarized by Hunt, 1975a; Hunt and Jacobson 1974a,b; and Jacobson and Hunt, 1973) have provided a coherent view of how positional identity is acquired, at least in amphibians. These experiments have addressed only the matter of polarity; this is tested by rotating the eye (to invert the dorsoventral axis) or transplanting it from left to right orbit (to invert the anteroposterior axis) at early development stages and seeing at what point the map changes from a situation (prior to stabilization) in which the retina accommodates for the reversal and maps normally on the tectum (Fig. 14-2B) to one in which, like Sperry's early experiments, the map is reversed (Fig. 14-2C).

Szekely (1954) found in newts that the anteroposterior axis becomes fixed before the dorsoventral axis and Jacobson (1968a) was able to confirm this for *Xenopus* where the anteroposterior axis is established at stage 30 (Nieuwkoop and Faber, 1956) and the dorsoventral axis at stage 31, a few hours later (at 20° tank temperature). This stabilization of polarity information occurs at a time when fewer than 500 of the total population of 100,000 ganglion cells are postmitotic (Jacobson, 1968b; Straznicky and Gaze, 1971) and when it appears that gap junctions (which allow free diffusion of small molecules between cells) between ganglion cells are being lost (Dixon and Cronly-Dillon, 1972).

Such correlation might suggest that as long as free diffusion of "polarity" molecules between cells is possible, then the polarity label expressed by single cells is modifiable. Once cells become separated from their neighbors, whatever polarity label they have at the time can no longer be altered. A similar explanation would hold if the positional information was determined in the pigment epithelium rather than the ganglion cells, since, as they become postmitotic, ganglion cells lose their association with pigment epithelial cells.

One question is whether there is in fact any evidence of retinal polarity, even in labile form, prior to this stage of stabilization to a fixed polarity. This was investigated (Hunt and Jacobson, 1973a) by removing an eye at stage 22, well before stabilization of polarity, maintaining it in tissue culture, and then replacing it in the orbit of an older animal. When tested later it would map on the tectum according to its orientation at the time of removal. Furthermore, if the eye had been rotated for six hours before removal from the stage-22 embryo, it would still map normally with respect to the orientation of the eye at the time of explantation, suggesting that the positional information in prestabilized eyes can be rapidly modified. Jacobson (1976b) found that if a stage 24–25 eye is placed in culture for four to six hours and then returned to the orbit, the eye polarity adjusts to that of the host. If, however, the eye is put in culture with an ionophore (X537A) which in other systems causes uncoupling of cells, the retinal polarity is no longer modifiable: if the eye is inverted, so is the retinotectal map. This again suggests that as long as intercellular communication is possible, the polarity is modifiable: as soon as it is broken, polarity becomes fixed. Further experiments (Hunt and Jacobson, 1972) showed that if an eye is taken from a prestabilization stage and transplanted to the body wall of another embryo, it adapts its polarity according to the orientation in which it is placed on the body wall. This implies that polarity information is not a unique property of the orbital tissue and can be provided by other parts of the body.

These results indicate that polarity in the eye becomes established very early and relates to the eye's position relative to the body polarity. Up to a particular stage of development, the eye polarity is modifiable by the context in which it is placed, but once stabilized, polarity information appears to be relatively difficult to alter.

Owing to the considerable technical difficulties, comparable studies on the development of polarity in the tectum are few. In one recent study (Chung and Cooke, 1975), the tectum in *Xenopus* was rotated at stages 21–24 or at stage 37. In both cases it was found that the retinal projection did not recognize the reversal unless the caudal diencephalon was included in the reversed tissue. The authors argue that an organizer of tectal polarity may reside in the caudal diencephalon. It would be presumed that at a late stage a polarity becomes established by the individual tectal cells which is independent of extrinsic influences. Such an expectation is supported by the results (to be described later) of tectal rotation or translocation in adult fish and amphibians.

### Development of radial identity

Radial identity has been approached in two ways: (1) by studying how retinal fragments, substituted for the whole eye early in development, map onto the tectum and (2) by forming "compound" eyes made from two halves which may be identical (e.g., double temporal or double nasal—Gaze et al., 1963; Straznicky et al., 1971) or different (e.g., left eye temporal and right eye nasal or temporal half and ventral half—Hunt and Jacobson, 1973b).

The retinal fragments are particularly interesting because they do one of two things: provide a projection which maps coherently over the entire surface of the tectum or provide a map which reduplicates itself like the compound eye shown in Figure 14-3 (Feldman and Gaze, 1975). Some even do both. McDonald (see Feldman and Gaze) has attempted to explain this by using the rule of distal transformation. He envisions that, for the eye, proximal positions represent the central retina composed of the first-formed cells, while distal positions are represented by the peripheral retina. If a fragment contains the central region, then, according to the rule, a whole retinal map would be formed, but if the central region is not included a reduplicated map would form. While this explanation seems reasonable, it does not easily explain a case in which both a normal and reduplicated map occur in the same animal.

The work on compound eyes that were initially constituted of two temporal halves or two nasal halves at stage 30–32 showed that each half projects to the whole tectum in overlapping fashion (Fig. 14-3). These studies were particularly important in that they provided a contrary result to what would have been expected from the rigid affinity between retina and tectal points proposed by

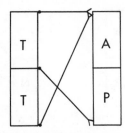 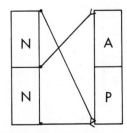

**Figure 14-3.** Mapping of retinas made of two temporal halves (TT) and two nasal halves (NN) on the tectum. (A) Anterior, (P) posterior.

Sperry. One criticism, that the region of tectum normally receiving temporal retina has overgrown while the other half has degenerated, was answered by uncrossing the chiasm and allowing the compound eye to grow instead into the previously normally innervated ipsilateral tectum (Straznicky et al., 1971). Again an overlapping map results. This pattern suggests that each half of the compound eye, despite the physical continuity between them, behaves as a whole with respect to its tectal projection and indeed it may even have a separate optic nerve. It suggests that the definition of the border of the field is extremely important as a signal to the cell of the cell's relative position in the sheet. Further studies on compound eyes using temporal and ventral halves (so that each has a common quadrant—ventral-temporal) show that they do not map such that this quadrant from each half overlaps in its projection, but instead behave as independent entities (Hunt and Jacobson, 1973b).

These results suggest that the positional label a cell acquires (as expressed in its projections) with respect to center and edge of the retina can be modifiable much later in development than polarity labels can. We will see that while there are very few cases in which polarity can be reversed by experimental manipulation in adult animals radial position is still susceptible to modification.

### Size disparity, spatial disparity, and mismatch experiments

These studies test how the map of the retina on the tectum is affected if an experimental manipulation is performed on one or the other. There are three principal experimental approaches. The first approach, size disparity, involves removal of part of the retina or tectum and seeing whether the remaining fragment maps with its partner as it did previously (Fig. 14-4B) or whether there is regulation such that the fragment becomes connected with the whole area of the intact partner (Fig. 14-4C). The second approach involves rotation or translocation of part of the tectum, or recombination of tectal fragments to see whether the mis-

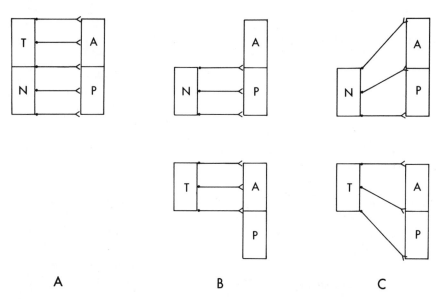

                A                        B                        C

**Figure 14-4.** Projection of retina to tectum (A) in normal animals, (B) in animals with either nasal (N) or temporal (T) retinal halves removed, with no regulation, (C) in animals with half the retina removed and regulation of the retinotectal projection. Anterior tectum—A; posterior tectum—P.

placed region connects with the usual retinal cells or whether the connections made in the misplaced region adjust to the context in which they find themselves. Finally, mismatch studies test if axons from one region of the retina can be forced to grow into an inappropriate remnant of the tectum.

Most of the relevant work has been done on adult goldfish and the frogs, *Rana* and *Xenopus*. More recently, however, comparable studies have been performed on the developing visual systems of chick, rat, and hamster; the results are in reasonably good accord with those obtained from the regeneration experiments. The experiments differ from those already described since in fish and frogs they concern restitution of a map in animals operated as adults while in mammals and chick the injury causing modified connections is at later stages of development than those considered so far. They are outlined below with a preliminary comment concerning technical considerations.

### Technical considerations

There are three ways in which the retinotectal map has been demonstrated and none is totally satisfactory. The most commonly used method is physiological. In most experiments on frogs and fish an electrode with a fairly large tip is lowered into the optic tectum. A stimulus is waved about within the field of view and the boundary of the region in space causing a response at the recording position is recorded. A number of problems associated with the use of the approach are enumerated below.

a. The electrode appears to record activity of preterminal axons: it does not demonstrate the exact location of synapses. It is generally assumed that selectivity does not change according to experimental conditions, but this has not been tested. Such a possibility is important because in some cases an altered map could result from the electrode recording activity in the axon further away from the synaptic terminals than normal, rather than from a reorganization in the distribution of the pathway. Apart from this, the minimum number of synapses or axons necessary to elicit a physiological response is not known.

b. The stimulus is a visual one and so the map might be more appropriately termed visuotopic rather than retinotopic (Hunt and Jacobson, 1974a). This is not merely splitting hairs, but is especially important if the retina has been directly lesioned, because the visual map could be distorted by mis-shaping of the eye and by damage to the lens. Any disturbance of the optics of the eye will modify the visuotopic map with no effect on the retinotopic map.

c. The methods of presentation of results are confusing, and somewhat misleading. Since the technique involves finding visual-field positions for tectal coordinates, maps are always presented of visual field relative to tectal position, even though (as in the case of tectal rotations) it might be logically presented the other way round. For this reason, a two-dimensional simplification of the experiments is given here, which, it is hoped, is more comprehensible. One further problem is that in most charts, the center of the visual field for the electrode position is given and not the full extent of the field. This gives the erroneous impression of a tighter specificity than what actually exists.

d. Because of the curvature of the tectum, most recordings are restricted to the map of the upper visual field.

Anatomical methods may be used to identify the projection area of the retina. Such methods are applied by injecting the eye with a tritiated amino acid which will be transported to the tectum only by axons of intact ganglion cells, or by removing the eye, precipitating degeneration of the axons of remaining ganglion cells. Subtotal lesions can be made in the retina to map the relation of the retina to the tectum. If the lesion in the retina is long-term, it is often very difficult (especially in a small eye) to assess accurately the area of the lesion. Studies using normal silver stains to identify absent projections resulting from lesions are not absolutely reliable. Retinotectal relationships may also be mapped

by allowing a short survival after a retinal lesion and charting the tectal degeneration it causes. There are two problems: first, it is possible to chart only one
position on each animal (although combined with autoradiography two or three
positions can be mapped—see Miller and Lund, 1975), and second, it is impossible to define the extent of the lesion accurately. not only are ganglion cells in
the area of damage destroyed, but axons from more peripherally located cells will
also be cut.

A third way of defining the retinotectal connection is behaviorally; usually
the investigator produces an orienting response to a stimulus presented within the
visual field. Although this approach provides good correlative data to anatomical
and physiological results, its weakness is that it presumes no compensation of
visumotor performance if the visual map is disturbed. One interesting recent
study (M.Y. Scott, 1975) involves heartbeat conditioning to visual stimuli,
which would seem to be a particularly good way to map a scotoma without
involving a guided motor response.

### Retinal lesions

The first size disparity studies (Attardi and Sperry, 1963) showed that if
retinal lesions are made in goldfish, the remaining retina does not expand its
projection on the tectum to occupy denervated areas. The results may be
criticized in that the only methods then available to the workers (normal silver
stains) are unsatisfactory and also that the survival time is rather short. Nevertheless, subsequent anatomical (Meyer, 1975) and physiological (Horder, 1971)
studies have shown little spread of temporal retina into the posterior tectum (Fig.
14-4B, bottom) although there does appear to be as shown in Fig. 14-4C, top)
spread of nasal retina into anterior tectum (through which the axons run on their
way to the posterior tectum).

In chick (Crossland et al., 1974) after quadrantic retinal lesions, even of the
upper temporal retina which denervates more anterior tectal regions, there is no
regulation of the projection from the remaining retina to innervate the whole
tectum (Fig. 14-4B, top). Similarly, lesions made in the retina of perinatal rats
and hamsters frequently result in areas of tectum receiving a minimal projection
from the lesioned eye (Lund and Lund, 1973, 1976; Frost and Schneider, 1975,
1977). While it is true that these areas become innervated by axons from the
other eye as well as from the visual cortex, even after removal of all these other
sources, there is still a gap in the retinal map (Frost and Schneider, 1975). The
gap corresponds in approximate position with the projection area of the lesion,
although it is extremely difficult to define correspondence of lesion size and gap

area. Such correspondence is important because it has been noticed (Lund and Lund, 1976) that the region adjacent to the gap may be innervated by anomalous retinal positions leading to the possibility that part of the gap has been filled by axons from intact cells in the lesioned eye. Anomalous mapping was also described for the aberrant uncrossed axons from the normal eye which fill the gap in the crossed projection from the lesioned eye (see Fig. 10-10). The gap is filled by axons arising from a large part of the normal retina and not just from the position homotypic to the lesion. Such a result argues against special affinity as a necessary determinant for establishing ordered patterns of retinotectal connections, and suggests that some measure of fiber ordering is important, although where and how this occurs remains to be determined.

While a confusing picture emerges concerning the effects of retinal lesions, it does appear that optic axons do not readily expand into a larger area of available tectal tissue.

### Tectal lesions (Fig. 14-5)

After partial tectal lesions in fish (e.g., Gaze and Sharma, 1970; Yoon 1971; Sharma 1972a,b; see Keating, 1976 and Meyer and Sperry, 1976 for further references), newborn hamsters (Schneider and Jhaveri, 1974; Schneider et al., 1975; Finlay et al., 1977), and young frogs (Jacobson and Levine, 1975a), the whole retina ultimately projects in a coherent fashion to the remaining tectum, compressing its projection into the available space (Fig. 14-5C). Compression occurs slowly if at all in older frogs (e.g., Straznicky, 1973), but can be accelerated by uncrossing the optic chiasm (Gaze and McDonald, unpublished). In goldfish, there is clear evidence that initially axons grow back to their normal

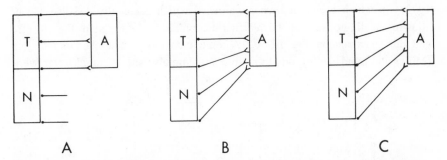

**Figure 14-5.** Effect of posterior tectal lesion on the retinotectal map. (A) Shortly after the lesion, (B) at an intermediate stage in regulation, and (C) at a later, presumably stable stage.

sites (Fig. 14-5A) and that later there is a gradual compression starting from the edge of the scotoma (Fig. 14-5B) and continuing to the periphery of the visual field (Fig. 14-5C). This has been shown well behaviorally using a cardiac conditioning response to visual stimuli (M.Y. Scott, 1975) and has also been demonstrated physiologically by recording several times from the same animal (Cook and Horder, 1974; Yoon, 1976b). One question of some dispute is that if the nerve is not crushed along with the tectal lesion, is there, together with the compression, some evidence of continued normal mapping? Several earlier authors in this field (e.g., Gaze and Sharma, 1970) found such mapping, although it is not mentioned in more recent studies. Studies showing both compression and normal mapping are interesting in that figures presented suggest pockets of tectum doing one or the other, rather than direct overlap between the two. It would be interesting to follow up these results both anatomically and physiologically, using recording methods with better resolution.

One question never addressed is what happens to the axons which have been deprived of their tectal area of termination in the period prior to compression? One would presume that they may cross to the opposite tectum, since, as we have already seen, such a crossed pathway does occur after tectal lesions. If so, what is the interrelation between the crossed projection and the compressed projection to the lesioned side? Are they axons of different cells or branches of the same axons? These are rather crucial questions and deserve attention. If branches of the same axon, do they therefore form a compressed map on one side and a normal map on the other?

Transverse slits have also been made across the tectum. In goldfish Yoon (1971) found that if these are permanent, the retinal map tested physiologically compresses on the rostral part of the tectum despite the physical presence of a caudal tectum unavailable to optic axons (Fig. 14-6A). Further work (Yoon, 1972a) in which a gelatin barrier was used which eventually dissolves away allowing healing of the slit showed first a compression of the map (Fig. 14-6B) and then an expansion (Fig. 14-6C). A comparable study undertaken on fetal rats (Miller and Lund, 1975) found that in many animals there is complete recovery

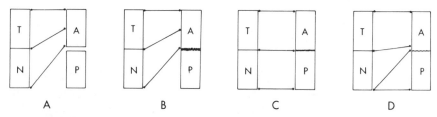

**Figure 14-6.** Results of tectal slits in goldfish and rats. See text for description.

after tectal slits, giving a result like Figure 14-6C. In some animals, however, a duplicate map of parts of the nasal retina was found, one position as normal and the other in front of the slit (Fig. 14-6D). The temporal retina does not show a duplicate map.

## Mismatching of retinal and tectal segments (Fig. 14-7)

Such mismatching has been investigated in two studies in goldfish. In one by Yoon (1972b), the temporal hemiretina and caudal tectum were removed; the optic nerve was left untouched. In three animals, the nasal retina formed a coherent map over the foreign rostral tectum. In the fourth, the map showed a mirror reversal of its caudal limit. In subsequent testing, the reversal disappeared and a normal map was found. Anatomical studies by Meyer (1975) showed that after nerve crush, nasal retina grows into foreign rostral tectum and temporal retina into foreign caudal tectum.

## Tectal rotation and tectal translocation

These two operations have been used to see whether the tectum is a passive receiver of the retinal input or whether the retinal axons "search out" points in the tectum even though these points are misplaced. The results from different experiments are quite discordant. Of the studies in which part of the tectum was rotated, those of Sharma and Gaze (1971), Yoon (1973, 1976a), and some of those of Jacobson and Levine (Levine and Jacobson, 1974; Jacobson and Levine, 1975a) show that the optic axons "recognize" the rotation (Fig. 14-8B). Such reversal occurs even when the part of the tectum has been reversed and the map,

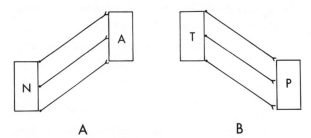

A                                            B

**Figure 14-7.** The effect of removing noncorresponding retinal and tectal regions on the distribution of axons from the remaining fragment.

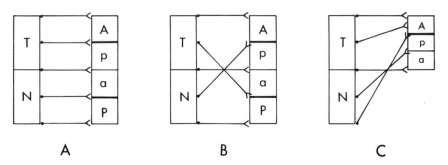

**Figure 14-8.** The effect of rotation of a fragment of tectum, ap. (A) The retinotectal map adjusts, (B) the retinotectal axons regrow to their original synaptic sites, (C) rotation and compression resulting in axons regrowing to their original sites but compressing as well.

normal and rotated, is compressed upon the remaining tectum (Fig. 14-8C, Yoon, 1977). On the other hand, other results of Jacobson and Levine indicate that optic axons organize in a coherent manner relative to one another, as if treating the tectal tissue as a substrate in which to end rather than a set of specifically programmed loci to which the axons have to relate (Fig. 14-8A).

Similar inconsistencies are encountered in experiments in which regions of tectum are translocated from right caudal to left rostral tectum (Jacobson, and Levine, 1975b, Sharma, 1975) or from caudal to rostral on the same tectum (Hope et al., 1976). Some animals form a continuous map across the tectum, while others map according to the position of origin of the transplant. Sharma found that the variation was a function of time and, like the compression studies, axons would initially map to a position appropriate for the position of origin of the tissue and then "regulate" to form a coherent single map across the tectum.

The easiest compromise in explaining these and other more recent results of similar design is to accept Sperry's original proposal that some form of special affinity guides axons to their target regions on the tectum, but that there is, in addition, an interaction between the fibers in which they attempt to maintain their normal contextual order. Such interaction may override the affinity matching if it occurs between the fibers in the transplant and host tectum. The variation may depend on the degree of contiguity existing between transplant and host.

## Summary

These results, contradictory though they sometimes appear to be, generally support the comments made earlier in this chapter, namely that polarity is determined early in development and that the relative position a cell occupies in the map is open to modification for a much longer time. The studies involving rotation of the eye and of the tectal fragments, as well as those concerning the initial regeneration patterns which follow partial tectal lesions, argue for the presence of special affinities which are important in matching appropriate retinal and tectal fragments. However, it might be supposed that these are not the only factors involved in guiding axons to their proper place on the map. Axons themselves appear to interact with one another and to adjust to some extent to the area of tissue available to them. Within that area, they have the capacity to form a coherent order among themselves. This capacity is particularly evident in cases of compression; it also appears to be the case in rats having local retinal lesions where an aberrant uncrossed pathway fills the deafferented gap, even though the map in the region of the gap may be topographically inappropriate.

Several points seem to be important with respect to proposed axonal interactions. One is the contextual relation of adjacent axons, which may perhaps be disturbed by establishing borders—in some cases, clearcut surgical divisions; in others, territories limited by certain pathway distributions. In addition, the actual territory on the map occupied by any single axon appears to be conditioned by interaction of the axon with its neighbors: in the compressed state all the axons crowd more tightly, not just the ones at the back of the tectum. Finally, the studies of partial retinal lesions show, in most cases, limited expansion across the map of optic axons from intact retina adjacent to the lesion. This implies that some constraint exists to limit the lateral expansion of an axon population in such cases.

It is possible in time that further factors, in addition to axon interactions and special affinities, may be identified which control the formation and maintenance of the retinotectal map, but even with these two, it becomes possible to account for certain of the contradictions by supposing that the particular conditions of the experiment emphasize one or the other. At the very least it would appear that the retinotectal map is not the product of a single regulating process.

## Mammalian Geniculocortical Map

In the relay of visual information to the primary visual cortex, area 17, there are two matters concerning mapping which are of particular interest. One involves the mapping of the input from two eyes in the monkey and the other, the map found in the Siamese cat.

### Cortical map in monkeys

In the lateral geniculate nucleus there is a series of laminae, each having a continuous map of the contralateral visual field, but each receiving input only from a single eye. As can be seen in Figure 14-9, the laminae supplied by each eye overlie one another such that a single point in space is represented in register traveling vertically from one lamina to another. In layer IV of the primary visual cortex, although there is a complete map from the geniculate laminae serving each eye. the geniculocortical inputs from each eye do not overlay one another, but instead are segregated into adjacent stripes 400 μm wide (Fig. 14-9; Hubel and Freeman, 1977; Hubel et al., 1977). This means that as one progresses tangentially across layer IV, there will be discontinuity in the map every 400 μm. This is illustrated in Figure 14-9. Points 1–6 on crossed and uncrossed laminae of the lateral geniculate nucleus respond to corresponding positions on the visual field. In layer IV of the visual cortex, these become segregated according to the plan indicated. Above and below layer IV the position in the field to which a unit responds varies considerably (even for adjacent neurons) such that although, as indicated in Figure 14-9, there is an ordered progression across the map in a gross sense, there is a local almost random progression.

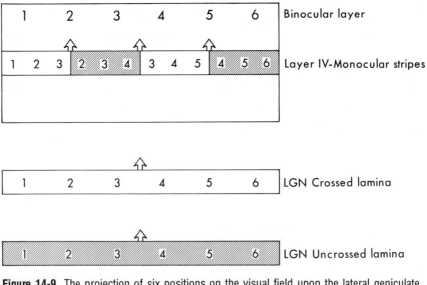

**Figure 14-9.** The projection of six positions on the visual field upon the lateral geniculate nucleus (LGN) and visual cortex (top block). Shaded LGN and cortical layer IV receive input from the ipsilateral eye, their unshaded areas from the contralateral eye.

One matter of interest concerns the development and stability of this pattern. Rakic (1976) has found that in fetal monkeys even after the input from the two eyes has become segregated in the lateral geniculate nucleus, little evidence exists for segregation at a cortical level. Segregation is gradually achieved over a period extending beyond the first month postnatal (Hubel et al., 1977). If one eye is removed shortly after birth, the cortical stripes which would have been served by that eye are narrower, while those served by the other eye are wider than normal. As we will see in the next chapter, the relative width of the stripes can also be changed by closing one eye shortly after birth. In all cases, however, the total width of left eye plus right eye stripe remains the same. Thus while there appears to be a stable progression of the visual map across the cortex, the extent to which each eye feeds into it can be altered by the input, which functions to regulate the amount of retraction of axon terminal area that the input served by each eye experiences.

## Siamese cats

We have seen already (Chapter 10) that in Siamese cats, cells in the retina lying temporal to the vertical midline representation project in anomalously large numbers to the contralateral side of the brain. Some of these end aberrantly in the principal laminae of the lateral geniculate nucleus. Further work has shown that associated with this additional crossed temporal input, the visual map on area 17 of the visual cortex is highly abnormal and that two distinct patterns of projection are noted. In one type, the "Midwestern" cat (Kaas and Guillery, 1973), as shown in Figure 14-10, lamina A maps in a normal manner in area 17, the vertical midline being represented at the junction between areas 17 and 18. The more lateral part of lamina A1 projects normally, as does the medial normal segment. The abnormal segment gives a topographic visual field map overlapping, in mirror image, the map from lamina A. This suggests that the abnormal segment of lamina A1 is projecting as normal to the cortex, and is unaffected by the error in the retinogeniculate pathway.

Such a conclusion is hard to reach in the other form of Siamese visual cortical map, the "Boston" cat (Hubel and Wiesel, 1971) shown in Figure 14-10. In this case, the border between areas 17 and 18 receives input from at least 20° into the ipsilateral visual field, and corresponds therefore with the projection from the most lateral part of the abnormal segment. The overall progression of the map is indicated in Figure 14-10. In addition, some mirror symmetric positions comparable to those found in the Midwestern cat are also recorded. Without suggesting some massive reorganization in the distribution of

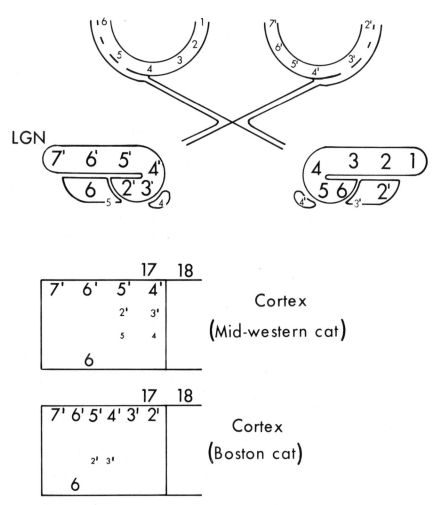

**Figure 14-10.** Map of retina on lateral geniculate nucleus (from Fig. 10-9) and on visual cortex of two varieties of Siamese cat. In cortex, size of numbers signifies frequency with which particular response pattern is found.

geniculate cells, it is hard to escape the conclusion that the organization of the geniculocortical pathway has been modified as a result of the aberration in the retinogeniculate pathway. Even stronger evidence of a secondary reorganization of a pathway is provided by the callosal connection. In normal animals, it is distributed at the border of areas 17 and 18, but in the Boston Siamese cat, it is more broadly spread, extending from the representation of the vertical midline in areas 17 and 18 and encompassing the abnormal ipsilateral field representation lying in between (Shatz, 1977). This raises the possibility that the distribution of a pathway in a map is determined not only by special relations which exist between axons and the cells on which they end, but is also dependent on the pattern of inputs from sensory receptors. A further example of this is provided by the secondary interconnections between the tecta of frogs, which adapt to aberrations introduced in the retinotectal map (Gaze et al., 1970; Keating, 1974).

### Cutaneous map

The skin of the trunk and limbs is innervated by the sensory processes of the cells whose somas lie in the dorsal root ganglia. Ganglia in each segment of the cord innervate a particular patch of skin and there is partial overlap between areas of innervation of ganglia of adjacent segments. A large number of studies, both clinical and experimental, conducted on a variety of vertebrates, have shown that if one sensory nerve is damaged, the adjacent nerves will sprout to occupy the area of skin formerly innervated by the damaged nerve. In the experimental studies, two major issues have emerged. One concerns the dynamic interaction between adjacent cutaneous nerves, and the other concerns the effects of peripheral innervation on central connections.

Several interesting studies have been carried out on the hindlimb innervation of salamanders. The aim is to define the nature of the interactions between adjacent sensory nerves and between sensory nerves and skin. In *Amblystoma tigrinum,* the tiger salamander, the hindlimb is innervated by three nerves, 15, 16, 17, as shown in Figure 14-11. If nerve 16 is cut, the sensory fields of 15 and 17 (tested by their response to light touch) expand to cover the area formerly occupied by nerve 16. Of particular interest is the finding that if, instead of cutting nerve 16, colchicine is applied to it for a short time and then washed off, nerves 15 and 17 again extend their sensory fields even though, as far as can be seen, nerve 16 has not degenerated and can still carry impulses (Aguilar et al., 1973). Colchicine, as we have seen in Chapter 3, blocks fast axoplasmic transport. The time course of expansion of the adjacent nerves is the same as after cutting nerve 16. If nerve 16 is cut and nerve 15 treated with colchicine, nerve 15 does not expand its territory. Further experiments (Cooper et al., 1977) have charted the numbers of mechanoreceptors in the salamander skin, their distribution, and their thresholds to stimulation. After colchicine treatment, there is no change in any of these properties, thus confirming the early experiments showing that there is no transitory degenerative stage which itself stimulates the sprouting, rather than the effect being related to a failure of fast transport systems. In addition, after cutting one nerve and allowing sprouting from adjacent nerves,

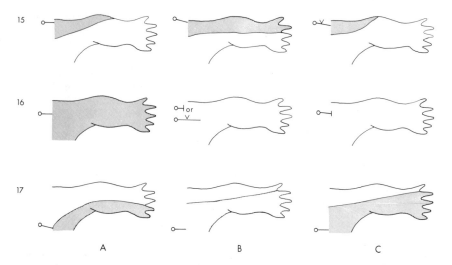

**Figure 14-11.** Innervation pattern of hindlimb of salamanders by nerves 15, 16, and 17. (A) Normal, (B) nerve 16 cut or colchicine administered (indicated by V). Colchicine administration results contrary to diagram in a normal receptive field for 16, but the same expansion of 15 or 17 occurs. (C) Nerve 15 given colchicine and nerve 16 cut.

the number of mechanoreceptors, their distribution, and their thresholds are restored to normal levels, suggesting that some regulatory behavior exists which controls these parameters.

The implication of this work is that sensory nerve processes are constantly adjusting their terminal distribution in relation to their neighbors. Sprouting plays a major part in this process and seems to be particularly dependent on the integrity of protein transport systems, rather than necessarily on the physical presence of axons or degeneration products. Two possibilities could be envisioned. In one, the nerve may secrete a substance from its terminals which has an inhibitory effect above certain concentrations to growth of other axons. Alternatively, axon growth could be inhibited by contact, as it is in culture (e.g., Dunn, 1971). The integrity of the surface molecules essential for this process may depend on a continuous stream of materials from the cell body. One slight problem with this study is that definition of sensory fields depends on physiological responses to light touch. It is conceivable that processes are already present but nonresponsive. If this were the case, such suppression of sensory responses would, perhaps, be more remarkable than the concept of sprouting, but since suppressive mechanisms have been indicated at neuromuscular junctions (see Chap. 13), they cannot be overlooked.

In another study, this time on the axolotl (Johnston et al., 1975), limb nerves were cut and allowed to regenerate, and it was found that while they reoccupied their approximate sensory field, individual fibers, instead of innervating single receptive fields, often innervated several. Whether this innervation became more restricted with time was not investigated. If the nerves on opposite sides of the limb were cut and crossed, they grew into the area to which they were redirected and innervated it. Thus, at least with respect to peripheral processes, sensory nerves appear to spread in the area to which they are directed and establish a field by interaction with neighboring nerves. Nerve–skin spatial specificity would appear to be relatively loosely programmed, although, of course, one does not know whether the skin may have some influence on the pattern of central connections made by the sensory neuron.

A second approach to cutaneous innervation would appear at first to contradict the idea of lack of spatial specificity between nerve and skin. Thus, Miner (1956) transplanted skin from the back to the belly and vice versa in frog tadpoles. These animals metamorphosed, and the frogs grew up with a patch of pale skin on their back and pigmented skin ventrally. Stimulation of this skin caused anomalous wiping responses to the position the skin would normally have occupied, except at the edge of the graft where normal reflexes were elicited. In two cases, high-threshold normal responses were obtained.

There are two possible explanations for the transposed responses: (1) the original nerves could have sought out the transplanted skin but retained their previous central connections, or (2) adaptation had occurred in the central connections of the nerves innervating the peripheral region to which the skin was transplanted. The first option requires a special affinity between regions of skin and certain nerves, something which would not be expected in view of the foregoing discussion. The second option requires that a region of skin modulate central connections.

This phenomenon has recently been reexamined (Jacobson and Baker, 1969; Baker and Jacobson, 1970; Baker, 1972). They have confirmed that if a large enough piece of skin is reversed in *Rana pipiens* tadpoles before stage 15 (Taylor and Kollros, 1946) then misdirected wiping responses result. They found a number of additional interesting facts: (1) From shortly after reversal of the skin up to several days after metamorphosis, normally directed responses were elicited, but then misdirected responses began to be elicited along with normal responses, and finally predominated. (2) Reversal of response did not occur if the area of skin transplanted was smaller than 40 mm$^2$. (3) Reversal did not occur if the transplantation was done later than stage 15. (4) In cases where no reversal occurred, nerves innervating adjacent normal skin crossed the boundaries of the transplant, but when reversal of response occurred, this did not happen.

The authors suggest that these findings, together with limited anatomical investigation of nerve growth patterns, support the concept that peripheral neural processes are not being redirected to find the patch of skin they normally innervate. This leads to the possibility that, providing neurons innervating the aberrantly placed skin patch do not also innervate the adjacent normal skin, their pattern of central connections may be modified. Such a possibility clearly needs further attention. It would seem important to confirm that the transposed skin is indeed being innervated by nerves which would normally innervate the region to which it has been transposed, and that this innervation does not change with time. One should also find out how broadly the central connections of individual nerves are spread and whether this changes if the peripheral field is modified; when and how the central connections develop; and how much degeneration (even transganglionic—see Westrum et al., 1976) occurs with skin transplantation. Furthermore, it would seem important to know whether the changes in central responses are anatomical or primarily physiological, comparing with those shown by Wall and his colleagues (see Chap. 15). It is unfortunate that the transplantation experiment has not been explored further with some of these points in mind, since, as it stands, it suggests a major mechanism—that the pattern of axonal connections of a neuron may be modified by the tissue to which its sensory processes are exposed. In this respect, it is reminiscent of the Siamese cat and of the frog intertectal pathway, where disturbance of a primary input appears to have major secondary consequences to other pathways.

In summary, the available evidence suggests that peripheral sensory neural processes interact with one another in the establishment and maintenance of sensory fields. The territory a neuron occupies seems to depend on the integrity of its neighbors and on their having an intact transport system. Sensory neurons exhibit a broad specificity as to the area of skin they innervate. The transplantation experiments raise the question of whether this is always true or whether the peripheral associations can in some way modify the central connections which seems more likely.

## Summary

The development of a topographically ordered connection between two regions depends first on the cells in each region acquiring a spatial identity which, at least with respect to retinal polarity, becomes fixed early in development. The mechanism whereby spatial identity is determined is still largely unknown, although it appears likely to arise from the establishment of a chemical gradient to which the cells in a sheet respond differentially. In the matching of the maps it appears that the cells of two sheets which have common spatial identity may exhibit a special affinity one to another, which plays a part in forming a coherent projection. However, there is good indication that this is not the only factor: interaction between axons appears equally important. Such interaction is evident not only in the retinotectal mapping but also in the geniculocortical projection of monkeys and in the cutaneous innervation patterns.

In the case of cutaneous innervation it appears that the relative spacing of axons in a map may be determined by materials released by the individual axons and it would be very interesting to know whether similar mechanisms occur in the central nervous system. A further point—which derives from the Siamese cat studies, studies of the intertectal pathway in frogs, as well as possibly the skin transplantation studies in frogs—is that the functional patterns a pathway transmits may also play a part in how it maps or how pathways interacting with it map. Such a mechanism has particular significance in cases where separate mapped projections converge on a region and have to be correlated one with another, as for example the intertectal pathway in frogs and the visual callosal pathway in mammals.

# VI

# THE ROLE OF FUNCTION
# IN DETERMINING CONNECTIONS

In terms of an animal's behavioral response to a stimulus, two extremes are encountered. In one, a single stimulus presented at a rigidly set time early in development will cause an invariant response to be elicited if that same stimulus is presented at any point throughout the life of the animal. This is the rigid imprinting behavior shown particularly well in birds. At the other extreme, the stimulus may be presented at any time in the animal's life, it must usually be presented a number of times, and the appropriate response may or may not always be elicited each time the particular stimulus is given after the initial training period. This is typical learned behavior. Between these extremes are various combinations of the two.

One of the most interesting questions, in view of the large amount of attention directed to it, concerns the development of response patterns in the visual system. These patterns, like imprinting behaviors, require an appropriate stimulus during a fixed period early in development—the so-called critical period or sensitive period—to mature normally and once this period is over, a change in response pattern is difficult to achieve. The principal differences from classical imprinting are that the sensitive period lasts a matter of weeks rather than hours, and the entraining stimulus requires more than one presentation. The great attraction of studying visual development, apart from its relative ease of manipulation, is that not only can the whole behavior of the animal be studied, but also the response of single cells can be investigated. As such, visual development provides a good system for assessing the role of function in determining a neuron's behavior, and may provide an insight into the more complex and less accessible mechanisms of memory and learning.

# 15

# ENVIRONMENTAL INFLUENCES ON
# VISUAL RESPONSIVENESS

A host of clinical studies have shown that disturbances of the optics of the eye during early life will cause a long-lasting impairment of vision, even though the optical defect may later be corrected and no pathological changes are evident in the retina (see for example Von Senden, 1960; Von Noorden, 1967). The general condition is referred to as amblyopia and affects as many as 6% of sample populations (Schapero, 1971). It may be broken down into four broad categories (Von Noorden, 1967).

1. Deprivation amblyopia (also called amblyopia of disuse, amblyopia ex anopsia), in which a visual image is prevented from falling upon the retina as a result of congenital or traumatic cataracts, corneal opacity, or clouding of the ocular media.

2. Ametropic amblyopia, in which as a result of uncorrected refractive errors early in childhood, there is a permanent reduction in acuity. This is particularly noticeable for astigmats (in whom the visual image is blurred in one orientation).

3. Anisometropic amblyopia, in which as a result of blurred vision in one eye, that eye has reduced acuity and tends to be suppressed by the good eye.

4. Strabismic amblyopia, in which as a result of a major misalignment of the eyes, the acuity in the deviant eye is reduced and vision in that eye is "suppressed" by the good eye.

Since in each of these cases little recovery of vision can be expected from correction of the optical defect in later life, the source of the problem is presumed to be in the central nervous system. The clinical studies present essentially two phenomena: one in which the amblyopia results from a total or partial deprivation of form vision and the other in which it arises as a result of an imbalance between the two eyes, the good eye actively suppressing the other. The latter may be termed "suppression amblyopia" and is best exemplified as a result of squint in which the visual image may be perfectly good on the disadvantaged retina. In other cases, such as a unilateral deprivation amblyopia or anisometropic amblyopia, the effects observed might be expected to result from both deprivation per se and competitive suppressive effects.

Earlier behavioral studies (reviewed by Riesen, 1966) clearly showed that amblyopia is not unique to man, but can be elicited in a variety of animals. In subsequent work major emphasis has been given to the cat and more recently to the monkey largely as a result of the pioneering work of Hubel and Wiesel (Hubel and Wiesel, 1963, 1965, 1970; Wiesel and Hubel, 1963a,b; 1965a,b; 1974). Discussion here will be centered around cat and monkey; research on other animals will be briefly reviewed separately. A variety of experimental rearing conditions imposed during a limited period of development have been shown to produce aberrations in performance which are difficult or impossible to reverse in later life. These include binocular deprivation (performed by suturing both eyelids closed, by the use of opaque contact lenses, or by rearing in the dark), monocular deprivation (suturing the eyelids of one eye), deprivation of a congruent image seen by the two eyes because of squint (induced by cutting some of the extraocular muscles), and partial visual deprivation, (in which certain parameters of vision such as stripes in all but one orientation are missing). The consequences of this controlled rearing have been analyzed behaviorally, anatomically, and physiologically, in the latter case by recording the patterns of action potentials (firing patterns) produced by individual cells in response to a variety of visual stimuli. In order to understand the nature of the physiological changes caused by the deprivation, it is obviously important to know how cells in normal animals respond, particularly in the visual cortex, where most of the major deprivation effects are encountered.

## Normal Response Patterns of the Primary Visual Cortex, Area 17

On the whole, cells of the visual cortex fire hardly at all, or at relatively low rates, unless an object is placed in a particular part of the field of view of the animal, this area being described as the receptive field of the cell. Even then the cell will only respond well if the object is presented in a certain manner; most cells of the cortex respond best to bars or edges moved across the field in a specific manner. They are particularly sensitive to the orientation of the bar and to its direction of movement. Usually the stimulus is more effective when presented through one eye than the other; when presented to both eyes simultaneously, the magnitude of the response varies with the binocular alignment.

The orientation specificity of a cell is said to be tightly tuned when the cell fires only to bars presented in a narrow range of orientations and broadly tuned if responses occur to a wider range of orientations. It is thought that the orientation selectivity is brought about in part, at least, by local inhibitory circuits within the cortex. Support for this proposal comes from the fact that iontophoretically applied bicuculline (an antagonist of GABA, an inhibitory transmitter in the cortex) reduces or abolishes orientation selectivity of cortical neurons (Sillito, 1975) and from intricate analyses of response properties of normal cortical neurons. If an electrode is moved perpendicularly through the cortex, the cells recorded from are likely to respond to the same orientation, irrespective of their depth in the cortex. On the other hand if the electrode is moved obliquely or tangentially through the cortex, there is generally a gradual shift in the optimum orientation to which the cell responds. This has led to the concept of perpendicularly disposed orientation columns formed of cells responding to the same orientation with adjacent columns of cells responding to slightly different orientations. Whether these really are discrete columns in the formal sense or are merely due to a gradual shift of optimal orientation response over the cortex is not altogether clear (Hubel and Wiesel, 1974).

While direction selectivity is easy to identify if a cell fires when a bar is moved across its receptive field in one direction but not in the opposite direction (without changing the orientation), some care must be exercised in dissociating orientation and direction selectivity, especially if the testing stimulus is a moving bar. This question is discussed by Pettigrew (1974) and Stryker and Sherk (1975).

The degree to which one or the other eye drives a cell is a matter of considerable importance in studies of visual deprivation. It was noted in early studies on normal visual cortical physiology that many cells were driven by both eyes, although not equally by each. Hubel and Wiesel (1962) devised a convenient scheme, depending on an arbitrary set of scales 1–7, for expressing the relative importance, or dominance, of each eye in firing the cell. Class 1 cells are driven only by the eye contralateral to the recording electrode. With increasing numbers, the cells are progressively influenced more by the ipsilateral and less by the contralateral eye, class 4 being influenced equally by both and 7 by only the ipsilateral eye. A typical ocular dominance histogram appears in Figure 15-1A. It is collected from a composite of cells in the cat cortex and this kind of pattern serves as a baseline for many experimental studies. There is an important qualification which must be made, however. It is clear that in certain layers (Stryker and Shatz, 1976) and perhaps regions of area 17 (Albus, 1975), there is a tendency toward a segregation of the inputs from the two eyes resulting in an ocular dominance histogram similar to that shown in Figure 15-1B. This local variation has two implications. First, it is important in any study involving manipulation of the visual environment that there should be control samples taken from the same layers and regions as in the experimental situations. Second, it is important to understand the local patterns of segregation of the influence of the two eyes within the cortex in order to explain the significance of the experimental studies.

The separation of inputs from each eye has been studied in most detail in the monkey, although the general principles appear to apply, albeit somewhat less neatly, in cats as well. In monkeys, the afferents from the lateral geniculate nucleus end in the intermediate layers of the cortex in parts of layer IV. Within this layer the afferents are segregated into stripes, adjacent stripes receiving input from one or the other eye (Hubel et al., 1977; Fig. 15-2). Physiological recording shows the cells in each ocular dominance stripe to be monocular, either class 1 or 7, and the cells at the border between two stripes to be binocularly influenced. A pooled dominance histogram for this layer would appear much as in Figure

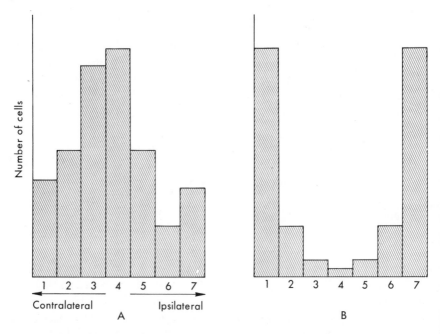

**Figure 15-1.** Normal ocular dominance histogram, following the convention of Hubel and Wiesel (see text). (A) A sample of cells taken from all layers of cat cortex (after Hubel and Wiesel, 1962); (B) an expected histogram recorded from cells of layer IV.

15-1B. Recordings from the layers above layer IV show less indication of monocular domination, although in the deeper layers it appears quite strong. In the cat, anatomical studies show a slightly less distinct separation of crossed and uncrossed inputs to layer IV, in that while uncrossed inputs are clearly segregated, the crossed pathway is continuous, but with markedly reduced density where the uncrossed fibers distribute. Physiological studies of layer IV (Stryker and Shatz, 1976) indicate that the cells are generally driven by only one eye and are clustered according to eye preference.

Binocularly driven cells respond best if stimuli are presented in discrete regions of the visual fields of both eyes. These regions may activate conjugate or disparate points on the two retinas, individual cortical cells responding optimally, therefore, to different degrees of binocular alignment (Nikara et al., 1968). The role that coordinated binocular vision plays in the development of this response specificity is clearly of interest.

Cells may be further classified according to their patterns of responses to these various stimuli. The most commonly used system is of simple, complex, and two classes of hypercomplex cell (Hubel and Wiesel, 1962). These are clearly defined cell types and as far as can be seen are not differentially affected in any of the deprivation manipulations. For this reason they will not be described further.

## Introduction to Experimental Studies

Of the properties described above, orientation, direction, ocular dominance, and binocular disparity can each be altered by disturbing the visual input. Such possibilities provide a splendid system in which to investigate how function can modify the behavior of a neuron. Of particular interest is to what extent the visual responsiveness of individual neurons develops independently of the visual input and to what extent neurons require that input to develop their normal functional patterns. Three possibilities have been proposed:

1. The properties develop independently of visual input, but require input for their maintenance, and degenerate or regress if that input is lacking (Fig. 15-3A).
2. The properties develop in some measure independently of visual input, but require further and continued specific afferent stimulation for their maturation and final stabilization. Inappropriate reinforcement may change the specificity of the cell (Fig. 15-3B).
3. The properties are not innate but develop gradually as appropriate entraining stimuli are presented (Fig. 15-3C).

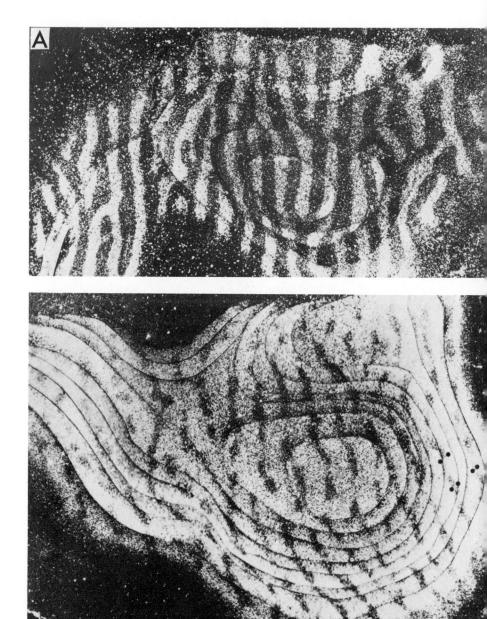

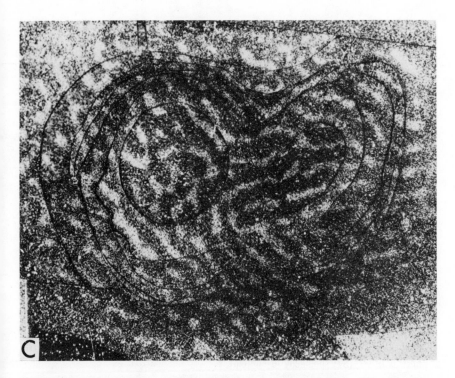

**Figure 15-2.** Ocular dominance columns in striate cortex of normal (A) (× 10) and long-term monocularly deprived (B,C) (×10 and × 9, respectively) rhesus monkeys. Each figure is a montage of serial dark-field autoradiographs of sections cut tangential to layer IVc. The monkeys were injected intraocularly with ³H-proline about two weeks before being killed and the labeled (light) bands correspond to ocular dominance columns for the injected eye. In (B) the open eye was injected and in (C) the closed eye (from Hubel et al., 1977, with permission).

Evidence has been marshalled to support each viewpoint. Before considering these and other matters associated with the role of function in the development of neural connections, a number of cautionary comments concerning the research in this area must first be made. Such comments are necessary because this is a highly attractive but difficult field in which to work and because some results by different groups of workers are completely contradictory.

The first problem concerns the state of the preparation, especially anesthetic levels and optics. Blood gas levels are very important in young animals where suboptimal oxygen levels result in increasing numbers of sluggish, erratic, or nonresponsive units (Sherk and Stryker, 1976). With respect to optics, young kittens, especially up to three weeks of age, have very cloudy ocular media, and it is quite unlikely that a discrete visual image can be projected upon the retina

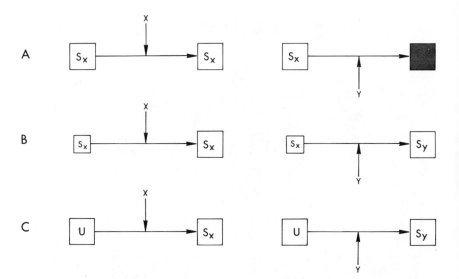

**Figure 15-3.** Three schema to account for the role of environment in the development of response specificity of visual cortical neurons. The first box in each column denotes the responsiveness prior to specific stimulation. ($S_x$) unit shows specificity to stimulus X; (U) uncommitted; (small box) specificity not secure. The arrow signifies the entraining stimulus, X or Y. The second box denotes the effect of entraining. (dark box) Degenerated or nonresponsive; (Sy) specificity to stimulus Y.

(Thorn et al., 1976). Mention, therefore, of imprecisely tuned responses in young kittens could be due as much to anesthetic and optical problems as to neurological shortcomings of the system.

A second problem concerns the comprehensiveness of the studies. Comprehensiveness is particularly a problem with the physiological studies associated with orientation effects; of numerous papers on the subject all but four are short reports or abstracts. Of these four, one is a series of short reports without full experimental detail and another suffers some technical shortcomings. Also, in many studies, the number of units per animal is often small or unstated and in some cases conclusions are drawn from experiments on single animals.

A third problem concerns the lack of controls: often control animals have not been investigated rigorously along with the deprived animals under the same experimental conditions. This is important because electrodes of different construction sample different populations of cells; also, variation in details of anesthetic conditions and stimulus presentation may modify the specificity of responses. Furthermore, a knowledge of the course taken by the electrode penetration is significant, especially since an electrode may run for a considerable distance in a single lamina. Unless there are control penetrations through the same layers, unwarranted conclusions may be drawn, especially with regard to the matter of ocular dominance.

Finally, significance is sometimes attached to the presence of regions of "silent" cortex along an electrode penetration running through grey matter. These can occur in normal animals; and in general while the absence of silent areas is of positive significance, their presence adds little weight to an argument.

With these reservations in mind, a series of separate experimental situations will be discussed, and will then be synthesized to extract the main points.

### Early Visual Responsiveness of the Cortex

In order to know whether response patterns develop independently of input, it is worth studying visually naive kittens, either as soon as the eyes open, or after a short time in the dark, so that the optical media of the eyes have cleared. This situation was first approached by Hubel and Wiesel (1963) recording from cortical cells of eight-day-old kittens. They found that while units were sluggish and fatigued early, orientation and directional specificity were present, as well as a normal pattern of ocular dominance. Buisseret and Imbert (1976), although unable to define the characteristics of units in eight-day-old kittens, found an essentially similar pattern in slightly older animals either normally reared or dark reared. Sherk and Stryker (1976), in a careful analysis of 22–29-day-old kittens reared up to that time with eyelids sutured, found good orientation and direction-ally selective responses. The matter of binocular alignment was not addressed in these reports.

Against these studies are those of Pettigrew (1974) who found that orientation specificity does not develop until between the fourth and fifth week postnat-ally, and that directional selectivity, only rarely found during the second week, develops during the third week. With continued binocular deprivation, however, some normal units were found. Pettigrew also studied binocular disparity by altering the position of a visual image on the nondominant eye using a prism, to see how this affected the rate of firing of the unit to a stimulus viewed by both eyes. In adults the prism-induced disparity produces a tightly tuned response, but in young animals the tuning is extremely sloppy until about 30 days postnatal. Blakemore and Mitchell (1973) confirmed Pettigrew's findings of an absence of oriented units in young animals, but subsequently Blakemore and Van Sluyters (1975) found some tightly tuned units even at nine days of age.

Given the several positive studies, it would appear that binocular responsiveness, orientation selectivity, and direction selectivity can all develop in cortical neurons independently of visual input. Pettigrew and Garey (1974) have shown, however, that by presenting a particular stimulus to an uncommitted neuron, its response properties can be biased toward those of the presenting stimulus. Because of this effect, it has been suggested that with the usual methods of testing in which the optimal stimulus is moved many times across the field of the neuron, uncommitted units are actually being entrained. Sherk and Stryker (1976) deliberately used random stimulus presentations in their study in order to avoid this problem, but they still found units with good specificity. It does appear, however, the binocular disparity develops very late, and deprivation studies (Pettigrew 1974) support the idea that visual input is important for its development.

Studies of synaptic development in the visual cortex of cats (Cragg 1972, 1975a) show that the major increase in synaptic numbers occurs between postnatal days 8 and 37, and that the increase in synapses within the deeper layers precedes that in the more superficial laminae. The data further suggest a drop in numbers of synapses per neuron somewhere between day 37 and adulthood. It is unfortunate that, due to the considerable technical difficulties involved, we know little of when individual groups of axons arising from particular populations of cells make their connections in the cortex. Anker and Cragg (1974) found that thalamic axons are already present in the cortex at birth and that callosal axons appear to arrive in the visual areas on about day 26 postnatal. Nothing is known beyond Cragg's work of the postnatal development of intracortical connectivity in the cat. This is particularly unfortunate because without this information, it is difficult to provide a morphological foundation for the physiological events occurring in the cortex as a result of deprivation in this animal.

## Deprivation of Pattern Vision

Such deprivation is achieved by restricting vision in one or both eyes by suturing the eyelids together at an early stage, by the use of opaque occluders, or (in the case of binocular deprivation only) by rearing in the dark. In two cases, deprivation is of pattern vision with a variable reduction in intensity, while in the third it involves total loss of all visual input. While some evidence exists to suggest that the amount of light incident upon the retina may be significant in the deficit observed, it is far from clearcut and indeed one study (Wilson et al., 1976) found that higher illumination levels to one of two pattern-deprived eyes conferred no advantage to that eye. Thus the important feature is pattern deprivation.

In most studies on kittens, deprivation is maintained from shortly after the time at which the eyes open naturally to at least four months postnatally. This period encompasses the critical period, defined by the fact that deprivation during this time leads to the most profound and long-lasting effects on central visual organization and responsiveness. The effects of deprivation are tested at the end of this period by allowing normal vision through the deprived eye or eyes. Most of the studies test responsiveness shortly after normal vision is restored, but some have also investigated the effects of prolonged periods of normal vision. The extent of the critical period in monkeys is still in dispute; in humans it depends to some degree on the tests being used.

In this section, we will consider first the general effects of deprivation throughout the visual system, then the deficits resulting from monocular or binocular deprivation in experimental animals. Separate consideration of monocular and binocular deprivation is justified since in the latter case there is a domination of the cortex by the nondeprived eye which puts the deprived eye at a competitive disadvantage. While there is some evidence of interocular interaction in binocular deprivation, it is clearly of less significance, and in general the effects of deprivation are ultimately less severe than in the monocular case.

### General effects of deprivation throughout the visual system

The *retina* of visually deprived cats shows no difference from normal in the types of ganglion cell recorded from and the frequency with which they are encountered (Sherman and Stone, 1973). A prolongation of the b-wave of the electroretinogram (attributed to glial cell activity—Dowling, 1970) has been reported for dark-reared cats (Cornwell, 1974), but no comparable anomaly was found in visually deprived children (Yinon and Auerbach, 1974). Histological studies of the retinas of normal and dark-reared monkeys (Hendrickson and Boothe, 1976) failed to show any qualitative differences, but recent work has reported fewer ganglion cells in deprived monkeys (Von Noorden et al., 1977). The major defects are found in the lateral geniculate nucleus, in the visual cortex, and superior colliculus.

In the *lateral geniculate nucleus,* only minor physiological changes have been noted (Eysel and Gaedt, 1971), but there are marked reductions in the size of neuronal cell bodies when stained with Nissl methods. These compare with the transneuronal atrophy observed after cutting the optic nerve.

In the *cerebral cortex* there are pronounced physiological aberrations; although anatomical changes have been identified, they are more subtle and have been given perhaps less attention than they deserve. Thus Coleman and Riesen

(1968) found that the stellate cells of layer IV of the visual cortex had fewer and shorter dendrites in dark-reared compared with normal cats. LeVay (1977) noted in the cells of layer IVC of deprived kittens a disruption of ribosomal aggregates, which might suggest a disturbance of protein synthesis in these cells. Cragg (1975b) found in cats an overall shrinkage of the visual cortex of about 9% after bilateral lid suture and this was almost the same as after eye removal. The total number of synapses in the visual cortex was 70% of normal after both eye removal and deprivation. The question of precisely which synapses were involved was not addressed. In monkey, there is a significant reduction in number of spines on the dendrites of a variety of cell types of the visual cortex after dark rearing (Boothe and Lund, 1976). Whether this is due to a failure of maturation or an excessive spine loss in development is not altogether clear at present.

Many of the superior collicular changes are dependent on the activity of the visual cortex and will be discussed separately.

### Monocular deprivation

Monocular deprivation demonstrates a form of competition amblyopia in which the deprived eye becomes dominated by the normal eye. This leads to a series of behavioral deficits demonstrable when the cat views the world with the deprived eye. The various studies on cats (Riesen, 1966 for a review of earlier work; Wiesel and Hubel, 1963b, 1965b; Dews and Wiesel, 1970; Ganz and Fitch, 1968; Ganz et al., 1972; Spear and Ganz, 1975; Rizzolatti and Tradari, 1971) describe a reasonably concordant set of results. In general, monocular deprivation maintained over the first three months of life is most damaging, and short periods of deprivation have a more severe effect if they occur earlier rather than later during that time. Immediately after the full period of deprivation the animal is essentially blind when viewing the world through the deprived eye or eyes. In the following weeks there is marked improvement, although permanent deficits still occur.

Both Ganz and Fitch (1968) and Dews and Wiesel (1970) found a reduction in acuity. In more recent work (Mitchell et al., 1977; Mitchell, personal communication) a modified jumping stand for testing acuity after periods of deprivation of up to four months has been used. The method allows a finer test of acuity to be made more quickly than before and, more important, makes it possible to follow the course of recovery after deprivation.

After monocular deprivation through the critical period there is a gradual recovery over a period of two months of acuity to about 2 cycles/degree (compared with a figure of 8–10 cycles/degree in normals). A startle response to fast approaching objects never develops and the animals show permanent difficulty in guiding their limbs to small objects (Ganz and Fitch, 1968; Dews and Wiesel, 1970). This led Ganz and Fitch to suggest that the deprivation may disturb some aspect of depth perception. It must be remembered, however, that profound deficits of visuomotor control may be produced by dissociating visual and sensory feedback systems in the presence of perfectly normal vision (Held, 1970). Ganz and Fitch found no recovery in pattern discrimination, although subsequent studies with more persistent or aggressive testing have been able to demonstrate ability to discriminate using the deprived eye, but very much more slowly than normal.

Rizzolatti and Tradari (1971) noted in their monocularly deprived cats that the animals adopted a strategy in which they scanned the stimulus, moving their head in the process. A possible reason for this comes from the visual perimetry studies of Sherman (1973). He tested the response of deprived cats compared with normals to visual stimuli presented in the peripheral field, while the cat is fixating straight ahead. In normal cats, each eye scans 135° of visual field; in monocularly deprived animals, the nondeprived eye behaves normally and the deprived eye responds only to objects in the monocular segment of the visual field. This last response can be elicited immediately after opening the deprived eye. The fact that earlier reports indicated monocularly deprived cats to be initially blind in one eye may be because testing was directed at the binocular part of the visual field. The scanning behavior mentioned by Rizzolatti and Tradari could be related to an attempt to fixate the visual image in the monocular field of view. An attempt by Spear and Ganz (1975) to address this question by making a lesion in the peripheral visual field representation of the cortex does not altogether resolve the issue.

One of the problems in interpreting the behavioral deficits is knowing which part of the visual system is involved—the primary or secondary visual cortices, the superior colliculus, or other areas—and whether indeed the deficit lies in the visual system at all, rather than in its interrelations with the motor systems concerned with effecting the response.

Physiological studies report that the deprived eye is quite ineffective in driving cells in the visual cortex, thus producing an ocular dominance histogram biased toward class 1 or 7 (Fig. 15-4), depending on whether the eye contralateral or ipsilateral to the recording electrode was deprived. In those few cases when units are driven by the deprived eye, they are highly abnormal. There is an asymmetry in the responses in that the deprived eye is more effective in firing units on the contralateral side of the brain than ipsilaterally.

Studies on monocularly deprived monkeys indicate behavioral (Von Noor-

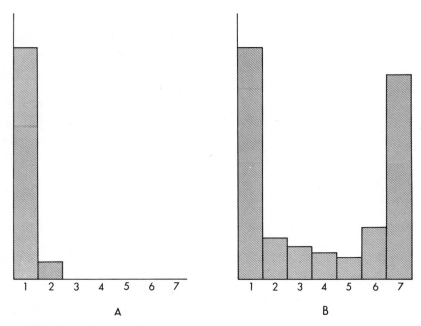

**Figure 15-4.** Ocular dominance histograms from cat cortex. (A) after the eye contralateral to the recorded cortex had been visually deprived through the critical period, and (B) after squint had been induced early in visual development (after Hubel and Wiesel, 1965).

den, 1973) and physiological (Hubel et al., 1977) deficits comparable to those experienced in similarly reared cats.

Anatomical investigation of the visual afferents to the cortex of monocularly deprived kittens shows that in layer IV, the ocular dominance stripes found in normal animals are still present, but the inputs from each eye are more strictly segregated than in normals (Shatz et al., 1975). In recordings from layer IV cells, reasonable numbers of cells are found to be driven by the deprived eye which have clearly definable receptive fields. Such cells are found in the stripes to which the deprived eye projects. The other layers show the usual bias toward the nondeprived eye (Stryker and Shatz, 1976). This result emphasizes the importance of laminar differences normally disregarded in overall ocular dominance histograms. In the monkey (Fig. 15-2), there is a pronounced reduction in the width of those stripes served by the deprived eye and an increase in those served by the normal eye. The total width of the two stripes is the same as in normal animals.

In the lateral geniculate nucleus of the cat, the cells in those layers receiving input from the deprived eye show a profound shrinkage in Nissl preparations.

Recent studies on the effects of monocular deprivation (Garey and Blakemore, 1977) indicate that the largest cells shrink more than the smaller ones. An early report showed a small proportion of physiologically sluggish cells (Wiesel and Hubel, 1963b). More recent work has shown a selective reduction in the number of Y cells encountered in a penetration. (Y cells are defined as ones driven by the Y input from the retina. They project to the visual cortical areas 17 and 18, with large, fast conducting axons. These contrast with X cells driven by X retinal input which have slower conducting axons projecting only to area 17.) These results are interpreted as indicating a selective loss in the Y cell population in conditions of deprivation (Sherman et al., 1972). However since the Y cells are the largest in the geniculate and since these cells show proportionately greater shrinkage in deprivation, the chance of encountering them is reduced. This shrinkage becomes of significance in view of Singer and Tretter's claim (1976) that the Y cells are unaffected by deprivation and that the main defects are a reduction in the efficiency of transmission through the lateral geniculate nucleus and possibly also of intracortical inhibition.

The reason for the cell shrinkage is not altogether clear. The Nissl-stained area of the cell reflects to some extent the amount of protein synthesis occurring in the cell. Since much of the synthetic activity of the cell is directed at maintaining itself, it is possible, as suggested in Chapter 3, that cell body size could relate to total cytoplasmic volume, and in particular to the volume of the axon and its ramifications. This relationship is of considerable importance when interpreting the results of Guillery and co-workers, Dürsteler et al. (1976), and others. Guillery and Stelzner (1970) found that after monocular deprivation cell shrinkage is pronounced in the binocular segment of the contralateral lamina A, but is absent from the monocular segment. This result suggests that an inhibitory influence is being exerted at some level by the nondeprived eye and that this influence is instrumental in causing the cell shrinkage, rather than the shrinkage being caused by a direct transneuronal effect from the deprived eye which would also involve the monocular segment. The site of the interaction could be by direct competition within the lateral geniculate nucleus or by competition between axonal fields in the cortex.

Support for the concept of competition was provided by an ingenious experiment (Guillery, 1972a) in which one eye was occluded at about one week postnatal and at the same time a lesion was made in a region of the nonoccluded eye. The lesion produces a patch of transneuronal atrophy in the appropriate segments of the lateral geniculate nucleus receiving input from that area. Thus in Figure 15-5 such atrophy would occur in lamina A1 of the ipsilateral nucleus. The cells in the corresponding part of lamina A fail to show the cell shrinkage

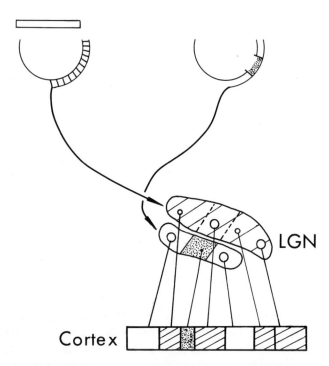

**Figure 15-5.** Diagram showing the combined effects of an early retinal lesion (dots) and deprivation (bar over eye and lines) through the critical period on cell-body size in the lateral geniculate nucleus of the cat. The hypothesized effect on the width of ocular dominance columns in the visual cortex is also shown. There is marked cell shrinkage in the segment of lamina A1 served by the lesioned area of retina. The deprivation causes shrinkage in segments of lamina A which lie adjacent to parts of A1 receiving input from intact retina; cells in lamina A lying in the monocular segment or in the segment adjacent to the part of A1 receiving from the lesion do not show significant shrinkage. It is presumed that the width of cortical ocular dominance stripes bears a direct relation to the cell-body size of LGN cells.

seen over the rest of the binocular region of that lamina. Behavioral studies show that the deprived eye is still functional over that segment although not with that particular test over the rest of its binocular field; and a recording from the visual cortex indicates that responses attributable to the deprived eye are found only in the region corresponding to the scotoma (Sherman et al., 1974).

These results are consistent in showing that a major result of monocular deprivation is dominance of the deprived eye by the normal one.

The study of overall cortical ocular dominance in monocular deprivation has provided an easy index (allowing for the cautions already expressed) for defining quite carefully the time period during which visual deprivation causes long-lasting effects and for investigating the question of reversibility. Hubel and Wiesel (1970) found that the sensitive period extends from the beginning of the fourth week postnatal to the end of the third month. Periods of deprivation after this time produce no long-lasting deficit. Deprivation for a short time (e.g., six days) early in the sensitive period has a major long-lasting effect on the ocular dominance patterns, while a similar period of deprivation performed later in the sensitive period is less significant. The associated behavioral studies of Dews and Wiesel (1970) already mentioned show a close parallel with the physiological experiments.

These studies indicate that there is a clearly defined period during which deprivation has a major effect on the establishment of normal ocular dominance patterns, and that deprivation during the early part of this period is more damaging than at later times. The period of sensitivity begins as the ocular media clear and continues through a period when visual acuity develops (Mitchell et al., 1976) and during which there is a massive increase in the number of synapses in the visual cortex (Cragg 1975a).

If, during the critical period, the deprived eye is exposed and the nondeprived eye is closed (termed reverse suturing), visual function and normal physiological responsiveness may be acquired by the newly exposed eye and at the same time may be lost by the newly deprived eye (Blakemore and Van Sluyters, 1974; Movshon, 1976a,b). At the time of reverse suturing, neuronal somata in the layers of the lateral geniculate nucleus served by the initially deprived eye show a marked shrinkage, while the cells in the other layers appear normal. After cross-suturing, these effects are reversed, the cells in one set of layers shrinking and those in the other set enlarging (Dürsteler et al., 1976).

The reversal is most readily demonstrated by cortical ocular dominance patterns as shown in Figure 15-6. As can be seen if performed at five weeks of age, reversal is quite rapid, showing the opposite bias by nine days, and proceeding to completion by 63 days. With cross-suturing at progressively later times, the reversal process becomes gradually slower and reversal is incomplete (Fig. 15-6) until animals cross-sutured after the end of the sensitive period fail to show any reversal in ocular dominance patterns. The reversal of behavioral responses (visual placing, visual following, visual cliff, visual startle response, optokinetic nystagmus, and maximum acuity) and of cell sizes in the lateral geniculate nucleus correlates remarkably well both in terms of rate and completeness with the physiological findings (Movshon, 1976b; Mitchell et al., 1977). Orientation columns are found for the previously seeing eye, but initially they are quite disorganized for the other eye, although they appear to become ordered with time. Associated with this is the finding that at these intermediate times, as the

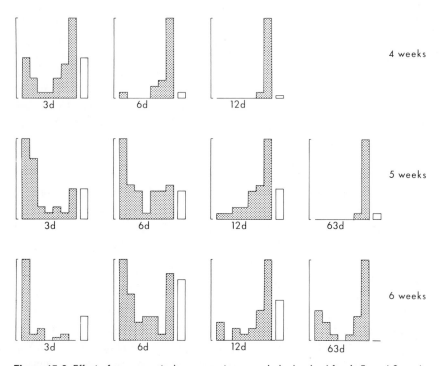

**Figure 15-6.** Effect of reverse suturing on a cat monocularly deprived for 4, 5, and 6 weeks on ocular dominance histograms recorded 3–63 days after. In all cases, the eye contralateral to the cortex being recorded from was deprived first and the ocular dominance histogram immediately after eye opening appeared as in Figure 15-4A. The gradual shift toward domination by this eye is shown. Visually nonresponsive units indicated by unshaded column (after Movshon, 1976a).

reversal process is in progress, half of the binocular units respond to different orientations with the two eyes, the disparity between optimal orientations being as much as 90°. This surprising result argues for a period in which connections are particularly plastic, being susceptible to a variety of imbalances imposed on the system.

Given the permanence of the physiological effects of monocular deprivation maintained through the critical period, two recent studies have offered a surprising result—that cells can be made to respond to the deprived eye in a comparatively normal manner. In one study, the normal eye was removed after five months of deprivation. When recorded immediately afterwards, the deprived eye was found to drive 31% of the cells identified compared with 6% when the normal eye was present (Kratz et al., 1976). Of these cells, a considerable proportion showed normal orientation and direction selectivity. These figures did not change significantly with age, and never attained the values characteristic of normal animals. These results would suggest that even though covered during testing, the normal eye of monocularly deprived kittens limits the efficacy of the deprived eye in driving cells and that on removing this inhibitory influence, the deprived eye shows a remarkable capacity for recovery, although it never regains its full functional capabilities. In an anatomical study of comparable design, the soma sizes of neurons of the deprived laminae of the lateral geniculate nucleus show some recovery toward normal if the optic nerve of the nondeprived eye is cut at the same time that vision is restored to the deprived eye. Such recovery was not found if the normal nerve was intact (Cragg et al., 1976).

In another set of studies (Duffy et al., 1976 a, b) it was found that units which are nonresponsive to stimulation of the deprived eye can be made to respond by intravenous injection if bicuculline or ammonium acetate, thus reducing inhibitory responses at all levels in the visual system. Each unit shows good specificity of response which corresponds in field position, field size, cell class, and orientation and direction specificity with that shown by the same unit driven by the nondeprived eye.

These two studies strongly suggest that in cases of monocular deprivation, the deprived eye still retains connections in the central visual system which allow it to drive the visual cortex in a comparatively normal manner. Sustained inhibitory inputs seem to prevent this from happening in most of the test situations.

### Binocular deprivation

The behavioral effects of binocular deprivation have been given somewhat less attention than those caused by monocular deprivation. A recent comprehensive study indicates that while an animal may initially be visually incapacitated after four or seven months of dark rearing, there is considerable recovery of visual function over time, as determined by a variety of tests (Van Hoff-van Duin, 1976). Mitchell (unpublished work) has found that, unlike monocularly deprived kittens, the visual acuity of binocularly deprived animals returns to normal levels over several months.

The main physiological effects of binocular deprivation are a reduction (or loss) of specifically responsive cells, and an increase in the number of cells which respond sluggishly, are poorly tuned, or are nonresponsive to visual stimuli. In addition the overall ocular dominance histogram tends to be flatter than normal with relatively fewer cells in categories 3–5. Details of the degree of loss of specific responsiveness vary widely between studies. Thus several reports indicate that in kittens reared with lid suture there is a reasonable percentage of units with normal or somewhat more broadly tuned orientation and directional selectivity (Hubel and Wiesel, 1965; Singer and Tretter, 1976; Kratz and Spear, 1976). Two other studies, by contrast (Pettigrew, 1974; Blakemore and Van Sluyters, 1975), found virtually no selective responses in bilaterally lid-sutured kittens while in two more studies on dark-reared kittens (Buisseret and Imbert 1976; Blakemore and Mitchell, 1973) there were very few selectively responsive cells. It is not clear at present to what degree the differences are due to differences in technique, interpretation, or the different ways of inducing deprivation.

In the lateral geniculate nucleus there is evidence of cell shrinkage at least throughout laminae A and A1. The amount of shrinkage seems to be less than after monocular deprivation.

In view of the suggestion that there is some type of interaction between the inputs from each eye even in binocularly deprived cats, the results of Kratz and Spear (1976) in which the authors reared kittens with one eye deprived and the other removed are not altogether surprising. They found that while cortical units in these cats were abnormal, more of them were responsive than in binocularly deprived animals and, of these, a higher percentage was orientation selective.

One of the more surprising findings in these studies on binocularly deprived cats is the good behavioral recovery from the deprivation. Physiological studies have not generally given the same impression, although most have been concerned with recording shortly after cessation of deprivation. However, several more recent reports do indicate that there is physiological recovery which correlates with the behavioral improvement.

A behavioral study on dark-reared monkeys (Regal et al., 1976) indicates that, while acuity levels do not improve beyond about half the normal figure, they are nevertheless considerably better than after monocular deprivation (Von Noorden, 1973). Although there is considerable variation between animals, there seems to be a continued deficit in general visuomotor responsiveness. Physiological studies of the visual cortex show that infant monkeys, after short periods of deprivation, give normal unitary response patterns both with respect to direction and orientation, and the sequence of change in orientation found in normal animals also occurs (Wiesel and Hubel, 1974). After longer periods of deprivation, the Wiesel and Hubel study implies that the situation is little changed, while in another study (Crawford et al., 1975) most units were difficult to classify or were nonresponsive. Both studies are preliminary comments based on single animals. In each, both young and old animals show a major loss of binocularly driven units, either compared to normal newborn or adult monkeys (which are quite similar).

## Squint and alternating deprivation

A common cause of competition amblyopia is misalignment of the eyes resulting from convergent and divergent squint. This condition has been mimicked in the laboratory by cutting the lateral or medial rectus muscles early in development. The principal finding in the cortex of cats treated in this way is the presence of areas driven only by one eye, or the other eye giving an ocular dominance histogram biased toward classes 1 and 7 (Fig. 15-4B). A similar phenomenon can be produced by alternating monocular deprivation by changing an opaque contact lens daily from one eye to the other during rearing (Hubel and Wiesel, 1965; Blake and Hirsch, 1975). This would suggest that coincident input from the two eyes is necessary in order to develop binocular response characteristics. Such a conclusion has been questioned by Maffei and Bisti (1976) who induced a squint in kittens and reared them in the dark. Rather than the ocular dominance histogram being as for a binocularly deprived cat as might be expected, it had two peaks, at 1 and 7, as for a strabismic cat. In other words, even without binocular interaction between the normal and deviated eyes, the strabismic eye exerts an effect which the authors suggest may come from feedback from the eye muscles.

It should be added that in humans, in contrast to most cat studies, amblyopia usually develops in the strabismic eye, and it has been reported that this is also the case for monkeys (Baker et al., 1974). The reason for the species difference is not clear, but it is possible that the amblyopia may be more pronounced in animals with a true fovea centralis.

## Partial visual environments

To achieve partial deprivation, animals are generally reared in the dark and for certain periods each day from about the third week of life onward for up to several months, they are exposed to an abnormal visual world. The first studies were done using stripes of particular orientation (Hirsch and Spinelli, 1970). In subsequent studies, kittens have been reared in environments without lines at all, with stripes moving in only one direction, and without movement (under conditions of stroboscopic illumination).

In the Hirsch-Spinelli studies (1970, 1971) kittens were fitted during exposure periods with goggles, which had lenses bringing a set of stripes into focus on the retina. One eye viewed a set of horizontal stripes, while the other viewed vertical stripes. From recordings from the cortex it was apparent that cells responded only if the stimulus was oriented close to the horizontal (when viewed with the eye which had seen horizontal stripes during the test period) or close to the vertical (when viewed with the other eye). Most cells responded only to one eye, but one example was found of a binocular unit, which responded to both orientations. The validity of the method of Hirsch and Spinelli, who used moving spots to find the optimal orientation to which a particular cell responds, has been questioned (e.g., Pettigrew, 1974), but other studies in which more acceptable stimulus presentation methods were used have confirmed the basic observations (Pettigrew et al., 1973, Stryker and Sherk, 1975). Pettigrew et al. (1973) also identified, in their one animal, distinct ocular dominance columns. Interestingly, as they moved from one to another in their tangential electrode track, the orientation which would drive the cells would also change from horizontal to vertical, and so on. Stryker and Sherk (1975) confirmed the original observation that, besides the cells responding to specific orientations, many others were hard to excite visually. Behavioral studies (Hirsch, 1972) show only a slight reduction in perceptual ability when the trained versus other orientations are tested. Using the same situation, Leventhal and Hirsch (1975) reared animals exposed to obliquely oriented stripes. When tested later, less than one-third of the units responded to the training orientation, the remainder preferring horizontal or vertical stripes. This surprising result might reflect an artifact of the rearing situation, but if it does not, it must relate to the basic mechanisms associated with the perception of oblique lines as opposed to horizontals or verticals.

Blakemore and Cooper (1970) devised a somewhat different rearing situation and put the kitten in a cylinder with horizontal or vertical stripes on the side and mirrors above and below to give the impression of a continuous stimulus pattern. One kitten reared in the horizontal environment showed cortical units responding only to the horizontal stripes; the cortical units in the other kitten, reared in a vertical environment, responded only to vertical stripes. Preliminary testing showed that the kittens appeared blind to orientations other than those to which they were trained, although longer-term acuity studies (Muir and Mitchell, 1973) showed at worst an 18–40% reduction in acuity in the untrained orientation. Further work (Blakemore and Mitchell, 1973) indicated that as little as 1 hour of exposure to vertical stripes on day 28 is sufficient to bias the unit responses toward the vertical when the kitten was tested after a further 6 weeks in the dark (Fig. 15-7). Rearing either simultaneously or alternately to vertical and horizontal stripes produces a high proportion of units responding to one or the other orientation. It is unfortunate that most of these important observations depend on individual experiments involving so few animals. This is particularly disturbing because Stryker and Sherk (1975), using exactly the same rearing system as that employed by Blakemore and colleagues, were unable to replicate their results, although as mentioned above Stryker and Sherk could confirm Hirsch and Spinelli's findings. Their negative result remains to be explained: certainly

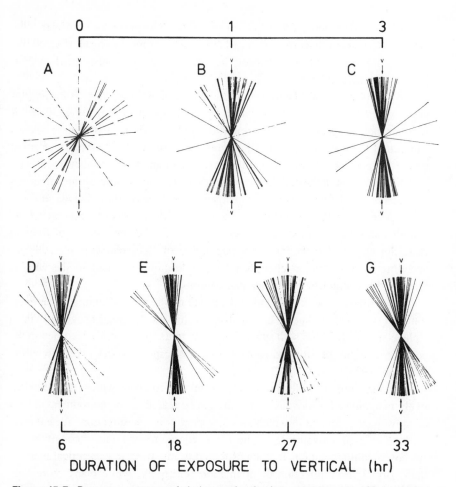

**Figure 15-7.** Response patterns of dark-reared animals exposed on day 28 to vertical stripes for different periods. Each line signifies one cortical unit responding selectively to a stripe of that orientation (from Blakemore and Mitchell, 1973, with permission).

they sampled larger populations than most other studies and did so in a manner designed to reduce subjective bias. It is possible that electrode design or some other detail of the preparation, in particular the conditions of rearing, is involved.

Another set of observations (Freeman et al., 1972; Freeman and Thibos, 1975a,b), on human astigmats, is consistent with the proposed correlation between orientation of rearing and neural patterns. Freeman and colleagues found that in some patients, even with careful optical correction, there was reduced visual acuity in the orientation of the astigmatism, and an associated depression in the cortical evoked response for the same orientation. The presumption was that the optical astigmatism, uncorrected in early life, had presented a world in which one orientation was blurred and as a result the nervous system had responded much as in the experimental cases.

An important consideration in these studies is whether there is dropout of units which do not become activated by their usual driving orientations or whether there is specific entraining of uncommitted cells or modification of cells committed to other orientations. Pettigrew and Garey (1974) found that short periods of presentation of a particular orientation can indeed develop specificity to that orientation in the cell being studied, but the permanence of this effect has not been tested. Unlike procedures in other studies, the conditioning was done in anesthetized animals. Interestingly, there was a delay between presenting the stimulus and the development of specificity and in the early stages of training there appeared a largely transitory population of "complementary" cells responding to orientations other than the conditioning orientation; the authors thought these might be inhibitory interneurons.

An extreme case of orientation deprivation is rearing an animal in an environment devoid of lines. It has been achieved by rearing kittens in a "planetarium" (a dome with spots on it). The units subsequently recorded in the cortex are highly sensitive to spots and do not respond to lines (Pettigrew and Freeman, 1973).

Another important feature of the visual environment which can be selectively manipulated is movement: kittens can be reared in an environment devoid of movement or in one in which all movement is in one direction. An environment devoid of movement can be achieved by rearing under conditions of stroboscopic illumination. In animals reared in this way, fewer cortical neurons than normal show direction or orientation selectivity, while many more than normal respond to strobe flashes (Cynader et al., 1973; Olson and Pettigrew, 1974).

Direction selectivity has been tested in both the superior colliculus (Vital-Durand and Jeannerod, 1974) and visual cortex (Tretter et al., 1975). In both cases the kittens showed a preferential response for stripes moving in the direction to which they had been exposed. Daw and Wyatt (1976) reared kittens in an environment in which stripes moved in one direction. At various times they reversed the direction of movement of the stripes and found that if this was done before five weeks of age, later units recorded in the cortex were responsive to movement of the reversed direction. After reversal later than five weeks, cortical cells did not show physiological reversal. This suggests a critical period for directional selectivity which ends at about 5 weeks.

In summary, despite a lot of superficial studies and a significant negative result, it would seem fair to say, given the total amount of work and variety of approaches, that units do tend to adopt orientations available to them during a critical period of development. Although there is suggestion of entraining of units during this time, the question of training versus loss of responsiveness has yet to be resolved. In addition to the orientation of the stimulus, it appears that movement, including the direction of movement in the entraining environment, is important for establishment of a physiologically normal cortex.

One question must be raised of all these experiments, however: for the larger part of each day (16 hours or more) during training, the kitten is in the dark and one wonders whether this contributes to the aberration. The final behavior may reflect a combination of selective rearing and dark rearing together, rather than selective rearing effect alone. Curiously, this matter has not been addressed.

### Superior colliculus

The significance of the effects of deprivation on the superior colliculus is difficult to assess since many of the results mirror events occurring in the cortex (Wickelgren and Sterling, 1969; Flandrin and Jeannerod 1975) and some of the aberrations are lost by chronic or acute decortication (Wickelgren and Sterling, 1969; Berman and Sterling, 1976). Hoffman and Sherman (1975) found the major deficit in the input to the colliculus to be related to that from the cortex rather than that directly from the retina. These findings would suggest that the deficits in the colliculus are largely secondary to those in the cortex.

### Animals other than cat and monkey

With few exceptions, studies on other animals have been confined to the effects of complete pattern deprivation. Much of this work has been summarized very carefully by Chow (1973). Here we will consider only the general points and the more recent developments.

In the *retina*, deprivation has been described as causing anything from no change to total degeneration. The retinal degeneration found in chimpanzees (Chow et al., 1957) seems inconsistent with the results of other studies and the possibility of some secondary factor must be entertained. Of the studies on smaller animals, comment has been made of slightly reduced levels of certain enzymes (Schimke, 1959; in rabbits) and of RNA in at least the ganglion-cell layers (Brattgard, 1952 in rabbits; Rasch et al., 1961 in rat, as well as monkey and chimpanzee). There is controversy as to whether or not the thickness of individual retinal layers is disturbed. Similarly, there is uncertainty as to whether deprivation causes changes in the proportion of certain synaptic patterns. While some studies have described such changes (Sosula and Glow, 1971; Fifkova, 1972a,b), a recent and carefully controlled experiments (Chernenko and West, 1976) failed to find any such difference.

In the appropriate regions or layers of the *lateral geniculate nucleus* of rabbit (Chow and Spear, 1974), squirrel (Guillery and Kaas, 1974), and dog (Sherman and Wilson, 1975) there is shrinkage of cell somas in Nissl-stained preparations after monocular deprivation. In the dog an increase in size of cells served by the nondeprived eye was also noted. Cell size has not been measured in the rat, but one report indicates a reduction in cell numbers after deprivation (Fifkova and Hassler, 1969).

In the *visual cortex* of a number of rodents, a reduction in the volume, particularly of the neuropil between cell bodies, has been reported (e.g., Gyllensten et al., 1965; Fifkova and Hassler, 1969). Structural studies have shown significant changes after both dark rearing and monocular deprivation, one result being fewer spines on apical dendrites of layer-V pyramidal cells (Valverde, 1967). The lower spine counts appear to be due to fewer spines being formed in early development and a subsequent reduction in number to lower levels than normal (see Fig. 11-3A) (Ruiz-Marcos and Valverde, 1969). In other deprivation studies on rats (Fifkova, 1970a,b), fewer synapses were observed in the cortex, especially in the upper layers; relatively more of the terminals contain round vesicles and fewer contain flattened vesicles than in normal animals. Dark rearing of rats also leads to a reduction in number of synaptic vesicles per terminal (Cragg, 1967) and a similar phenomenon has also been noted in rabbits (Vrensen and de Groot, 1974). While a reduced acuity has been reported for dark-reared rabbits (Vrensen and de Groot, 1975), in general the rabbit's visual system appears rather more resilient to environmental manipulation than most other experimental animals so far studied. Rats, by contrast, show quite severe functional deficits after deprivation. (See Tees, 1976 for a detailed discussion of the nature of the deficits in rats; and Daniels and Pettigrew, 1976, for a comment on the failure to show deprivation effects in rabbits.)

Light microscopic studies of the *superior colliculus* report little or no change in the thickness of the superficial layers, but one electron microscopic study (Lund and Lund, 1972) has reported changes in the relative proportion of terminals containing round over those containing flattened synaptic vesicles. This observation is tentatively interpreted as being the result of a failure of normal development of a set of terminals containing flattened vesicles (perhaps inhibitory synapses) at a late stage in synaptic maturation of the superior colliculus.

One final study worth mention showed that units recorded from the forebrain visual center of owls are also susceptible to deprivation effects (Pettigrew and Konishi, 1976). This system has a particular advantage in that, unlike mammalian visual connections, the pathway from the analogue of the visual cortex to the analogue of the lateral geniculate nucleus runs a separate course from the "geniculocortical" pathway. This permits investigation of the role of the "corticogeniculate" feedback circuit in determining patterns of responsiveness in the "cortex" without interfering directly with "geniculocortical" axons.

In summary, the results from other species appear, for the most part, in good accord with those from cats and monkeys. It is unfortunate that more has not been done with animals such as the rat for which there is an enormous amount of normal behavioral data, some normal physiology, and the opportunity for relatively straightforward anatomical investigations. In using the rat the problem of small numbers of animals, which plagues much of the cat and monkey work, can be overcome with less difficulty.

## Other Examples of Functional Dependency of Brain Organization

There is good indication in the auditory system that comparable events may occur. Thus rats deprived of auditory input during development subsequently show marked behavioral deficits in responsiveness to auditory inputs (Tees 1967a,b). Furthermore, rearing in a particular pattern of sound increases the responsiveness of units in the inferior colliculus to the rearing pattern and reduces the responsiveness to background noise (Clopton and Winfield, 1976). Adult animals maintained in the same environment showed no such pattern. The olfactory system also may require a functioning input since if the nares are blocked early in development, the olfactory bulbs are smaller (Meisami, 1976).

An interesting parallel is provided by work on the preoptic area of the hypothalamus of rats (Raisman and Field, 1971, 1973). This area is of particular interest because its synaptic patterns show sexual dimorphism: in the female there are more synapses upon spines (from sources other than the amygdaloid nucleus) than in the male. These differences have some light-microscopic correlates which suggest that they are not general throughout the brain. If males are castrated on postnatal day 4, the spine synapse counts lie within the female range, but this is not the case if castration is performed on day 7. Conversely, females injected with testosterone on day 4, but not on day 16, have spine synapses counts in the male range. In view of the correlated physiological effects associated with the testosterone injections, the authors suggest that the different connectivity relates to the ability of the female to maintain cyclic patterns of gonadotrophin release or behavioral estrus cycles. It is interesting that the spine synapse proportions appear to depend on the presence of adequate testosterone levels during a critical period of early development and in this respect parallel the visual deprivation studies.

A further area of direct relevance concerns the role played by the environment in the overall development of the cortex. Rats reared in "enriched" environments (with all sorts of objects and playthings introduced in their cage) compared with ones reared in "deprived" environments, have a thicker cortex, an increase in a number of metabolic indices, and an increase in the number of higher-order dendritic branches of certain cell types (Diamond et al., 1966; Bennett et al., 1964; see Greenough, 1976 for recent comprehensive review). These effects appear to be irreversible. Correlative electron microscopic studies suggest associated changes in size of synaptic appositions and in synaptic numbers; however, particularly with regard to the latter finding, sufficient account of other variables has not been taken. These results suggest that the concept of input over a critical period of development being essential for normal brain maturation extends beyond primary sensory events and may be carried to higher-order functions associated with the general behavior patterns of the animal. The clinical studies on the effects of lesions on language ability (Lenneberg, 1967) indicate that localized injury to language centers before a certain age does not disturb the subsequent acquisition of language ability, but at later times causes irreversible deficits. This again suggests an early flexibility in the organization of a central function and a subsequent polarization over the period during which the functional capacity of the region is maturing.

## Discussion

The various studies reviewed here indicate that deprivation and distortion of the visual environment during a limited period of development have a profound effect on the responsiveness of neurons of the visual cortex and superior colliculus (where the effects are largely secondary to the cortical events). Associated with this are distinct, although rather less dramatic than expected, behavioral changes. Anatomical studies of the cortex give indication of reduced connectivity (as judged from spine changes and synapse counts) which in some cases mimics quite closely the effects of cutting the optic nerve.

Despite the vast amount of data, we are still unable to provide concise answers to the two crucial questions: (1) how does a functioning input imprint itself indelibly upon a particular set of neurons and (2) why should this be necessary? Failure to answer the first question is due partly to the fact that the basic mechanisms whereby a neuron normally acquires its characteristic patterns of responsiveness have still to be elucidated. Rather than unraveling normal physiology, the study of deprivation effects appears to have added a new dimension to the problem. In addition, the subject is still being approached at a relatively superficial level, partly because the field is a new one involving all the difficulties associated with sampling large, heterogeneous populations of cells, but also becuase there tends to be a lack of regard for the various pitfalls in experimental procedure mentioned earlier in this chapter.

These problems are well illustrated in the consideration of ocular dominance. The ocular dominance index is used very commonly in deprivation studies, yet only recently has it been realized that even in normal animals the ocular dominance histogram varies significantly according to the region and layer from which recordings are being taken. In addition, cells in normal cats which by regular testing appear to respond to stimulation of only one eye can with more sophisticated testing be shown to be responsive to the other eye also (Kato et al., 1977). One wonders how similar testing may apply to monocularly deprived animals, especially since cells nonresponsive to the deprived eye can be made responsive by drugs which abolish inhibitory circuits and by lesions. In the case of eye dominance, it would appear that a nonresponsive cell receives input from the ineffective eye and can be driven by it under special conditions. These various points make interpretation of one of the simplest indices used in deprivation studies rather less straightforward than is generally acknowledged.

Despite the reservations expressed throughout this chapter, it is possible to offer some suggestions for the manner in which abnormal response patterns are generated. First it is clear that, with the exception of disparity responses, characteristic features of visual cortical neurons are expressed by at least some cells even prior to visual exposure, or in the total absence of it. Second, as optically good vision becomes available to an animal, so do the neurons become especially sensitive to various parameters of the visual environment. Once this period of susceptibility is over, reversibility or recovery of responsiveness is minimal except in special cases. These cases indicate that failure to respond does not mean an absence of connections, but rather an imbalance or suppression of the ineffective connections. This leads to two possibilities. The active input may stimulate the formation of synapses, perhaps by some sort of trophic behavior of the kind proposed elsewhere (Chap. 12), while relatively fewer synapses are formed by the inactive input. Such a possibility has some support from the imbalanced geniculocortical pathway, particularly in monkeys, after monocular deprivation. The other possibility is that there is a pruning or loss of some synapses during development and that the less active synapses are particularly susceptible. Reduction in numbers of afferents to individual muscle fibers has been observed through a critical stage of development (see, for example, Brown et al., 1976). In the visual cortex, spine numbers (an index of one type of synapse) are reduced in normal development and this process is more extreme after visual deprivation (Boothe and Lund, 1976). In neuromuscular development, various mechanisms of suppression of ineffective synapses have been proposed (see Chap. 13) and it is possible that similar mechanisms also apply in deprivation states.

The sites of the effects of visual deprivation have yet to be elucidated. Certainly the distribution of geniculocortical axons can be affected. However, since some of the spine changes (in monkey, at least) involve cells not receiving a direct thalamic input, and since manipulations abolishing intracortical inhibitory mechanisms reverse deprivation effects, it may be presumed that part of the disturbance is due to intracortical mechanisms. The further possibility, that mechanisms within the lateral geniculate nucleus are also responsible for some of the changes, should not be overlooked either. It is perhaps simplistic to look for one site at which deprivation has an effect; it is quite likely, for example, that at every point at which some form of interaction occurs between the inputs from the two eyes, a normal balance of relative influence is set up and this can be disturbed by an asymmetry of the two inputs.

Two key features of the normal visual cortex are the ordered change in orientation preference and the presence of stripes of cortex driven completely or preferentially by one or the other eye. Although orientation specificity exists very early, it can be disrupted by training so that cells which should have responded to an orientation other than the entraining orientation cease to do so. Whether they develop a permanent new specificity as a result of the entraining is uncertain. The complex patterns of orientation response demonstrated by Movshon (1976a) during the course of accommodating to reverse suturing attest, nevertheless, to the labile nature of the cells' responses during the critical period.

Ocular dominance is more problematic, largely because of the laminar and regional differences in normal animals, which were not appreciated when many of the experimental studies were conducted. In the normal animal, layer-IV cells of the cat show a strong tendency for ocular separation, in that they fall into classes 1 or 7 by traditional testing. This is not so for the other layers. However, in some conditions, particularly when different images are presented to both eyes (as in squint, alternating monocular deprivation, or when different orientations are presented to each eye), it appears that the ocular separation is found throughout the layers and is not restricted to layer IV. In this case the basic pattern is emphasized rather than modified. The fact that, even in monocular deprivation, the cells of layer IV maintain their ocular dominance patterns points to the existence of some basic organizational constraints which cannot easily be overridden by imbalance of the visual input, as opposed to others like orientation preference which appear more labile.

Why the functioning of afferents should be so important in reinforcing or biasing the basic response patterns of individual cells is not clear. In the visual system, where normal functioning depends so heavily on the coordination of the inputs from the two eyes, there is good reason to have an inbuilt mechanism which will correct or compensate for misalignment of the eyes or poor optics in one eye. If misalignment is minor, some degree of central correction might be expected; if it is major, then it would be clearly advantageous to suppress the input from one eye to avoid the confusions associated with seeing two different fields of view at once. Similarly, if the image on one retina was poorly formed, this too might be suppressed to avoid interference with the good eye.

It is difficult to apply a similar rationale to the effects associated with partial visual deprivation as well as to comparable phenomena in the auditory system, to endocrine controls of synaptology, and to environmental effects on cortical development. It may perhaps be better to look for a fundamental mechanism which is important for the development of all parts of the nervous system, although possibly more critical to some than others.

It may be a reasonable premise that there is not enough genetic material to program the nervous system to the level of precision necessary for optimal functioning. Brindley (1969), for example, has estimated that a "quasi-random" network of $10^{10}$ forebrain neurons each making $10^3$ connections would require that all or most of the genetic information in the zygote be directed to organizing the forebrain. This is clearly preposterous and in fact is an overestimate in that it does not take into account the nonrandom nature of neural development or various combinatorial effects of genes which should reduce the number of bits of information necessary. Nevertheless, taking these into account, an unreasonably large figure might still be expected, especially since safeguards have to be built in to account for problems such as vascular anomalies in development, temporary metabolic disturbances, and the associated local cell death which might arise from them.

Thus while a genetic program might lay down the basic organization of the nervous system, some degree of secondary regulation might be necessary to correct for minor misalignments and incongruities between the component parts. Since a primary requirement is that the system functions, the best control is the function itself. In a gross behavioral sense, a similar mechanism might operate to program an animal's response to its environment such that stimuli experienced frequently in early stages of development become the ones to which it responds most effectively later. The extreme case of this might be the imprinting behavior of birds. In mammals, it could be proposed that a similar sort of imprinting occurs and that adolescent play is purposely directed toward reinforcing response patterns important to the animal's later behavior. In this context the work on enriched and deprived environments has an obvious relation.

Thus the basic biological need for functional modulation of brain organization may stem from a requirement to align a crudely made system so that it functions satisfactorily. It may depend on quite a simple pattern of events requiring that active inputs be "facilitated" and inactive ones be "suppressed" either by virtue of the relative numbers of synapses they make or by physiological mechanisms. Such behavior, while generally applicable, may be especially exploited in the visual system where exact alignment of the eyes may be somewhat precarious. It may also account for the development of specific behavior patterns.

# VII

# THE ADULT BRAIN

We have so far considered the plasticity of connections occurring during development, but equally interesting is what happens once a synaptic connection is formed. Is it irrevocably fixed once established, or can it be broken and reformed with a different partner? An additional question is whether synapses can be added in the brain once the period of synapse formation has passed.

In this section we will review evidence of plasticity in adult brains with two separate questions in mind. The first is whether synaptic mobility occurs in the mature mammalian brain, either normally or in response to injury. The second is what chance an axon has of growing back to its original terminal site if it has been damaged along its course. Consideration of both points is important in the matter of recovery of function after injury to the adult brain. We have already seen that neurons of the mammalian central nervous system are generated during only a limited period of early development. They cannot be replaced by a new wave of cell division if they are damaged after the generative period has ended. Thus if local injury occurs due to trauma, vascular damage, or disease, recovery is limited by the degree to which regeneration of damaged neural processes is possible and by the capacity of uninjured neurons to form new synaptic contacts or to use their existing ones more effectively in compensation for the lost connections. Synaptic mobility in adult brains is also important when cellular models of memory and learning are considered. If these models require that synaptic connections be made or broken, then one prerequisite is that the making and breaking of synapses must be demonstrated in adult brains.

# 16

# SYNAPTIC MOBILITY

In the peripheral nervous system, damage to a nerve is likely to result in sprouting of adjacent, intact nerves which invade the denervated area. This phenomenon is termed collateral sprouting. The earlier work in which its presence was defined is well reviewed by Edds (1953), and we have already discussed more recent studies in peripheral nervous systems as well as the identification of its existence in the developing mammalian central nervous system.

The possibility that collateral sprouting may also occur in the adult central nervous system was first suggested by the pioneering experiments of Liu and Chambers (1958) and has been subsequently confirmed by a number of studies in a variety of systems. In general, work concerning the existence, specificity, or randomness of collateral sprouting in the adult mammalian central nervous system after injury has progressed slowly, largely due to the limited nature of the changes occurring, and to the technical difficulties in demonstrating them unequivocally and characterizing them successfully. Whether synaptic mobility occurs normally in the adult is even more difficult to investigate, and while one would like to know whether it exists as a possible basis for memory and behavioral changes, there is virtually no evidence to show the presence of synaptic changes in adult animals which are not secondary to death of some neural processes.

Liu and Chambers (1958) used a modification of the then newly developed Nauta Gygax method for demonstrating degenerating axons in the spinal cord of cats in two experiments. The study depends on the fact that each sensory nerve root normally distributes to a limited number of segments of the spinal cord. In their studies, they were attempting to extend this distribution experimentally. In the first experiment (Fig. 16-1A), they cut one pyramidal tract, thus depriving the deeper part of the dorsal horn of the spinal cord of that side of a major input throughout its rostrocaudal extent. Three or four years later, when residual de-

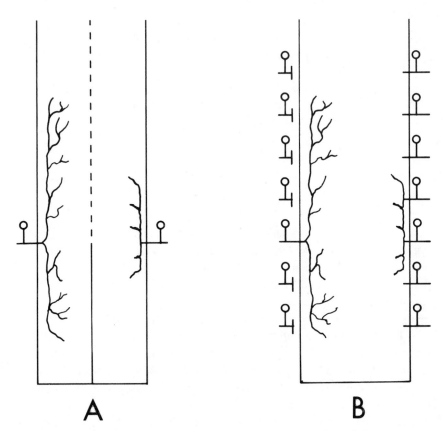

**Figure 16-1.** Diagram of spinal cord of adult cat. In (A) the descending pathways have been cut on the left side and dorsal root afferents in the lumbar region spread further rostrally and caudally than on the control side. In (B) all but one dorsal root is cut and this too extends its terminal distribution.

generation from the first lesion had long since disappeared, they cut one dorsal root (at C8 level) on both the previously lesioned and on the normal sides and found that the distribution of fibers from the dorsal root was denser and more extensive on the lesioned side. In the second experiment they cut all but one of the dorsal roots on one side (Fig. 16-1B). They tested the projection of that one segment more than 280 days later on the partially deafferented and on the control side. They found that the nerve root on the experimental side projected 3–6 segments more rostral than the control root and appeared to project more densely to the areas of normal projection. The authors have attempted to relate the collateral sprouting to the spasticity which occurs after cord lesions (McCouch et al., 1958).

The conclusion drawn was that intact tracts give off collateral sprouts to innervate neurons which have lost synaptic contacts as a result of damage to some of their afferents. There are two notes of caution which must be added. First, subsequent work has shown that the degenerative method they used for demonstrating the axon distribution is quite refractory for small and unmyelinated axons and terminals. This possibility was raised by Liu and Chambers who wondered whether small axons, unstainable in the controls, may have become fatter and stainable in the experimentals, i.e., they may have normally projected to the regions of apparently increased projection, but the method would not show them. The second point is that if an area is deafferented, it is likely to shrink, due to removal of axons and terminals, as well as possibly to transneuronal changes. Even if another pathway to the region does not increase its density of projection, it will appear denser simply because the same number of terminals now occupies less space. Despite these questions, these interesting experiments have stimulated a considerable number of further studies.

In more recent work Murray and colleagues have investigated further the question of sprouting in the adult cat's spinal cord and have attempted to correlate this with behavioral changes (Murray and Goldberger, 1974; Goldberger and Murray, 1974). In the first study they transected the spinal cord on one side between levels T12 and L1, sparing the dorsal columns. Initially the hindlimb on the lesioned side is quite useless and most reflexes are depressed. In time the cat is able to use the limb fairly normally, although with testing it appears that reflexes, such as tactile placing, which depend on descending pathways, never recover. Anatomical studies of the projection of a dorsal root caudal to the cut, tested 10 months later with degeneration methods or at various survival times with autoradiography, show a heavier projection on the experimental compared with the control side, mainly to areas to which the dorsal root normally projects.

In the second series of experiments all the dorsal roots of the lumbosacral region of one side were cut. This resulted initially in a completely useless hindlimb on that side. In time the limb was held up so that the cat walked efficiently on three legs; finally it participated to some extent in walking. By about one week after the lesion, many reflexes showed an exaggerated response (This did not occur if the cord was also hemisected at a more rostral level.) Anatomical studies showed that the descending pathways in this case appear to sprout in response to the absent dorsal root projection. There are two notable points. First, some of the heavy sprouting is into regions such as Clarke's column, which normally have little or no projection from supraspinal sources, but do have a dorsal root input. Thus while the possibility that some of the increased density of projection could be due to shrinkage, it would be unlikely to be so in this region. The second point is that while the more dorsal layers (especially lamina III) of the dorsal horn of the spinal cord receive a massive input from the dorsal root, removal of this input by the first lesion is not sufficient to stimulate sprouting of nearby descending axons. In other words, there is some limitation to the extent of sprouting that can occur. The authors suggest that the sprouting may indeed provide the basis in each case for the recovery of reflexes and posture of the partially deafferented limb, and while there are other alternatives (e.g., denervation supersensitivity), this proposal has some merit.

Other studies, such as that of Goodman and Horel (1966) and Cunningham (1972), have suggested the presence of sprouting in the visual system of adult rats after removal of one input to a region. Absolute certainty in identifying sprouting is hampered by the limitations of light microscopic methods, particularly of quantitation of degeneration and of detailed assessment of shrinkage after lesions, and clearly other approaches should be used to confirm and extend these findings.

Björklund and his colleagues (see review by Moore et al., 1974 for a summary) have done a number of studies showing that the monoaminergic (catecholamine fluorescent) systems of the brain appear to sprout after a variety of insults. Since most of these studies relate more closely to the matter of regeneration, discussion of them will be reserved for Chapter 17. One study (Stenevi et al., 1972) is relevant here. After visual cortex lesions in adult rats, the sparse catecholamine fluorescence normally present in the dorsal lateral geniculate nucleus becomes more intense, due apparently to a sprouting in response to the retrograde degenerative reaction of cells in the nucleus which project to the cortex. The result is a little surprising in that the lesion effectively causes a net *reduction* in synaptic sites, not an increase, as would occur in the other studies described here. Whether the increased fluorescence is the result of sprouting or of increased fluorescence of axons already present which are unable to release transmitter is difficult to resolve.

These pathways can also be demonstrated by injecting tritiated amino acids into the nuclei of origin of the fluorescent axons. In one such study Pickel et al. (1974) showed that the locus coeruleus projects both to the cerebellum and forebrain. Damage to the cerebellum results in an increased label in the forebrain, apparently due to axonal sprouting. The presumption is that individual axons send branches to both regions and that damage to one branch causes a reactive response in the other.

One negative note should be added and that stems from the work of Kerr (1972) on the potential convergence of trigeminal and dorsal root afferents in cats. These two inputs overlap in their terminal distribution in the first cervical segment and in part of the caudal medulla. Therefore, it might be expected (and indeed predicted as an explanation for recurrence of certain neuralgias occurring after facial surgery) that following lesions of the trigeminal nerve, there would be a sprouting of dorsal root afferents to occupy the denervated sites. None was found using degenerative techniques; there even was a reduction, if anything, in the density of degeneration on the previously lesioned side compared with normal. Apart from the fact that the survival times after the second lesion (6–13 days) are rather long and some (perhaps significant) degeneration may have already disappeared, it is difficult to see any flaw in these results. It is unfortunate that the work is often quoted as evidence that sprouting does not occur in the adult mammalian brain. Clearly the results are relative to only one system and show that there does not appear to be sprouting of dorsal root afferents into trigeminal terminal regions, much as Goldberger and Murray failed to find descending afferents sprouting into deafferented lamina III of the cord. Further, the possibility exists that other pathways not tested in these studies may sprout into the deafferented region.

As is obvious from the discussion so far, although light microscopic methods can be used to suggest that collateral sprouting may occur, they are not on their own totally satisfactory for demonstrating the presence or absence of synaptic change. A number of systems have been studied with the electron microscope, some in conjunction with light microscopic and/or physiological correlative work, and these perhaps provide the beginnings of a more satisfactory attempt to assess the full impact and limitations of collateral sprouting. They will be dealt with separately.

### 1. Septal nuclei

These nuclei are situated in the forebrain. They are relatively easy to locate in rats, where they are quite prominent and are notable in that they receive two main inputs. One of these, arriving from the hippocampus by way of the fimbria, is cholinergic and has terminals containing small vesicles; it synapses on spines and shafts of dendrites. The other arrives by way of the medial forebrain bundle from nuclei scattered along its course and is aminergic. It forms synapses on cell bodies as well as dendrites. If a fimbria is cut, thus depriving the septal nuclei of hippocampal afferents, there is a noticeable increase in the number of intact terminals making multiple contacts in individual sections. This might suggest that there has been a sprouting of undamaged terminals already in the area to occupy synaptic sites vacated by degenerated hippocampal afferents (Raisman, 1969). Studies using fluorescence microscopy (Moore et al., 1971) are consistent with this in showing an increased fluorescence of the medial forebrain bundle fibers within the septal nuclei after cutting the fimbria.

Further support for this idea of takeover of synaptic sites came from the opposite experiment in which the medial forebrain bundle was cut. After allowing sufficient time for degeneration to disappear, a second lesion was made, this time involving the fimbria so that its distribution could be tested. One notable feature is that the hippocampal afferents now make a significant number of axosomatic synapses, normally a characteristic of axons of the medial forebrain bundle. Additional quantitative data added further support to the concept of synaptic plasticity (Raisman and Field, 1973). The authors found, for example, that half the synapses on dendritic spines degenerate within a few days of cutting the fimbria, but that with time the proportion of spine synapses made by non-degenerating terminals is restored to its preoperative figure. Study of short-term changes indicated as described in Chapter 3, that astrocytes remove degenerating terminals, leaving postsynaptic sites which appear then to become secondarily occupied by other terminals. An ingenious attempt was made to see whether reinnervation of deafferented sites is specific or is from the closest terminal. At short times after the lesion, the synaptic pattern formed by the nearest normal terminal to each degenerating axospinous synapse was recorded. In a majority of cases, this is with a dendritic shaft. If proximity were the factor determining which terminal takes over the vacated site, there should be a lot of double contacts, making one synapse with a dendritic shaft (the original contact) and one with a dendritic spine (the new contact). Instead the predominant form of double contact is with two spines. This result is quite provocative in that it suggests that sprouting may be selective, and that spine synapses are more amenable to reorganization than those on other parts of the dendrite.

## 2. Superior colliculus, upper layers

In Chapter 10 we saw that if the afferents from one eye to the superior colliculus are restricted early in development, there is considerable reorganization of the remaining optic afferents, which does not occur in older animals, where at the most there may be a slight sprouting of the normal uncrossed pathway within the optic fiber layer (Goodman et al., 1973). Even the presence of this sprouting might be questioned since it was shown by degenerative methods in a place where in adult rats residual degeneration from the first lesion can persist almost indefinitely; it would clearly be worthwhile to repeat the study using autoradiographic methods. Despite the absence of extensive plasticity in the termination of retinal afferents, there is indication of considerable mobility of synaptic connections after optic deafferentation in the adult which is seen with the electron microscope (Lund and Lund, 1971a,b). The changes can only be appreciated after characterization of the synaptic patterns in the normal superior colliculus and after quantitative studies.

Terminals of retinal ganglion-cell axons contain round vesicles and pale mitochondria (Fig. 16-2a). They more commonly make asymmetric synaptic junctions. They account for about 75% of all terminals in the area containing round vesicles and are notable in that they are the major (if not only) round-vesicle-containing population which forms synapses with other terminals (serial synapses as shown in Fig. 3-5). There is a substantial proportion of terminals (39%) containing flattened vesicles and the majority of these make symmetric synaptic contacts (Fig. 16-2b).

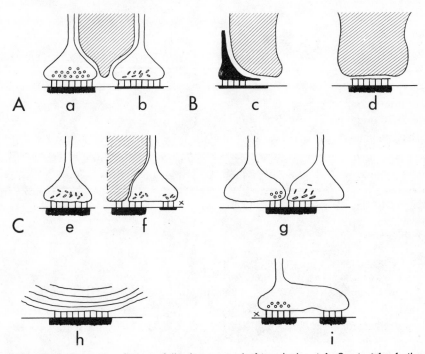

**Figure 16-2**. Sequence of events following removal of terminal a at A. See text for further description.

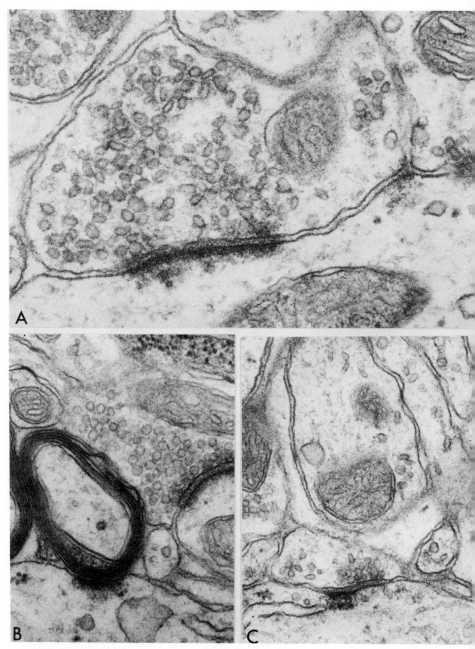

**Figure 16-3.** Electron micrographs of abnormal synaptic associations in superior colliculus. See text for description. (A) ×60,530, (B) ×52,700, (C) ×34,800.

If one or both eyes are removed, there is a rapid degeneration of optic terminals. These become removed by astrocytes from the postsynaptic site (Fig. 16-2,c,d): the majority of vacant postsynaptic sites appear to be shortlived and various anomalous associations are recognized:

1. Whereas in normal animals only 2% or less of the terminals containing flattened vesicles occupied asymmetric synaptic junctions, more than 10% do so within two weeks of eye removal (Fig. 16-2,e; Fig. 16-3A), and some of these are presynaptic dendrites.

2. At short times in particular, it is not uncommon to see a synaptic specialization occupied by two presynaptic profiles. Most commonly one is an astrocyte, and the other a terminal that usually but not always contains flattened vesicles (Fig. 16-2, f). Sometimes, there are two terminals, each with the same type of synaptic vesicle (round or flattened) or with different types occupying one asymmetric synaptic specialization (Fig. 16-2,g).

3. Myelin lamellae are seen occupying postsynaptic specializations (Fig. 16-2,h; Fig. 16-3B).

4. Some of the synaptic associations show an absence of vesicles immediately adjacent to the presynaptic membrane, which might suggest that these contacts are not functional (Fig. 16-2,i, Fig. 16-3C).

The results show first that a bare postsynaptic site can be reoccupied by a wide variety of other profiles in the area, including other axon terminals. But are these new associations functional? Some probably are not, because of the absence of synaptic vesicles, but even a normal-looking synapse may not be functioning normally (Mark et al., 1972; Model et al., 1971) especially with mismatching of synaptic vesicle type and postsynaptic specialization type. It is possible that the main function of the plasticity shown in the colliculus is to silence hyperactive postsynaptic sites, since in another situation the presence of bare postsynaptic sites correlates with that of denervation hypersensitivity (Kjerulf et al., 1973). Any new synaptic associations could well be secondary to this purpose. While this viewpoint may be unduly negative, it is one which should not be overlooked unless there is attendant evidence of associated functional plasticity. For this reason it is felt that studies of local sprouting using the electron microscope alone should be interpreted with caution with respect to their functional significance.

### 3. Other studies involving only cytological methods

Beyond these electron microscopic studies are several others which have demonstrated qualitatively the existence of synaptic mobility in the spinal trigeminal nucleus (Westrum and Black, 1971), the ventral cochlear nucleus

(Gentschev and Sotelo, 1973), and dorsal column nuclei (Rustioni and Sotelo, 1974). Only in the last are synaptic vesicles shown aggregated adjacent to the abnormal contact site, and the question of whether they are of any functional significance in terms of synaptic transmission must be raised.

In a series of studies Bernstein and Bernstein have shown that anomalous synaptic patterns suggestive of reorganization occur adjacent to spinal cord transection in rats and monkeys (J. Bernstein and M. Bernstein, 1971, 1973; M. Bernstein and J. Bernstein, 1973). Counts of boutons on motor neurons stained by the Rasmussen method indicate a drop in numbers soon after the lesion followed by a gradual return to normal, suggesting that axons separated from their normal terminal site may sprout proximal to the lesion to innervate partially deafferented neurons.

One final experiment should be mentioned (Ralston and Chow, 1973) since it involves a somewhat different plan. The visual cortex of rabbits was removed, resulting in retrograde degeneration of the projection cells in the lateral geniculate nucleus. This deprives the retinal ganglion cells of one of their targets, although interneurons on which they also end still persist. Ralston and Chow, in a quantitative study, found that the optic terminals did not die but formed more serial synaptic contacts with interneurons. In addition to this, the interneurons tended to synapse with one another more frequently.

### 4. Red nucleus

The work on the red nucleus is of interest in that there is a physiological correlate for the anatomical studies (Fig. 16-4). It is a region in which there is a convergence of inputs from the sensorimotor cortex and nucleus interpositus of the cerebellum. Axons of the sensorimotor cortex normally end on distal dendrites, while the axons from the nucleus interpositus end on cell bodies (Nakamura et al., 1974). After the nucleus interpositus is damaged, all degenerating axosomatic synapses disappear within 24 days. However, there is a significant population of axosomatic synapses, covering between 20% and 60% (average 25%) of the soma surface, which degenerates if the sensorimotor cortex is then ablated. This compares with a normal axosomatic covering of 60% from the nucleus interpositus. Thus the sensorimotor terminals have extended their distribution more proximally on the cell, but still retain distal terminations. Such distribution should modify the response properties of the neuron considerably, as has been shown by Tsukahara et al. (1974; 1975) who recorded intracellularly from the large cells of the nucleus. They found that, characteristic of distal terminations, sensorimotor stimulation produces a slow rise time in the mem-

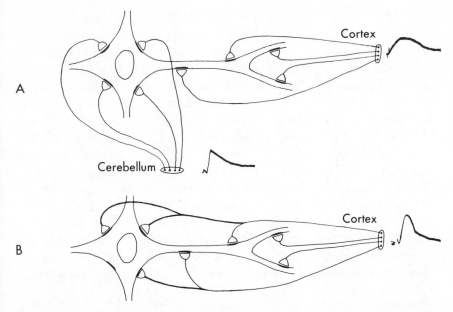

**Figure 16-4.** (A) Diagram of distribution of two principal inputs to cells of the red nucleus and intracellular physiological recordings produced by stimulation of each input. (B) Anatomical sprouting and associated physiological change of the cortical input after removal of cerebellar afferents.

brane potential changes recorded intracellularly from the soma and this input is relatively less sensitive to membrane potential displacement at the soma than somatic synapses would be. Cerebellar stimulation, by contrast, produces EPSP's with fast rise time, which are more sensitive to membrane potential changes. This is to be expected for a pathway making axosomatic synapses. Two weeks after interrupting the cerebellar input, the EPSP's from sensorimotor cortex stimulation (1) have a fast rise time and (2) are more sensitive to membrane potential displacement. Both factors are indicative of synapses closer to or on the cell body.

### 5. Hippocampal formation

The sprouting behavior in the red nucleus resembles that already described for the developing dentate gyrus of the hippocampal formation after early lesions (Chap. 13) in that terminals spread along the dendrite of a particular cell type. Studies by Cotman, Lynch, and colleagues, and by Zimmer, have shown that many of the synaptic changes they described after making lesions in young animals also occur, although to a lesser degree, after lesions in adults. Thus after lesions of the entorhinal cortex, the commissural (Lynch et al., 1973a&b) and association (Zimmer, 1973a; Lynch et al., 1976) inputs spread more distally along the dentate granule-cell dendrites into the zone formerly occupied by entorhinal afferents. They do not extend the whole way, however. For example, Lynch et al. (1976) found that the association afferents normally occupy the proximal 26% of the dendrites; after entorhinal lesions (unilateral or bilateral), they come to occupy 35–38% of the dendritic length (after correcting for shrinkage). It is possible that septal afferents may block complete sprouting; Lynch et al. (1976) suggest that the products of degeneration are instrumental in this.

After making unilateral entorhinal lesions, Steward et al. (1976) found a significant increase in the normally sparse crossed projection from the remaining entorhinal cortex to the dentate gyrus. Using autoradiographic methods, they found a sixfold increase in the density of the pathway. Normally this pathway cannot be demonstrated by recording gross potentials, but in the sprouted condi-

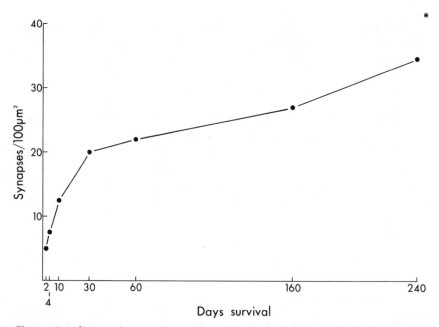

**Figure 16-5.** Changes in synaptic numbers with time in the dentate gyrus after entorhinal lesions made in adult rats. The asterisk signifies the normal density of synapses adjusted upward to take into account the shrinkage in the experimental animals (after Matthews et al., 1976).

tion its presence is clearly demonstrable. Matthews et al. (1976) studied the course of synaptic sprouting of this pathway, and, as can be seen from Figure 16-5, there is a massive initial increase in number of sprouted synapses followed by a slower increase to a number close to the normal density; but when the figures are corrected for shrinkage the amount is about 80% of normal. The authors found, as described for other regions, a transitory increase in vacated postsynaptic sites and a permanent increase in the number of terminals making multiple contacts.

### 6. Somatosensory pathways

A series of physiological studies by Wall and colleagues have shown that changes accompany partial lesions along the course of the somatosensory pathway, such that if a portion of the map of the body represented in a particular nucleus is denervated, then the deafferented region will then start responding to adjacent parts of the body map. The gracile nucleus in the medulla receives afferents from the lower limb and caudal trunk, while the adjacent cuneate nucleus receives from the more rostral body region. This map is retained in the next relay—the nucleus ventralis posterior. When the gracile nucleus was removed in adult rats (Wall and Eggers, 1971) there was an expansion of the trunk representation into the deafferented hindlimb area of nucleus ventralis posterior within three days.

Further and more detailed studies have concentrated on the plasticity of the spinal projection into nucleus gracilis in cats. In these, dorsal roots caudal to L3 (with the exception of L7 or S1) were cut on one side. Recording up to 12 hours after this, Millar et al. (1976) found a large number of cells in nucleus gracilis to be, as anticipated, hyperactive and nonresponsive to somatic stimuli. What may not have been expected, however, was that there were cells within the area of representation of the hindlimb which now responded to stimulation of the trunk, and yet others which responded to stimulation of two widely separated regions—one on the trunk and another on the heel, for example. These abnormal responses habituated easily and required flicking of the hair rather than simple displacement. With time (up to six months) the number of nonresponsive units is reduced and more inappropriate responses—wrong position and double position units—appear, and these tend to fire more securely than they did immediately after the lesion. The acute results suggest that the plasticity does not require synaptic sprouting, although it is possible that sprouting may contribute to the continuing changes with time; instead, some sort of physiological plasticity may occur.

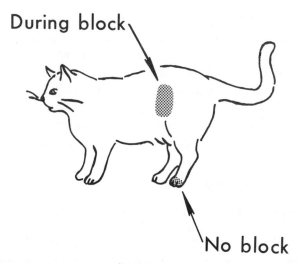

**Figure 16-6.** Receptive field recorded in the dorsal column nucleus with and without cold block of the spinal cord (after Dostrovsky et al., 1976).

In a subsequent study (Dostrovsky et al., 1976) the lumbar and sacral dorsal roots were cut on one side. Immediately, regions of the dorsal column nuclei, which had responded to stimulation of the foot, now responded to abdominal stimulation and the latencies were sufficiently short that monosynaptic activation would appear likely. Further, if instead of cutting roots a cold block was applied to the cord at L4, a unit which responded to stimulation of the toe might change its receptive field to a position on the abdomen (Fig. 16-6). Similar short-term changes are also reported to occur in the spinal cord.

To interpret these results it is presumed (and this is borne out by anatomical studies) that each dorsal root segment normally projects more broadly in the nucleus gracilis (and cuneatus) than the area from which physiological responses are elicited. If part of the projection is removed, the input to a cell which previously may not have driven it, can now do so directly. Such mechanism might also be involved in the earlier experiments of Wall and Egger (1971) rather than axonal sprouting, since three days is a rather short time for sprouting to occur across 1 mm, the distance across the relevant part of nucleus ventralis posterior.

Two further studies described in Chapter 15, each involving the permanence of the effects of monocular deprivation through the sensitive period in cats, show a comparable situation. In one, the failure of the deprived eye to drive units in the superior colliculus was negated by removal of the visual cortex (Berman and

Sterling, 1976). In the other (Kratz et al., 1976), the failure of the deprived eye to drive many cortical units was partially reversed by removal of the good eye.

These results raise two important points. One, introduced at the beginning, is that absolute precision in connections is not necessarily a requirement for transmitting a precise map through the brain. There may be spatially inappropriate connections which are normally ineffective in driving cells. The second point is that actual changes in synaptic connections are not always a prerequisite for functional plasticity. A change in balance of inputs due to loss of one component may be sufficient.

### 7. Plasticity occurring without lesions

Evidence for this is extremely limited. Without ways of labeling previously formed and newly formed synapses, it is impossible to say whether there is continued formation of new synapses in adult brains. Although all the aberrations found after lesions can eventually be found, albeit in low percentages, after an extension electron microscopic survey in a comparable region of a normal brain, this does not mean that synaptic takeover is necessarily occurring normally. One is never sure of how much degeneration may have occurred in the course of an animal's life due to trauma, vascular disorders, dietary inadequacies, and other factors. If such factors occurred, then the occasional atypical contacts could simply be of pathological origin. One intriguing study by Sotelo and Palay (1971) on the lateral vestibular nucleus of normal rats found a series of organelles occurring in terminals which they suggest might relate to growth or other plastic changes, but a direct correlation could not be made.

One point which should be mentioned here is the apparent ease with which terminals may be lifted from chromatolytic neurons and returned again once regeneration had occurred (Chap. 3). In the superior cervical ganglion, such synapse stripping is prevented by the presence of nerve growth factor (Chap. 4). There would seem to be no reason why similar trophic materials might not control the synaptic input of neurons in the adult central nervous system, and why such mechanisms should not extend to problems such as the endocrine control of behavior.

### Discussion

One major conclusion to be drawn from this set of experiments is that a considerable degree of synaptic reorganization occurs in the adult central nervous system after specific lesions. Such reorganization differs from that seen after comparable lesions in younger animals in being restricted, for the most part if not exclusively, to axons or terminals already in the region. This means that in some cases it can only be demonstrated satisfactorily using quantitative electron microscopy—a technique which, apart from being tedious, has many interpretive problems.

Several comments may be made about these experiments:

1. The sprouted terminals can change the physiology of the cells upon which they end; there is correlative work to suggest that they may be involved in behavioral changes. This means that much experimental psychology which tests "residual" function after lesions must be assessed in light of the potential for sprouting after any lesion in an adult brain. (It should be added in parentheses that the majority of the sprouting studies have been on the rat and the brainstem and cord of the cat. While one study has shown suggestion of synaptic mobility in primates, it would seem dangerous to generalize from smaller animals until some parallel experiments have been performed on monkeys.)

2. We still know very little about whether a deafferented synaptic site may be taken over by only certain kinds of terminals. The work of Raisman (1969) and of Lund and Lund (1971a,b) appears to show that postsynaptic sites normally in contact with terminals using one transmitter substance can be taken over by other presynaptic processes using a different transmitter substance. It is unfortunate that no physiological correlation has been provided to see whether these "mismatched" appositions are functionally effective. In this sense the occupation of denervated sites might seem random—after all even myelin lamellae may occupy them. However, Raisman and Field's (1973) quantitative data suggest that there may be some selectivity in the reinnervation process. This needs further investigation.

3. There are obvious limitations to the degree of sprouting; it is not necessarily into the whole deafferented region, as found by both Goldberger and Murray and Lynch et al. A number of possibilities suggest themselves. Adult axons may have intrinsic limited sprouting capacities. The number of local denervated sites is higher in adults than in young animals and perhaps sprouting is blocked once a certain number of contacts are made. Other possibilities include the limitations imposed by particular affinity mismatches, by the products of degeneration, or by other pathways not tested in the experiment also taking over denervated contacts.

4. Most of the emphasis has been on sprouting occurring as takeover of denervated postsynaptic sites. This is mainly because this mechanism can be demonstrated relatively easily. The alternative, of formation of new sites, is one which has been demonstrated in neuromuscular regeneration, but so far, due to technical difficulties, has not been shown in the central nervous system. It should also be added that emphasis has all been directed at asymmetric synaptic sites since, once it has lost its presynaptic component, a denervated postsynaptic symmetric specialization would be hard to iden-

tify and also since most long projection pathways—the ones that can be lesioned—make asymmetric contacts.

5. Given that synaptic replacement can occur after major disruptions, the possibility remains open that it can occur normally in nonpathological conditions, although as yet it has not yet been demonstrated unequivocally. The question of *de novo* synapse formation in adult brains remains to be shown and its presence must be demonstrated before one can safely expect synaptic mobility to play a significant role in memory and behavior.

6. Finally, the work of Wall and colleagues points to the fact that response plasticity can occur without any attendant synaptic sprouting, and while synaptic mobility may reinforce the changes, they are not an essential prerequisite for them. It might be anticipated that changes of this kind will be recognized more generally and that the whole question of what controls the efficacy of a synapse once it is made should be given equal consideration with what controls the making of the synapse in the first place. Indeed, realization that precise relationships demonstrated physiologically can be altered so easily and rapidly must cast some doubts on the use of physiological techniques for mapping specific connections between neurons and certainly on making statements regarding the anatomical precision of connection patterns.

# 17

# REGENERATION

Satisfactory regeneration in the strict sense requires that a damaged axon must regrow from the point at which it has been injured to reform its original patterns of connections, and that the normal function associated with it must be restored. Such regeneration should be distinguished (although often it is not) from the synaptic plasticity described in the previous chapter. We have already seen that in the central nervous system of adult fish and amphibians, cut axons can sufficiently regrow and reestablish their original connections to function in a relatively normal manner. Adult birds show no capacity for regeneration of interrupted central connections and while, as we have already seen, tracts can grow across sites of injury in the developing mammalian central nervous system, this capacity is lost in the adult brain with very few exceptions. It is lost despite the fact that axons (even ones originating from cell bodies situated within the central nervous system) do regenerate outside the confines of the brain and spinal cord and reinnervate skin, glands, and muscles.

From the clinical viewpoint, a failure of regeneration of cut axons within the human brain represents a major obstacle to treating a large number of cases, particularly patients with strokes and traumatic injuries. For this reason several different approaches have been adopted to attempt to identify the causes of this failure of regeneration. Each will be reviewed separately, and a synthesis attempted at the end of the chapter. While it will become clear that no immediate answer is provided as yet, it may be hoped that the recent major impetus of research on regeneration, in which a wide variety of approaches have been adopted, may offer some solutions which can in time be applied to human cases. The earlier literature in this field has been well reviewed by Ramon y Cajal (1928), Windle (1956), and Clemente (1964).

## Peripheral nerve regeneration

After injury, peripheral nerves of mammals, as of other vertebrates, show good regenerative capabilities. The specificity of the reinnervation has been discussed in a number of places throughout this book. Factors expediting optimal regeneration have received considerable attention, particularly as a result of the large number of peripheral nerve injuries caused by war wounds. Reviews of this work are provided by Guth (1956) and in the references quoted by McQuarrie and Grafstein (1973). Axons regenerate from the cut stump, growing at a rate of about 100 $\mu$m/hour after an initial delay. Although regeneration of neuromuscular connections is usually not complete, it can be improved by reducing the scar tissue to a minimum. In some cases the outgrowing axons grow back on themselves and form a whorl of fibers at the site of the cut. This is a neuroma and curiously elicits the response from the cell body that would be produced normally when the axon had reached its terminal region. (See Cragg, 1970 for a discussion of this matter.) The Schwann cells which formerly enveloped the nerve distal to the cut form cords of cytoplasm which seem to act as a guide for the regenerating axons. It appears, since there are more axons distal to the cut than proximal, that individual axons branch to supply several processes which then continue growing. Once synaptic connections are reestablished, the number of axons distal to the cut becomes reduced, presumably as a result of a dying back of those branches which fail to make appropriate connections. The process of overproduction of axons followed by a secondary loss parallels to some extent the pattern of development of neuromuscular innervation and possibly of central connections (see Chap. 13).

The anabolic reactions in the cell body have attracted considerable interest (see Liebermann, 1971). After the initial chromatolysis which follows axotomy, there is a massive proliferation of the granular endoplasmic reticulum and an associated increase in protein manufacture. The amount of protein transported along the axon is increased, but whether the rate of transport also increases is not absolutely certain. If the axon is cut a second time after this anabolic reaction has reached its peak, there is less delay in the new regenerative outgrowth (McQuarrie and Grafstein, 1973), indicating that the early response to injury involves a "priming reaction" of the cell body. Once this has been achieved it can be maintained by a second lesion. Of obvious importance is the nature of the priming reaction, whether it is a positive feedback from the injured axon, a loss of a repressive effect normally exerted by the axon terminal, or an excess of a circulating stimulatory agent within the neuron. In each case, retrograde axonal transport is a necessary means whereby the signal that conditions at the end of the axon are abnormal may reach the cell body.

### Regeneration in the central nervous system of nonmammalian vertebrates

It has been known for many years that central pathways in some nonmammalian vertebrates undergo regeneration and this fact has been exploited in many studies directed at the question of neural specificity (see for example Chaps. 9, 14). Only in the past few years has concerted attention been given to the reasons why central tracts should regenerate in these animals and not in mammals.

Regenerating optic pathways appear to grow through the region of the scar in a disorganized fashion (Sperry, 1944), but the axons reestablish a topographic order in the tract and enter the tectum in a comparatively normal manner (Horder, 1974a,b). Most of those axons destined for the medial part of the tectum run in the medial brachium, and most of those for the lateral tectum run in the lateral brachium. Some fibers enter by the wrong route but, despite this, terminate in the appropriate region. Once they reach the optic tectum, the optic axons first distribute in a disorganized manner and an ordered topographic map gradually emerges over time (Gaze and Jacobson, 1963; Gaze and Keating, 1970). The resolution of the map, judged by behavioral acuity tests in goldfish, is on average about 80% of normal (Weiler, 1966). In several electron microscopic studies of regeneration of lizard tails (Egar and Singer, 1972) and the optic nerve of tadpole (Reier and Webster, 1974) newt (Turner and Singer, 1974), and goldfish (Murray, 1976) the course of recovery after injury has been investigated. In each case emphasis is given to the close relation of the ependymoglial cells (much like radial glial cells of developing mammals, see Chap. 7) to the regenerating axons, and the suggestion is made that these cells play a major role in providing a guide for the regrowing pathways by making gutters, troughs, and tunnels. One further point is the finding that regenerating optic axons contain considerable numbers of dense granular vesicles and that these disappear once the axons reach their site of termination (Murray, 1976). Similar vesicles have also been described in regenerating peripheral nerves, but it is unclear what their function may be.

### Response to damage of the immature mammalian central nervous system

Axons in the developing mammalian central nervous system can grow through damaged regions and may even form their normal patterns of connections once they have done so. We have already seen, for example, that retinal axons will grow through a slit made across the superior colliculus and terminate in the usual position caudal to the slit. Similarly, even after total tectal lesions made in neonatal hamsters or fetal rats, axons grow for considerable distances over areas of injury before finding a region in which to terminate. It has been suggested a number of times that the capacity for growth across a lesion shown in young animals is only revealed if a scar does not develop. There are, however, plenty of examples where there is no reparative growth after lesions in neonatal rodents which leave little obvious scar reaction (e.g., Miller and Lund, 1975; Prendergast and Stelzner, 1976). It would seem more likely that the growth is related to some property of the axons themselves, but exactly how this happens is difficult to determine since the state of maturation of the damaged pathway is often not known. It is not clear whether the growth is due to regeneration (that is, the growth of cut axons which had already made connections), the growth of cut axons which had not yet made synaptic connections, or the growth of axons which had yet to reach the area of injury at the time it was made. These distinctions are important to make, since the state of maturity already achieved may have a major bearing on what growth is possible.

One interesting extension of this approach is the experiment in which the region containing the locus coeruleus (a central source of monoaminergic fibers) was transplanted from fetal rats into the area of the hippocampus of adult animals (Björklund et al., 1976). It readily innervated the hippocampus in a quite normal manner. In a second study fetal spinal cord was transplanted to the spinal cord region of adult rats, but although it differentiated there, it never formed connections with the host nervous system (Thuline and Bunge, 1972).

### Regeneration occurring in the adult mammalian central nervous system, without further treatment

As was indicated at the beginning of this chapter the capacity for regeneration of the mammalian central nervous system is severely limited. The only well documented example is that of the pathway from the hypothalamus to the pituitary gland (hypothalamo–hypophyseal tract) studied by Adams et al. (1969, 1971) and Beck et al. (1969). The tract is composed of large, unmyelinated axons which transmit neurosecretory granules synthesized in the cell body, to the terminal region where they are released into the vascular system. If the pituitary stalk (in which these axons run) is cut in adult ferrets, new fibers start growing across the cut within two weeks and the posterior pituitary is fully reinnervated within one year. Some of the regenerating axons grow back into the pars distalis of the anterior pituitary; others innervate an ectopic infundibular process. That the axons reestablish normal functional relations has been studied by comparing groups of animals in which regeneration is allowed and in which it is prevented by placing a block at the time of cutting. In the latter group urinary volume is increased from normal, testes are smaller, the vulva is shrunken, and the coat condition is poor. In the animals in which regeneration is allowed, these same events are noted initially, but from three months onward, the normal condition is restored; subsequently, four of five female animals mated satisfactorily. The authors attributed the success of regeneration to minimal trauma in surgery and to the fact that the cut ends tend to push together.

Björklund and his colleagues have used fluorescence methods to demonstrate the response of particular aminergic pathways to various disruptions of their course in normal rat brains (see Moore et al., 1974, for a review of their primary work). The advantage of the approach is that the aminergic system represents only a small part of the total neuropil and individual axons can be resolved against an unstained background. If the spinal cord is crushed, it is apparent that within several days fluorescent axons begin to grow into the region of damage. More significant are the several studies in which peripheral tissues, in particular iris muscle, are transplanted into the brain and thereby interrupt the course of certain pathways. Within a few days, aminergic axons containing norepinephrine begin to grow toward the transplants and within two weeks are ramifying in it and apparently forming synapses within it. Thus a tissue which normally would be innervated by peripheral adrenergic axons has become innervated instead by central adrenergic pathways. It should be noted that other aminergic systems, notably 5-hydroxytryptamine and indoleamine fibers, do not appear to invade the transplants.

While these studies do not demonstrate restoration of an original condition, they do indicate that aminergic pathways have the capacity to sprout after injury in a manner not shared by most other axonal systems. It would be interesting to know whether after a single transection of a major central aminergic pathway, such as the medial forebrain bundle, its axons would in time grow back to the regions they formerly innervated.

Other studies show some evidence of tissue repair after minimal lesions of the central nervous system. The most interesting of these by Rose et al. (1960), concerned the effect of heavy-ionizing monoenergetic particles delivered by a cyclotron upon the rabbit cortex. The resulting lesions appeared as flat disks 100–200 $\mu$m wide in which there was a total loss of neuronal cell bodies and stainable neuronal processes. Little vascular leakage was reported, but within seven weeks of the lesion a marked invasion of neuroglia into the area was noted. Subsequently, a stainable fiber plexus was found in the damaged area and dendrites could be traced into the region from neuronal somata situated deep to the lesion. The fiber plexus in the area hardly appears normal, but this is not surprising since the lesioned area never acquires a new population of neuronal cell bodies. The extent of new growth is difficult to assess because, while the stainability with light microscopy of the neural processes in the lesioned area is lost, further electron microscopic studies (Kruger and Hamori, 1970) indicated that the lesion does not, in fact, destroy all neural processes. As a result it is impossible to determine the degree to which the apparent sprouting is due to undamaged axon processes recovering or developing the capacity to be stained by the methods used, as opposed to real sprouting of damaged processes. Several other studies have indicated a capability for tissue repair after minimal surgical damage (e.g., Estable-Puig et al., 1964; Estable-Puig and Estable-Puig, 1972; Marks, 1972). It must be emphasized that although these studies show tissue reconstruction, they do not provide direct evidence of neuronal regeneration.

### Attempts to promote central nervous regeneration by reducing the amount of scarring

Many workers have noted that after a lesion axons can be followed into the region of the injury, but are apparently unable to penetrate the scar tissue formed by astrocytes at the site of injury. It would seem logical, therefore, that if the severity of the scar could be reduced, regeneration might be successful.

This approach was pursued for some years by Windle and his colleagues (see Windle, 1956; Clemente, 1964) following the serendipitous discovery (Chambers et al., 1949) that a preparation of fever-producing bacteria (Piromen), if administered intravenously or intraperitoneally after spinal cord section, had the effect of "loosening" the scar tissue, and nerve fibers were then recognized growing through the region of injury. Scott and Clemente (1955) found that a transected cat cord treated with Piromen showed less indication of retrograde degeneration of cell bodies whose axons passed through the cut than similarly operated but untreated cord. They also stimulated the cord above the cut and were able to record evoked responses up to 30 mm below the lesion in treated animals at times of up to 17 months after the lesion. In a further study, Liu and Scott (1958) tested ascending tracts and found that in treated animals the dorsal spinocerebellar tract had grown through the region of the lesion and extended for 25 mm more rostrally. The tracts were tested both physiologically and by using a degenerative technique (a modification of the Nauta method) after making a second lesion below the original cut. Survival times of 270–330 days were allowed between the first and second lesions.

These studies are of particular significance since they show that, besides crossing the lesioned area, axons are capable of growing for short distances in the regular central nervous tissue. The disappointing thing is that even having apparently solved the barrier problem associated with the scar, complete regeneration with restoration of original synaptic connections still fails to occur.

In a parallel study (Clemente, 1958), the facial nerve of cats was cut and the proximal stump directed into the cortex through a hole in the skull. After treatment with Piromen, ACTH, or desoxycorticosterone, the scar reaction appeared reduced and axons were followed up to 5 mm from the nerve stump into the brain tissue. Whether they formed functional contacts was never determined. A more recent study following growth of axons across a knife-cut lesion made in the forebrain of rats indicated that it too was enhanced by ACTH and by triiodothyronine (Fertig et al., 1971). The study suffers from several problems: the source and degree of penetration of the regenerating fibers were not investigated, nor was any attempt made to quantify the extent of the sprouting. Further work on the spinal cord of rats (Matinian and Windle, 1975) has indicated that if after midthoracic transection, the animals are treated with a mixture of proteases as well as with Piromen, there is anatomical and functional recovery of the deafferented spinal cord. It is unfortunate that the techniques used to assess both the regeneration and the functional recovery were relatively crude and it remains to be seen whether they can be shown with more sophisticated methods.

### Effect of chemical stimulants of nerve growth on central nervous regeneration

Nerve growth factor (NGF), as described earlier in this book, plays a role in maintenance of certain cells, and in sprouting of their axons. Scott and Liu (1964) thought that it promoted the growth of dorsal column axons across a lesion. However, the placing of the lesions was not well controlled, since they were made through the dura and the conduction velocities of the regenerating axons were rather high both for the diameter of the axons and the state of regeneration. Saunders (1972), who noted these problems, found a lack of effect of NGF on peripheral nerve regeneration after nerve crush, but this could be because insufficient NGF diffused from the injection site to the site where it was effective; and so this result too is equivocal.

More recently Björklund and colleagues investigated the effects of NGF on central (and peripheral) aminergic pathways (Björklund and Stenevi, 1971; Björklund et al., 1974). In the central nervous system, they studied, as before, the sprouting into transplants inserted in the course of the medial forebrain bundle. This sprouting is enhanced if NGF is given within seven days. The degree of sprouting is dose dependent, being higher the closer the injection of NGF is to the cell bodies of origin of the sprouting axons. Sprouting is prevented by administration of antisera to NGF along with the NGF. These interesting results are, of course, subject to the usual reservations which must be made in this type of work. In particular, there is the problem of whether NGF actually stimulates transmitter production and therefore enhances the fluorescence of fibers not previously seen rather than causing actual sprouting of axons. This reservation is all the more pertinent in view of the increased numbers of fluorescing axons shown to be present in normal tissue with the more recent glyoxylic acid method compared with the technique used in the earlier studies (Lindvall et al., 1974). A further problem is that of quantitation. This is extremely difficult to achieve in such material, but it would seem essential to attempt quantitation, especially since authors find considerable variation among animals.

### Failure of central regeneration may be the result of an autoimmune response

This proposal has been offered by two independent groups (Feringa et al., 1973, 1974; Berry and Riches, 1974). It derives from several observations:

1. There are a number of clearly identified and several suspected autoimmune diseases of the central nervous system which lead to neural degeneration.
2. The blood–brain barrier may normally prevent antibody-producing cells from entering the central nervous system, and, since this barrier is damaged in injury, such cells would have access to "foreign" protein.
3. Regeneration in nonmammalian vertebrates occurs in animals in which immune systems are slow to react.
4. Growth across lesioned areas in young rodents occurs prior to the development of immunological competence.
5. The corticosteroids which have been used to prevent scar formation also show immunosuppressive properties and may therefore function in this manner.

Feringa et al. (1973) tested the hypothesis by cutting the spinal cord in 75 rats of five groups: (1) control adult rats, (2) neonatal rats (before the immunological systems have developed), (3) immunotolerant rats (given intraperitoneal injections of adult spinal cord), (4) rats treated with anti-rat lymphocyte serum, and (5) rats treated with immunosuppressive drugs (azathioprine and levathyroxine). In only three rats (one each from groups 3–5) was there suggestion of axons crossing the lesion. These might have been axons left intact at the time of surgery rather than regenerating axons. Therefore, it would appear that immunological factors may not on their own play a major role in blocking central nervous regeneration—at least in the rat spinal cord.

## Discussion

Despite the variety of approaches to the question of neural regeneration, no clear primary reason for its failure in adult mammals emerges. We will consider the various possible reasons separately.

1. *The glial scar blocks regeneration.* Support for this proposal comes from several sets of experiments. The scar reaction is less impressive in young mammals where growth across a lesion sometimes occurs. One report has suggested that the glial scar is less pronounced in fish compared with mammals (Phelps, 1969). In addition, various chemical treatments which prevent an astrocytic scar from forming in cats result in growth of axons across lesions. However, despite this growth, the axons do not appear to continue growing back to reach their former terminal sites. The few cases where such growth has been claimed deserve further investigation.

2. *Autoimmune responses are involved.* While an interesting and logically reasonable idea, the experimental studies of Feringa and colleagues are not encouraging.

3. *Once an axon has completed its growth phase and made synaptic connections, it is no longer capable of reverting to its embryonic state and achieving new growth of more than local proportions.* This proposal explains the difference between fish and amphibians on the one hand and birds and mammals on the other, by arguing that one group retains the embryonic ability for continued axonal growth while the other does not. This explanation is not altogether satisfactory when the difference in regeneration capacity between axons of the central and peripheral nervous system of mammals is considered.

4. *Regeneration is only possible if there is appropriate feedback from the terminal area to maintain the cell and promote growth.* We have already seen that NGF promotes sprouting of neurons sensitive to it. Further, in the superior cervical ganglion, the transport of NGF from the region of axon terminals back to the cell body functions to maintain the cell and the retrograde degenerative effects of axotomy may be prevented by local application of NGF to the ganglion. In other peripheral ganglia, such as the ciliary ganglion which is insensitive to NGF, it might be anticipated that some other maintenance factor is involved. This could also be so in the central

nervous system, where the effects of NGF are equivocal, or at least restricted in scope. The rapid retrograde degeneration of thalamic cells after the terminal parts of their axons are removed by cortical lesions could well be due to removal of some such factor. It is obviously of considerable importance to know (a) whether such growth-promoting factors do indeed exist; (b) what their specificity may be (i.e., with respect to basic brain divisions, local regions, or cell types); and (c) whether they could be used to maintain cells and promote sprouting, as NGF does.

5. *Cut axons sprout proximal to the site of injury and this occupation of aberrant synaptic sites prevents any stimulus for growth.* Adult neurons certainly do show axonal sprouting proximal to lesions. As suggested in development, axonal growth may continue until a certain critical number of contacts are formed and the feedback from the region of the synapse is instrumental in preventing further growth. It is possible that this condition is maintained into adulthood and the sprouting phenomena described in the previous chapter are, in fact, deleterious to the progress of the axonal regeneration. There are two crucial experiments which can test this hypothesis. First, the capacity for proximal sprouting in fish should be studied in the same systems in which they have been shown to occur in mammals. This has not been done. Second, the capacity for regeneration can be tested in central fiber tracts which run for some distance in the absence of synaptic neuropil. This has been done and there is evidence in rat after optic nerve section (Sperry, quoted by Windle, 1956) and from the studies of optic chiasm or optic tract section in cats, monkeys, and humans that regeneration does not occur, there being instead retrograde degenerative changes of retinal ganglion cells.

6. *Neuroglia play an important role in axonal growth.* This could relate in two ways to regeneration. First, they (and indeed other tissue components) may provide guides for axonal growth. The Schwann cells of the peripheral nervous system appear to function in this manner. Since the radial glia seem to disappear in some regions of the mammalian brain during maturation, it is possible that regenerating axons no longer have the proper guides to direct them to their appropriate terminal area. The continued presence of ependymoglial cells in adult frog and fish brains, and their possible role in regeneration reactions, is at least consistent with this view. A second role proposed for glia is that of "trophic" cells which function to maintain a neuron in the way NGF-producing cells do. It is possible that the glia may even produce a specific central nervous system cell maintenance factor; perhaps there are insufficient cells in a region of regeneration to provide enough of this "trophic substance."

This section is highly speculative, largely because no clear pattern emerges from the available studies. It is quite possible that some or all of the suggested reasons for failure of regeneration in the mammalian central nervous system are relevant and that may be the crux of the problem. Most experiments are directed at testing one hypothesis. For example, autoimmune factors could be important, but if cell maintenance factors are not also given and glial guides not provided, regeneration might still fail, even though the immunity difficulties have been eliminated. In particular, the disappointing limited regeneration found by Windle and his colleagues *after* they had managed to get the axons to grow across a region of damage would seem to indicate the involvement of multiple factors. One final point is that many of the experiments investigating factors concerned with central regeneration have involved transection of the spinal cord. While of direct clinical significance in view of the large number of traumatic cord lesions which occur, especially in young people (Kurtzke, 1975), it is perhaps the wrong place to start experiments in view of the secondary problems associated with the lesions. These include the general health of the animal as well as the mechanical problems of maintaining the cut ends of the cord in apposition.

# VIII
# SUMMARY

As has been emphasized throughout this book, the final specific patterns of connections made by a neuron are not the product of a single event, but are the result of a continued interaction throughout all stages of development between the neuron and the environment in which it grows. Even when it is mature, a neuron is still susceptible to the vicissitudes of its surroundings.

One of the direct results of a continued dependency of a neuron on its surroundings is that considerable variation from the normal pattern might be expected. We have seen that this is the case between genetically pure populations: albino animals have major anomalies in their central visual pathways and several other examples were given in which slight variations in the distribution of certain pathways occur between different strains of mice. The extent to which the exact patterns of connections of a particular neuron in a vertebrate brain might vary from one animal to another in a genetically homogeneous population can only be guessed. In the case of the primary optic connections of the water flea, described in Chapter 1, there is some variability in the numbers of synapses formed by an identified neuron, even on opposite sides of the same brain, and greater variation might be expected in the vertebrate brain. Nevertheless, to achieve the level of acuity shown by the human visual system through the several synapses involved, considerable precision must occur somewhere. Does this mean that there must be anatomically precise connections therefore? The results of Wall and colleagues (Chap. 16), in which functionally precise patterns can be extracted from relatively diffuse anatomical connections, would suggest that such precise connections need not necessarily be the case.

We are still no further, therefore, in answering the question—how specific is specific?—asked in Chapter 1, despite the fact that this is such a fundamental point. What we do know is that the ordered patterns which we can define with our current and relatively crude techniques can be materially altered by a variety of approaches and that the resulting aberrations are more marked if the perturbation occurs during development rather than later in life. A point of interest in this work is that, however extreme the disorder, there is no indication that a group of neurons will behave in a totally random fashion. In this sense, the results of any experimental procedure will be to modify or remove certain constraints which limit some part of the growth of a neuron. Other limiting factors will still be present and the neuron may continue to respond to these as best it can. Recall, for example, the studies in which axons for one reason or another are made to end on the wrong side of the brain (Chap. 10). Despite this, they still end in the appropriate regions and attempt to form a coherent topographic map. In this case the particular constraints concerned occur in a sequential manner, the problem of laterality being faced by the axon before that of ordering of terminal growth patterns.

But may not any single event be subject to a variety of simultaneous constraints and controls? This would seem possible in the mapped projection of retinal axons on the optic tectum. It is difficult in the light of certain experiments to argue against some form of quite stable positional labels, but the modifiability occurring in other experiments might suggest that some measure of fiber ordering is also involved. Such multiple controls of a single event might in normal circumstances serve to sharpen the precision of the map beyond that which may be possible using only one controlling system, and in particular would serve as a check in eliminating aberrations which might result if a particular prerequisite for the successful functioning of one control system was absent. One of the indications of multiple controls is that with slightly different experimental details or designs, a different deficit or disordered connection pattern might be expected. This would explain some of the contradictory findings of the retinotectal mapping literature (Chap. 14).

Of considerable importance in understanding how a normal brain develops and where things can go wrong are the questions of what controls exist, what their nature is, when and how they operate, and what backups are there to compensate for local failures. To discuss this further, let us divide the life history of a neuron into four basic phases of development, according to the nature and permanence of the effects of the surroundings. These are:

1. A period extending up to early postmitotic times.
2. A period continuing from immediately after final mitosis to the time of establishment of connections.
3. A period beginning immediately after the establishment of connections and continuing throughout adult life. The characteristic events of this period are most evident early on.

4. A period partially overlapping period (3) in which the neuron acquires its functional identity.

Each will be discussed separately.

### Period 1

During this time, the neuron acquires the information necessary to express its basic identifying features. The protein synthetic machinery of the cell, presumably controlled by its genetic mechanisms, is strongly affected by the cell's surroundings. There are two major events which are imprinted upon the cell at this stage: a set of instructions specifying certain aspects of its intrinsic growth, and a set of instructions which detail its further interactions with its environment. Of those involved in intrinsic growth, the evolution of cell shape is clearly important. It has been shown that although cell shape, particularly dendritic character, can be somewhat modified by late events, the basic identifying features of a neuron are unmistakeable whatever the environment during development. This has been shown rather well for the Purkinje cell of the variously experimentally manipulated cerebella discussed in Chapter 11, but this is not a unique example. It is difficult to know how and when this feature becomes instilled into a cell. One problem is that we do not know what is responsible for defining cell shape. Is it the configuration of microtubules or neurofilaments? If so, how is their growth regulated to produce in one case a pyramidal neuron with its huge apical dendrite and in another, a symmetrically ordered stellate neuron? What defines the presence of dendritic spines, such a distinct characterizing feature of so many cell types? These usually contain neither microtubules nor neurofilaments. To these fundamental questions we have no answers. As to when a neuron acquires the instructions concerned with defining cell shape, all that can be said is that it is likely to be early in development simply because aberrations which affect the early postmitotic behavior of a neuron, in particular its migratory patterns, have little effect on the major character of the cell, even though orientation of the dendrites may be clearly disturbed. One obviously important question not answered is whether an individual stem cell may produce a series of different cell types or whether one cell will produce all the pyramidal class and another, the stellate class, and so on.

A second aspect of intrinsic control of growth concerns the possibility of a limited amount of growth and a preprogramming for cell death if a certain set of conditions is not met. The first point, that a neuron makes and can maintain only a certain volume of cytoplasm, is an inherent precondition of Schneider's theory of conservation of terminal space although, as we saw in Chapter 13, there are other interpretations of his experiments and enough exceptions exist to question the general validity of the theory. Despite this, the idea of a controlled finite cell volume may be important as it relates to such matters as the control of cell-body size. If somal size does, as has been suggested, reflect total cytoplasmic volume, what then is the primary control? Certainly in some cases (the visual deprivation literature provides a good example) cell-body size has clear extrinsic regulators, but it is not clear whether these are primary controls or whether they are secondary effects.

Programmed cell death is obviously an important way by which inappropriately connected or redundant neurons may be eliminated. In several examples, it has been shown that the neuron becomes particularly vulnerable during a limited time course of its development, notably the period during which it is making (or in some cases receiving) its major synaptic connections. This sensitive period can be defined rather well for neurons which are responsive to NGF (Chap. 12). While the trigger for maintaining the cell may be an external event (this is discussed for period 3 below), it would appear that at some point in development an intrinsic time-locked sensitive period is built into the neuron.

The extent to which timing is an essential feature in building up an organized and specifically connected nervous system is not altogether clear. Obviously the timing of cell generation relates to the ordered distributions of cells of the cerebral cortex, but ordered cell layering is found in other regions such as the midbrain without correlating with a timed sequence of cell generation (Chap. 7). It is possible that timing of axonal outgrowth may relate to its laterality of distribution (in the visual system) or to certain competitive disadvantages for synaptic sites (as has been proposed in the dentate gyrus of the hippocampal formation), but definitive proof is lacking. Further, the experiment in which an optic nerve was prevented from reaching the optic tectum of *Xenopus* until normal development had finished and then connected normally would argue (in that particular case) that timing of ingrowth is irrelevant to normal functioning.

The other built-in features a neuron may acquire during this first developmental period concern its direct interactions with the environment into which it is about to grow and mature. For the most part such interactions that fall in the general category of recognition processes may be attributed to the development of a unique set of membrane proteins. That differences do occur in membrane proteins from one neuron to another has been shown (Chap. 8). These may serve a number of functions: guiding the specific migration of the cell to its final location from which it develops its processes, guiding to some extent the disposition of both dendrites and axon, and, for the axon in particular, dictating its interaction with the substrate, its competitive or (in the case of fasciculation) its attractive interaction with other axons, and finally its matching with a particular synaptic partner. The fact that for a single cell the growing axon may follow different cues from those used by the migrating cell body and by the growing dendrites suggests that the membrane proteins, if indeed they are important in these events, are different for different parts of the cell. Whether they are different for any one part of a cell at different stages of development can only be surmised. What is important to realize is that the recognition process, however it works, is operative not only at the point at which an axon finds its appropriate synaptic site, but also at all points at which a migrating neuron and its growing axon tip or dendritic processes interact with their surroundings. The timing of acquisition of membrane "recognition" features has been addressed only with respect to those associated with positional identity, and it appears that for the retinal ganglion cells this becomes stabilized shortly after their final mitosis, although it exists much earlier in an unstable form.

### Period 2

During this period, extending from the early postmitotic period until the inception of synapse formation, the neuron expresses its identity acquired during period 1 by the interaction of its cell body and processes with their surroundings. As such the surroundings may have little or no permanent effect on the central synthetic machinery or controlling mechanisms of the cell, but instead exert their influences primarily through the various membrane interactions. The pattern of interactions shown by a neuron at this point would seem to depend, therefore, on the nature and distribution of proteins in the membrane, the main role played by the cell-body region being in the supply of appropriately programmed membrane proteins. With such a mechanism, feedback would not seem to be necessary.

During this period, the cell migrates, the axon and dendrites grow, and finally synapses are formed. It is a time when most of the plasticities of organization discussed throughout the book become expressed—largely by misposition of the cell body, maldisposition of the dendrites, and a variety of aberrations of axonal distribution. While there are examples of cell death resulting from aberrations occurring during this stage (the degeneration of granule cells in weaver mutant mice is one), the more usual response to abnormal conditions is for the neuron to follow a modified pattern of development, rather than to die.

### Period 3

This period is a time of consolidation, of enhancement of "good" connections, and of elimination of "bad" ones. It is a time when transneuronal influences between cells are particularly important and a time when the external influences exert their effects largely through the central machinery of the cell, turning it off completely, or altering levels of enzyme and other protein synthesis. It is during the early part of this period that cell death is a major feature, the factors preventing it apparently being the formation of synapses and/or the presence of maintenance factors like NGF produced by the target region and perhaps even by the target neuron. The whole area of trophic factors is one of great interest and clearly one in which significant progress might be expected in the next few years. It is also apparent that some degree of cell maintenance is afforded simply by the process of appropriate stimulation. Whether this is a direct effect of stimulation or is associated with the release of trophic substances such as adenosine is not altogether clear.

### Period 4

During this period, a neuron develops its characteristic functional identity. It overlaps period 3 in time, and some of the mechanisms involved are those of period 3, as well as perhaps those of period 2. However, it is apparent that a stable response pattern of a single neuron is not totally attributable to a series of hard-wired connections, but is also in part due to the development by the neuron of an "ability" to select from a wide variety of connections a certain pattern of inputs and to maintain this "choice" once selected and once beyond a critical period. The emphasis on maintenance of a physiologically stable pattern in a potentially dynamic state is supported by the several studies in which long-term deficits in visual responsiveness after selective rearing can be partially reversed (Chap. 15), and by the altered patterns of responsiveness in the somatosensory system as a result of cooling or local lesions (Chap. 16). This suggests that some of the selectivity of responsiveness shown by a neuron depends on its own intrinsic behavior, on the biophysical parameters involved in the preferential weighting of certain inputs, and of actual suppression of others.

In summary, therefore, the complex neural patterns which make up the vertebrate brain are largely the result of a set of relatively simple basic mechanisms, which, with the exception of those outlined in period 4, are shared by most other cells of the body. A lot of the work in developmental neurobiology, particularly in mammals, has simply involved the essential task of collecting data to find what factors are important at the different stages of development, and to study model systems for certain specific features. From this work a number of important issues have been identified. These include the ways in which controls are exerted upon the genome, the ways genetic information becomes translated into the characteristic features of the neuron, the mechanisms associated with intercellular recognition, the role and mechanism of cell maintenance factors, the mechanisms of programmed cell death, and the mechanisms whereby a neuron may selectively suppress the efficacy of particular inputs. All seem ripe for fruitful investigation in the next decade. From such studies, it may be hoped that some progress will be made toward solving the fundamental applied problems facing neuroscientists—the prevention and treatment of mental retardation conditions, abnormal developmental patterns, and degenerative states of the central nervous system—and providing insight into the basic mechanisms underlying learning and behavior.

# REFERENCES

Adams, J.H., P.M. Daniel, and M.M.L. Pritchard 1969 Degeneration and regeneration of hypothalamic nerve fibers in the neurohypophysis after pituitary stalk section in the ferret. J. Comp. Neurol., *135* : 121–144.

Adams, J.H., P.M. Daniel, and M.M.L. Pritchard 1971 Changes in the hypothalamus associated with regeneration of the hypothalamo-neurohypophysial tract after pituitary stalk section in the ferret. J. Comp. Neurol., *142* : 109–124.

Aguilar, C.E., M.A. Bisby, E. Cooper, and J. Diamond 1973 Evidence that axoplasmic transport of trophic factors is involved in the regulation of peripheral nerve fields in salamanders. J. Physiol., Lond., *234* : 449–464.

Albuquerque, E.X., J.E. Warwick, J.R. Tasse, and F.M. Sansone 1972 Effects of vinblastine or colchicine on neural regulation of the fast and slow skeletal muscles of the rat. Exp. Neurol., *37* : 607–634.

Albus, K. 1975 Predominance of monocularly driven cells in the projection area of the central visual field in cat's striate cortex. Brain Res., *89* : 341–347.

Altman, J. 1970 Postnatal neurogenesis and the problem of neural plasticity. *In* Developmental Neurobiology, W.A. Himwich (Ed.), Charles C. Thomas, Springfield, Ill., pp. 197–237.

Altman, J. 1971 Coated vesicles and synaptogenesis. A developmental study in the cerebellar cortex of the rat. Brain Res., *30* : 311–322.

Altman, J. 1973a Experimental reorganization of the cerebellar cortex. III. Regeneration of the external germinal layer and granule cell ectopia. J. Comp. Neurol., *149* : 153–180.

Altman, J. 1973b Experimental reorganization of the cerebellar cortex. IV. Parallel fiber reorientation following regeneration of the external germinal layer. J. Comp. Neurol., *149* : 181–192.

Altman, J., and W.J. Anderson 1969 Early effects of X-irradiation of the cerebellum in infant rats : decimation and reconstitution of the external granular layer. Exp. Neurol., *24* : 196–216.

Altman, J., and W.J. Anderson 1972 Experimental reorganization of the cerebellar cortex. I. Morphological effects of elimination of all microneurons with prolonged X-irradiation started at birth. J. Comp. Neurol., *146* : 355–406.

Altman, J., and W.J. Anderson 1973 Experimental reorganization of the cerebellar cortex. II. Effects of elimination of most microneurons with prolonged X-irradiation started at five days. J. Comp. Neurol., *149* : 123–152.

Anderson, M. J., and M. W. Cohen 1977 Nerve-induced and spontaneous redistribution of acetylcholine receptors on cultured muscle cells. J. Physiol., Lond., *268* : 757–774.

Anderson, M.J., M.W. Cohen, and E. Zorychta 1977 Effects of innervation on the distribution of acetylcholine receptors on cultured muscle cells. J. Physiol., Lond., *268* : 731–756.

Angevine, J.B. Jr. 1970 Time of neuron origin in the diencephalon of the mouse. An autoradiographic study. J. Comp. Neurol., *139* : 129–188.

Angevine, J.B. Jr., and R.L. Sidman 1961 Autoradiographic study of cell migration during histogenesis of cerebral cortex in the mouse. Nature, Lond., *192* : 766–768.

Anker, R.L., and B.G. Cragg 1974 Development of the extrinsic connections of the visual cortex in the cat. J. Comp. Neurol., *154* : 29–42.

Armstrong, J., K.C. Richardson, and J. Z. Young 1956 Staining neural end-feet and mitochondria after postchroming and carbowax embedding. Stain Technol., *31* : 263–270.

Attardi, D.G., and R.W. Sperry 1963 Preferential selection of central pathways of regenerating optic fibers. Exp. Neurol., *7* : 46–64.

Baisinger, J., R.D. Lund, and B. Miller 1977 Aberrant retinothalamic projections resulting from unilateral tectal lesions made in fetal and neonatal rats. Exp. Neurol., *54* : 369–382.

Baker, F.J., P. Grigg, and G.K. Von Noorden 1974 Effects of visual deprivation and strabismus on the response of neurons in the visual cortex of the monkey, including studies on the striate and prestriate cortex in the normal animal. Brain Res., *66* : 185–208.

Baker, R.E. 1972 Biochemical specification versus specific regrowth in the innervation of skin grafts in anurans. Nature, Lond., *236* : 235–237.

Baker, R.E. and M. Jacobson 1970 Development of reflexes from skin grafts in *Rana pipiens* : influence of size and position of grafts. Develop. Biol., *22* : 476–494.

Balsamo, J., J. McDonough, and J. Lilien 1976 Retinal-tectal connections in the embryonic chick : evidence for regionally specific cell surface components which mimic the pattern of innervation. Develop. Biol., *49* : 338–346.

Banker, G., L. Churchill, and C.W. Cotman 1974 Proteins of the postsynaptic density. J. Cell Biol., *63* : 456–465.

Barber, R.P., J.E. Vaughn, R.E. Wimer, and C.C. Wimer 1974 Genetically-associated variations in the distribution of dentate granule cell synapses upon the pyramidal cell dendrites in mouse hippocampus. J. Comp. Neurol., *156* : 417–434.

Barbera, A.J. 1975 Adhesive recognition between developing retinal cells and the optic tecta of the chick embryo. Develop. Biol., *46* : 167–191.

Barbera, A.J., R.B. Marchase, and S. Roth 1973 Adhesive recognition and retinotectal specificity. Proc. Nat. Acad. Sci., Wash., *70* : 2482–2486.

Barondes, S.H. 1975 Towards a Molecular Basis of Neuronal Recognition. *In* The Nervous System, D.B. Tower (Ed.), Vol. 1 : The basic Neurosciences, Raven Press, New York, pp. 129–136.

Barron, K.D. 1975 Ultrastructural changes in dendrites of central neurons during axon reaction. Adv. Neurol., *12* : 381–399.

Beazley, L.D. 1975 Factors determining decussation at the optic chiasma by developing retinotectal fibers in *Xenopus*. Exp. Brain Res., *23* : 491–504.

Beazley, L., M.J. Keating, and R.M. Gaze 1972 The appearance, during development, of responses in the optic tectum following visual stimulation of the ipsilateral eye in *Xenopus laevis*. Vis. Res., *12* : 407–410.

Beck, E., P.M. Daniel, and M.M.L. Pritchard 1969. Regeneration of hypothalamic nerve fibers in the goat. Neuroendoc., *5* : 161–182.

Bennett, E.L., M.C. Diamond, D. Krech, and M.R. Rosenzweig 1964 Chemical and anatomical plasticity of brain. Science, *146* : 610–619.

Bennett, M.R., and A.G. Pettigrew 1974 The formation of synapses in striated muscle during development. J. Physiol., Lond., *241* : 515–545.

Bennett, M.R., and A.G. Pettigrew 1976 The formation of neuromuscular synapses. Cold Spr. Harb. Symp. Quant. Biol., *40* : 409–424.

Berman, N., and P. Sterling 1976 Cortical suppression of the retino-collicular pathway in the monocularly deprived cat. J. Physiol., Lond., *255* : 263–273.

Bernstein, J.J., and M.E. Bernstein 1971 Axonal regeneration and formation of synapses proximal to the site of the lesion following hemisection of the spinal cord. Exp. Neurol., *30* : 336–351.

Bernstein, J.J., and M.E. Bernstein 1973 Neuronal alteration and reinnervation following axonal regeneration and sprouting in mammalian spinal cord. Brain. Behav. Evol., *8* : 135–161.

Bernstein, M.E., and J.J. Bernstein 1973 Regeneration of axons and synaptic complex formation rostral to the site of hemisection in the spinal cord of the monkey. Int. J. Neurosci., *5* : 15–26.

Berry, M., and P. Bradley 1976 The growth of the dendritic trees of Purkinje cells in irradiated agranular cerebellar cortex. Brain Res., *116* : 361–387.

Berry, M., and A.C. Riches 1974 An immunological approach to regeneration in the central nervous system. Brit. Med. Bull., *30* : 135–140.

Berry M., and A.W. Rogers 1965 The migration of neuroblasts in the developing cerebral cortex. J. Anat., *99* : 691–709.

Bignami, A., and D. Dahl 1973 Differentiation of astrocytes in the cerebellar cortex and the pyramidal tracts of the newborn rat. An immunofluorescence study with antibodies to a protein specific to astrocytes. Brain Res., *49* : 393–402.

Bignami A., and D. Dahl 1974 The development of Bergmann glia in mutant mice with cerebellar malformations: reeler, staggerer and weaver. Immunofluorescence study with antibodies to the glial fibrillary acidic protein. J. Comp. Neurol., *155* : 219–230.

Bignami, A., E.F. Eng, D. Dahl, and C.T. Uyeda 1972 Localization of the glial fibrillary acidic protein in astrocytes by immunofluorescence. Brain Res., *43* : 429–435.

Björklund A., B. Bjerre, and U. Stenevi 1974 Has nerve growth factor a role in the regeneration of central and peripheral catecholamine neurons? *In* Dynamics of Degeneration and Growth in Neurons, K. Fuxe, L. Olson, and Y. Lotterman (Eds.), Pergamon Press, Oxford and New York, pp. 389–409.

Björklund, A., and U. Stenevi 1971 Growth of central catecholamine neurones into smooth muscle grafts in the rat mesencephalon. Brain Res., *31* : 1–20.

Björklund, A., U. Stenevi, and N.A. Svendgaard 1976 Growth of transplanted monoaminergic neurones into the adult hippocampus along the perforant path. Nature, Lond., *262* : 787–790.

Black, I.B., I.A. Hendry, and L.L. Iversen 1971 Transynaptic regulation of growth and development of adrenergic neurons in a mouse sympathetic ganglion. Brain Res., *34* : 229–240.

Black, I.B., I.A. Hendry, and L.L. Iversen 1972 Effects of surgical decentralization and nerve growth factor on the maturation of adrenergic neurons in a mouse sympathetic ganglion. J. Neurochem., *19* : 1367–1377.

Blake, R., and H.V.B. Hirsch 1975 Deficits in binocular depth perception in cats after alternating monocular deprivation. Science, *190* : 1114–1116.

Blakemore, C., and G.F. Cooper 1970 Development of the brain depends on the visual environment. Nature, Lond., *228* : 477–478.

Blakemore, C., and D.E. Mitchell 1973 Environmental modification of the visual cortex and the neural basis of learning and memory. Nature, Lond., *241* : 467–468.

Blakemore, C., and R.C. Van Sluyters 1974 Reversal of the physiological effects of monocular deprivation in kittens: further evidence for a sensitive period. J. Physiol., Lond., *237* : 195–216.

Blakemore, C., and R.C. Van Sluyters 1975 Innate and environmental factors in the development of the kitten's visual cortex. J. Physiol., Lond., *248* : 663–716.

Blinzinger, K., and G.W. Kreutzberg 1968 Displacement of synaptic terminals from regenerating motoneurons by microglial cells. Z. Zellforsch. Mikr. Anat., *85* : 145–157.

Boothe, R.G., and J.S. Lund 1976 A quantitative study of pyramidal cell spine density in developing visual cortex of normal and dark reared macaque monkeys. Neurosci. Abstr., *2* : 1103.

Borges, S., and M. Berry 1976 Preferential orientation of stellate cell dendrites in the visual cortex of the dark-reared rat. Brain Res., *112* : 141–147.

Boycott, B.B., and H. Wässle 1974 The morphological types of ganglion cells of the domestic cat's retina. J. Physiol., Lond., *240* : 397–419.

Brattgard, S.O. 1952 The importance of adequate stimulation for the chemical composition of retinal ganglion cells during early post-natal development. Acta Radiolog., Suppl., *96*.

Bray, D. 1970 Surface movements during the growth of single explanted neurons. Proc. Nat. Acad. Sci., Wash., *65* : 905–910.

Bray, D. 1973 Branching patterns of individual sympathetic neurons in culture. J. Cell Biol., *56* : 702–712.

Bray, D., and M. Bunge 1973 The growth cone in neurite extension. *In* Locomotion of Tissue Cells, Ciba Foundation Symposium 14. W. Porter and D.W. Fitzsimmons (Eds.), Elsevier, Amsterdam, pp. 195–209.

Brindley, G.S. 1969 Nerve net models of plausible size that perform many simple learning tasks. Proc. Roy. Soc. Lond. B, *174* : 173–191.

Brodal, A. 1969 Neurological Anatomy in relation to Clinical Medicine, Oxford Univ. Press, New York.

Brown, M.C., J.K.S. Jansen, and D. Van Essen 1976 Polyneuronal innervation of skeletal muscle in new-born rats and its elimination during maturation. J. Physiol.,

Lond., *261* : 387–422.

Brückner, G., V. Mares, and D. Biesold 1976 Neurogenesis in the visual system of the rat. An autoradiographic investigation. J. Comp. Neurol., *166* : 245–256.

Buisseret, P., and M. Imbert 1976 Visual cortical cells: their developmental properties in normal and dark reared kittens. J. Physiol., Lond., *255* : 511–525.

Buller, A.J., J.C. Eccles, and R.M. Eccles 1960 Differentiation of fast and slow muscles in the cat hind limb. J. Physiol., Lond., *150* : 399–416.

Buller, A.J., W.F.H.M. Mommaerts, and K. Seraydarian 1969 Enzymic properties of myosin in fast and slow twitch muscles of the cat following cross-innervation. J. Physiol., Lond., *205* : 581–597.

Bunge, M.B. 1973 Fine structure of nerve fibers and growth cones of isolated sympathetic neurons in culture. J. Cell Biol., *56* : 713–735.

Bunge, M.B. 1977. Initial endocytosis of peroxidase or ferritin by growth cones of cultured nerve cells. J. Neurocytol., *6* : 407–439.

Bunt, A.H., A.E. Hendrickson, J.S. Lund, R.D. Lund, and A.F. Fuchs 1975 Monkey retinal ganglion cells: morphometric analysis and tracing of axonal projections with a consideration of the peroxidase technique. J. Comp. Neurol., *164* : 265–286.

Bunt, A.H., R.H. Haschke, R.D. Lund, and D.F. Calkins 1976 Factors affecting retrograde transport of horseradish peroxidase in the visual system. Brain Res., *102* : 152–155.

Caviness, V.S. Jr. 1977 The reeler mutant mouse: a genetic experiment in developing mammalian cortex. Neuroscience Symposia Series *2* : 27–46.

Caviness, V.S. Jr., and R.L. Sidman 1973 Time of origin of corresponding cell classes in the cerebral cortex of normal and reeler mutant mice : an autoradiographic analysis. J. Comp. Neurol., *148* : 141–152.

Ceccarelli, B., W.P. Hurlbut, and A. Mauro 1972 Depletion of vesicles from frog neuromuscular junctions by prolonged tetanic stimulation. J. Cell Biol., *54* : 30–38.

Chambers, W.W., C.Y. Liu, and C.N. Liu 1956 A modification of the Nauta technique for staining of degenerating axons in the central nervous system. Anat. Rec., *124* : 391–392.

Chambers, W.W., H. Koenig, R. Koenig, and W.F. Windle 1949 Site of action in the central nervous system of a bacterial pyrogen. Amer. J. Physiol., *159* : 209–216.

Chan-Palay, V. 1972 Arrested granule cells and their synapses with mossy fibers in the molecular layer of the cerebellar cortex. Z. Anat. Entwickl.-Gesch., *139* : 11–20.

Chernenko, G.A., and R.W. West 1976 A re-examination of anatomical plasticity in the rat retina. J. Comp. Neurol., *167* : 49–62.

Chiappinelli, V., E. Giacobini, G. Pilar, and H. Uchimura 1976 Induction of cholinergic enzymes in chick ciliary ganglion and iris muscle cells during synapse formation. J. Physiol., Lond., *257* : 749–766.

Chow, K.L. 1973 Neuronal changes in the visual system following visual deprivation. *In* Handbook of Sensory Physiology *7/3* pt.A: 599–627.

Chow, K.L., L.W. Mathers, and P.D. Spear 1973 Spreading of uncrossed retinal projection in superior colliculus of neonatally enucleated rabbits. J. Comp. Neurol., *151* : 307–322.

Chow, K.L., A.H. Riesen, and F.W. Newell 1957 Degeneration of retinal ganglion cells in infant chimpanzees reared in darkness. J. Comp. Neurol., *107* : 27–42.

Chow, K.L., and P.D. Spear 1974 Morphological and functional effects of visual depriva-tion on the rabbit visual system. Exp. Neurol., *42* : 429–447.

Chung, S.H. 1974 In search of rules for nerve connections. Cell, *3*: 201–205.

Chung, S.H., and J. Cooke 1975 Polarity of structure and of ordered nerve connections in the developing amphibian brain. Nature, Lond., *258* : 126–132.

Chung, S.H., R.M. Gaze, and R.V. Stirling 1973 Abnormal visual function in *Xenopus* following stroboscopic illumination. Nature, New Biol., *246* : 186–189.

Chung, S.H., M.J. Keating, and T.V.P. Bliss 1974 Functional synaptic relations during the development of the retino-tectal projection in amphibians. Proc. Roy. Soc. Lond. B, *187* : 449–459.

Clark, W.E. LeGros 1940 Neuronal differentiation in implanted foetal cortical tissue. J. Neurol. Psychiat., *3* : 263–272.

Clark, A.W., W.P. Hurlbut, and A. Mauro 1972 Changes in the fine structure of the neuromuscular junction of the frog caused by black widow spider venom. J. Cell Biol., *52* : 1–14.

Clarke, P.G.H., and W.M. Cowan 1975 Ectopic neurons and aberrant connections during neural development. Proc. Nat. Acad. Sci., Wash., *72* : 4455–4458.

Clarke, P.G.H., and W.M. Cowan 1976 The development of the isthmo-optic tract in the chick, with special reference to the occurrence and correction of developmental errors in the location and connections of isthmo-optic neurons. J. Comp. Neurol., *167* : 143–164.

Clarke, P.G.H., L.A. Rogers, and W.M. Cowan 1976 The time of origin and the pattern of survival of neurons in the isthmo-optic nucleus of the chick. J. Comp. Neurol., *167* : 125–142.

Clavert, A. 1974 Incidences des anomalies de la fente colobomique sur la migration des fibres optiques. Arch. Opht. (Paris), *34* : 215–224.

Cleland, B.G., and W.R. Levick 1974a Brisk and sluggish concentrically organized ganglion cells in the cat's retina. J. Physiol., Lond., *240* : 421–456.

Cleland, B.G., and W.R. Levick 1974b Properties of rarely encountered types of ganglion cells in the cat's retina and an overall classification. J. Physiol., Lond., *240* : 457–492.

Clemente, C. 1958 The regeneration of peripheral nerves inserted into the cerebral cortex and the healing of cerebral lesions. J. Comp. Neurol., *109* : 123–151.

Clemente, C.D. 1964 Regeneration in the vertebrate central nervous system. Int. Rev. Neurobiol., *6* : 257–301.

Clopton, B.M., and J.A. Winfield 1976 Effect of early exposure to patterned sound on unit activity in rat inferior colliculus. J. Neurophysiol., *39* : 1081–1089.

Close, R.I. 1972 Dynamic properties of mammalian skeletal muscles. Physiol. Rev., *52* : 129–197.

Cohen, M.W. 1972 The development of neuromuscular connections in the presence of D-tubocurarine. Brain Res., *41* : 457–465.

Coleman, P.D., and A.H. Riesen 1968 Environmental effects on cortical dendritic fields. I. Rearing in the dark. Amer. J. Anat., *102* : 363–374.

Colonnier, M. 1964 The tangential organization of the visual cortex. J. Anat. Lond., *98* : 327–344.

Colonnier, M. 1968 Synaptic patterns on different cell types in the different laminae of the cat visual cortex. An electron microscope study. Brain Res., *9* : 268–287.

Constantine-Paton, M., and R.P. Capranica 1975 Central projection of optic tract from translocated eyes in the leopard frog (*Rana pipiens*). Science, *189* : 480–482.

Constantine-Paton, M., and R.R. Capranica 1976a Axonal guidance of developing optic nerves in the frog. I. Anatomy of the projection from transplanted eye primordia. J. Comp. Neurol., *170* : 17–32.

Constantine-Paton, M., and R.R. Capranica 1976b. Axonal guidance of developing optic nerves in the frog. II. Electro-physiological studies of the projection from transplanted eye primordia. J. Comp. Neurol., *170* : 33–52.

Cooke, J.E., and T.J. Horder 1974 Interactions between optic fibers in their regeneration to specific sites in the goldfish tectum. J. Physiol., Lond., *241* : 89P–90P.

Cooper, E., J. Diamond, and C. Turner 1977 The effects of nerve section and of colchicine treatment on the density of mechanosensory nerve endings in salamander skin. J. Physiol., Lond., *264* : 725–749.

Cornwell, A.C. 1974 Electroretinographic responses following monocular visual deprivation in kittens. Vis. Res., *14* : 1223–1227.

Costa, E., A. Guidotti, and I. Hanbauer 1974 Do cyclic nucleotides promote the transsynaptic induction of tyrosine hydroxylase? Life Sci., *14* : 1169–1188.

Cotman, C.W., D.A. Matthews, D. Taylor, and G. Lynch 1973 Synaptic rearrangement in the dentate gyrus: histochemical evidence of adjustments after lesions in immature and adult rats. Proc. Nat. Acad. Sci., Wash., *70* : 3473–3477.

Cowan, W.M. 1970 Anterograde and retrograde transneuronal degeneration in the central and peripheral neurons system. *In* Contemporary Research Methods in Neuroanatomy. W.J.H. Nauta and S.O.E. Ebbesson (Eds.), Springer-Verlag, New York and Heidelberg, Germany, pp. 217–251.

Cowan, W.M., and W. Cuénod (Eds.) 1975 The Use of Axonal Transport for Studies of Neuronal Connectivity. Elsevier, Amsterdam.

Cragg, B.G. 1967 Changes in visual cortex on first exposure of rats to light. Nature, Lond., *215* : 251–253.

Cragg, B.G. 1970 What is the signal for chromatolysis? Brain Res., *23* : 1–21.

Cragg, B.G. 1972 The development of synapses in cat visual cortex. Invest. Ophth., *11* : 377–385.

Cragg, B.G. 1975a The development of synapses in the visual system of the cat. J. Comp. Neurol., *160* : 147–166.

Cragg, B.G. 1975b The development of synapses in kitten visual cortex during visual deprivation. Exp. Neurol., *46* : 445–451.

Cragg, B.G., R. Anker, and Y.K. Wan 1976 The effect of age on the reversibility of cellular atrophy in the LGN of the cat following monocular deprivation : a test of two hypotheses about cell growth. J. Comp. Neurol., *168* : 345–354.

Crawford, M.L.J., R. Blake, S.J. Cool, and G.K. Von Noorden 1975 Physiological consequences of unilateral and bilateral eye closure in macaque monkeys : some further observations. Brain Res., *84* : 150–154.

Creel, D., C.J. Witkop, and R.A. King 1974 Asymmetric visually evoked potentials in human albinos : evidence for visual system anomalies. Invest. Ophth., *13* : 430–440.

Crepel, F., and J. Mariani 1976 Multiple innervation of Purkinje cells by climbing fibers in the cerebellum of the weaver mutant mouse. J. Neurobiol., *7* : 579–582.

Crepel, F., J. Mariani, and N. Delhaye-Bouchaud 1976 Evidence for a multiple innervation of Purkinje cells by climbing fibers in the immature rat cerebellum. J. Neurobiol., 7 : 567–578.

Crick, F. 1970 Diffusion in embryogenesis. Nature, Lond., 225 : 420–422.

Croop, J., and H. Holtzer 1975 Response of myogenic and fibrogenic cells to cytochalasin B and to colcemid. 1. Light microscope observations. J. Cell Biol., 65 : 271–285.

Crossland, W.J., W.M. Cowan, and L.A. Rogers 1975 Studies on the development of the chick optic tectum. IV. An autoradiographic study of the development of retinotectal connections. Brain Res., 91 : 1–23.

Crossland, W.J., W.M. Cowan, L.A. Rogers, and J.P. Kelly 1974 The specification of the retino-tectal projection in the chick. J. Comp. Neurol., 155 : 127–164.

Cunningham, T.J. 1972 Sprouting of the optic projection after lesions. Anat. Rec., 172 : 289.

Cunningham, T.J. 1976 Early eye removal produces excessive bilateral branching in the rat : application of cobalt filling method. Science : 194 : 857–859.

Currie, J., and W.M. Cowan 1974 Evidence for the late development of the uncrossed retinothalamic projections in the frog, Rana pipiens. Brain Res. 71 : 133–139.

Cynader, M., N. Berman, and A. Hein 1973 Cats reared in stroboscopic illumination: Effects on receptive fields in visual cortex. Proc. Nat. Acad. Sci., Wash., 70 : 1353–1354.

Daniels, J.D., and J.D. Pettigrew 1976 Development of neuronal responses in the visual system of cats. In Studies on the development of behavior and the nervous system. Vol. 3, Neural and behavioral specificity, G. Gottlieb (Ed.), Academic Press, New York, pp. 196–232.

Das, G.D. 1974. Transplantation of embryonic neural tissue in the mammalian brain. I. Growth and differentiation of neuroblasts from various regions of the embryonic brain in the cerebellum of neonate rats. T.I.T.J. Life Sci., 4 : 93–123.

Das, G.D. 1975 Differentiation of dendrites in the transplanted neuroblasts in the mammalian brain. Adv. Neurol., 12 : 181–199.

Das, G.D., and J. Altman 1971 Transplanted precursors of nerve cells : their fate in the cerebellums of young rats. Science, 173 : 637–638.

Das, G.D., G.L. Lammert, and J.P. McAllister 1974 Contact guidance and migrating cells in the developing cerebellum. Brain Res., 69 : 13–29.

Daw, N.W., and H.J. Wyatt 1976 Kittens reared in a unidirectional environment: evidence for a critical period. J. Physiol., Lond., 257 : 155–170.

del Cerro, M.P., and R.S. Snider 1968 Studies on the developing cerebellum. Ultrastructure of the growth cones. J. Comp. Neurol., 133 : 341–362.

Del Cerro, M.P., and J.R. Swarz 1976 Prenatal development of Bergmann glial fibres in rodent cerebellum. J. Neurocytol., 5 : 669–676.

DeLong, G.R. 1970 Histogenesis of fetal mouse isocortex and hippocampus in reaggregating cell culture. Develop. Biol., 22 : 563–583.

DeLong, G.R., and A.J. Coulombre 1967 The specificity of retinotectal connections studied by retinal grafts onto the optic tectum in chick embryos. Develop. Biol., 16 : 513–531.

DeLong, G.R., and R.L. Sidman 1962 Effects of eye removal at birth on histogenesis on the mouse superior colliculus: an autoradiographic analysis with tritiated thymidine.

J. Comp. Neurol., *118* : 205–223.

DeLong, G.R., and R.L. Sidman 1970 Alignment defect of reaggregating cells in cultures of developing brains of reeler mutant mice. Develop. Biol., *22* : 584–600.

Detwiler, S.R. 1943 Reversal of the medulla in *Amblystoma* embryos. J. Exp. Zool., *94* : 169–179.

Detwiler, W.R. 1951 Structural and functional adjustments following reversal of the embryonic medulla in Amblystoma. J. Exp. Zool., *116* : 431–446.

Devor, M. 1975 Neuroplasticity in the sparing or deterioration of function after early olfactory tract lesions. Science, *190* : 998–1000.

Devor, M. 1976 Neuroplasticity in the rearrangement of olfactory tract fibers after neonatal transection in hamsters. J. Comp. Neurol., *166* : 49–72.

Devor, M., and G.E. Schneider 1975 Neuroanatomical plasticity : The principle of conservation of total axonal aborization. *In* Aspects of neural plasticity/Plasticite nerveuse, F. Vital-Durand and M. Jeannerod (Eds.). INSERM, Vol. 43, pp. 191–200.

Devreotes, P.N., and D.M. Fambrough 1976 Turnover of acetylcholine receptors in skeletal muscle. Cold Spr. Harb. Symp. Quant. Biol., *40* : 237–251.

Dews, P.B., and T.N. Wiesel 1970 Consequence of monocular deprivation on visual behavior in kittens. J. Physiol., Lond., *206* : 437–455.

Diamond, M.C., F. Law, H. Rhodes, B. Lindner, M.R. Rosenzweig, D. Krech, and E.L. Bennett 1966 Increases in cortical depth and glia numbers in rats subjected to enriched environment. J. Comp. Neurol., *128* : 117–126.

Dixon, J.S., and J.R. Cronly-Dillon 1972 The fine structure of the developing retina in *Xenopus Laevis*. J. Embryol. Exp. Morphol., *28* : 659–666.

Dostrovsky, J.O., J. Millar, and P.D. Wall 1976 The immediate shift of afferent drive of dorsal column nucleus cells following deafferentation : a comparison of acute and chronic deafferentation in gracile nucleus and spinal cord. Exp. Neurol., *52* : 480–495.

Dowling, J.E. 1970 Organization of vertebrate retinas. Invest. Ophth., *9* 655–680.

Droz, B. 1975 Synthetic machinery and axoplasmic transport : Maintenance of neuronal connectivity. *In* The Nervous System, Donald B. Tower (Ed.), Vol. 1, The Basic Neurosciences. Raven Press, New York, pp. 111–127.

Droz, B., H.L. Koenig, and L. Di Giamberardino 1973 Axonal migration of protein and glycoprotein to nerve endings. I. Radioautographic analysis of the renewal of protein in nerve endings of chicken ciliary ganglion after intracerebral injection of [$^3$H] lysine. Brain Res., *60* : 93–127.

Droz, B., and C.P. Leblond 1963 Axonal migration of proteins in the central nervous system and peripheral nerves as shown by autoradiography. J. Comp. Neurol., *121* : 325–346.

Droz, B., A. Rambourg, and H.L. Koenig 1975 The smooth endoplasmic reticulum : structure and role in the renewal of axonal membrane and synaptic vesicles by fast axonal transport. Brain Res., *93* : 1–13.

Duffy, F.H., J.L. Burchfiel, and S.R. Snodgrass 1976a Ammonium acetate reversal of experimental amblyopia. Neurosc. Abst., *2* : 1109.

Duffy, F.H., S.R. Snodgrass J.L. Burchfiel, and J.L. Conway 1976b Bicuculline reversal of deprivation amblyopia in the cat. Nature, Lond., *260* : 256–257.

Dunn, E.H. 1917 Primary and secondary findings in a series of attempts to transplant cerebral cortex in the albino rat. J. Comp. Neurol., *27* : 565–582.

Dunn, G.A. 1971 Mutual contact inhibition of extension of chick sensory nerve fibers *in vitro*. J. Comp. Neurol., *143* : 491–508.

Dürsteler, M.R., L.J. Garey, and J.A. Movshon 1976 Reversal of the morphological effects of monocular deprivation in the kitten's lateral geniculate nucleus. J. Physiol., Lond., *261* : pp. 189–210.

Ebbott, S., and I.A. Hendry 1978 Retrograde transport of nerve growth factor in the rat central nervous system. Brain Res., *139* : 160–163.

Eccles, J.C. 1970 Neurogenesis and morphogenesis in the cerebellar cortex. Proc. Nat. Acad. Sci., Wash., *66* : 294–301.

Eccles, J.C., R.M. Eccles, and A. Lundberg 1958 The action potentials of the alpha motoneurones supplying fast and slow muscles. J. Physiol., Lond., *142* : 275–291.

Edds, M.V. Jr. 1953 Collateral nerve regeneration. Quart. Rev. Biol., *28* : 260–276.

Edidin, M., and D. Fambrough 1973 Fluidity of the surface of cultured muscle fibers. Rapid lateral diffusion of marked surface antigens. J. Cell Biol., *57* : 27–37.

Egar, M., and M. Singer 1972 The role of ependyma in spinal cord regeneration in the urodele, *Triturus*. Exp. Neur., *37* : 422–430.

Estable-Puig, J.F., and R.F. de Estable-Puig 1972 Paralesional reparative remyelination after chronic local cold injury of the cerebral cortex. Exp. Neur., *35* : 239–253.

Estable-Puig, J.F., R.F. de Estable-Puig, C. Tobias, and W. Haymaker 1964 Degeneration and regeneration of myelinated fibers in the cerebral and cerebellar cortex following damage from ionizing particle radiation. Acta Neuropath., *4* : 175–190.

Eysel, U.T., and C. Gaedt 1971 Maintained activity in the lateral geniculate body of the cat and the effects of visual deprivation. Pflügers Arch., *327* : 68–81.

Feldman, J.D., and R.M. Gaze 1975 The development of half-eyes in *Xenopus* tadpoles. J. Comp. Neurol., *162* : 13–22.

Feldman, J.D., R.M. Gaze, and M.J. Keating 1971 Delayed innervation of the optic tectum during development in *Xenopus laevis*. Exp. Brain Res., *14* : 16–23.

Feldman, M.L., and C. Dowd 1975 Loss of dendritic spines in aging cerebral cortex. Anat. Embryol., *148* : 279–301.

Feringa, E.R., G.G. Gurden, W. Strodel, W. Chandler, and J. Knake 1973 Descending spinal motor tract regeneration after spinal cord transection. Neurology, *23* : 599–608.

Feringa, E.R., J.S. Wendt, and R.D. Johnson 1974 Immunosuppressive treatment to enhance spinal cord regeneration in rats. Neurology, *24* : 287–293.

Ferreira-Berutti, P. 1951 Experimental deflection of the course of the optic nerve in the chick embryo. Proc. Soc. Exp. Biol. Med., *76* : 302–303.

Fertig, A., J.A. Kiernan, and S.S. Seyan 1971 Enhancement of axonal regeneration in the brain of the rat by corticotrophin and triiodothyronine. Exp. Neurol., *33* : 372–385.

Fex, S., and B. Sonesson 1970 Histochemical observation after implantation of a 'fast' nerve into an innervated mammalian 'slow' skeletal muscle. Acta Anat., *77* : 1–10.

Fifkova, E. 1970a The effect of monocular deprivation on the synaptic contacts of the visual cortex. J. Neurobiol., *1* : 285–294.

Fifkova, E. 1970b Changes of axosomatic synapses in the visual cortex of monocularly deprived rats. J. Neurobiol., *2* : 61–71.

Fifkova, E. 1972a Effect of light and visual deprivation on the retina. Exp. Neurol., *35* : 450–457.

Fifkova, E. 1972b Effect of visual deprivation and light on synapses of the inner plexiform layer. Exp. Neurol., *35* : 458–467.

Fifkova, E., and R. Hassler 1969 Quantitative morphological changes in visual centers in rats after unilateral deprivation. J. Comp. Neurol., *135* : 167–178.

Fink. R.P., and L. Heimer 1967 Two methods for selective silver impregnation of degenerating axons and their synaptic endings in the central nervous system. Brain Res., *4* : 369–374.

Finlay, B.L., K.G. Wilson, and G.E. Schneider 1978 Anamolous ipsilateral retinal projections in Syrian hamsters with early lesions : topography and single-unit response properties. J. Comp. Neurol. (in press).

Flandrin, J.M., and M. Jeannerod 1975 Superior colliculus : Environmental influence on the development of directional responses in the kitten. Brain Res., *89* : 348–352.

Frank, E., and J.K.S. Jansen 1976 Interaction between foreign and original nerves innervating gill muscles in fish. J. Neurophysiol., *39* : 84–90.

Frazier, W.A., L.F. Boyd, and R.A. Bradshaw 1973 Interaction of nerve growth factor with surface membranes : biological competence of insolubilized nerve growth factor. Proc. Nat. Acad. Sci., Wash., *70* : 2931–2935.

Freeman, R.D., D.E. Mitchell, and M. Millodot 1972 A neural effect of partial visual deprivation in humans. Science, 175 : 1384–1386.

Freeman, R.D., and L.N. Thibos 1975a Contrast sensitivity in humans with abnormal visual experience. J. Physiol., Lond., *247* : 687–710.

Freeman, R.D., and L.N. Thibos 1975b Visual evoked responses in humans with abnormal visual experience. J. Physiol., Lond., *247* : 711–724.

Fricke, R., and W.M. Cowan, 1977. An autoradiographic study of the development of the entorhinal and commissural afferents to the dentate gyrus of the rat. J. Comp. Neurol., *173* : 231–250.

Frost, D.O., and V.S. Caviness 1974 Pattern of thalamic projections to the neocortex of normal and reeler mutant mice. Proc. Soc. Neurosci., 4th annual meeting, p. 217.

Frost, D.O., and G.E. Schneider 1975 Plasticity of retinotectal projections in newborn Syrian hamsters. Neurosci. Abstr., *1* : 495.

Frost, D.O., and G.E. Schneider 1977 Plasticity of retinofugal projections after partial lesions of the retina in newborn Syrian hamsters. J. Comp. Neurol. (in press).

Fujita, S. 1962 Kinetics of cellular proliferation. Exp. Cell Res. 28 : 52–60.

Fujita, S., M. Horii, T. Tanimura, and H. Nishimura 1964 $H^3$ - thymidine autoradiographic studies on cytokinetic responses to X-ray irradiation and to thio-TEPA in the neural tube of mouse embryos. Anat. Rec., *149* : 37–48.

Fukuda, Y., and J. Stone 1974 Retinal distribution and central projections of Y−, X−, and W− cells of the cat's retina. J. Neurophysiol., *37* : 749–772.

Fuller, J.L., and R.E. Wimer 1973 Behavioral genetics. *In* Comparative Psychology : A Modern Survey, D.A. Dewsbury and D.A. Rethlingshafer (Eds.), New York, McGraw-Hill, pp. 197–237.

Furshpan, E.J., P.R. MacLeish, P.H. O'Lague, and D.D. Potter 1976 Chemical transmission between rat sympathetic neurons and cardiac myocytes developing in microcultures : evidence for cholinergic, adrenergic, and dual-function neurons. Proc.

Nat. Acad. Sci., Wash., *73* : 4225–4229.

Ganz L., and M. Fitch 1968 The effect of visual deprivation on perceptual behavior. Exp. Neurol., *22* : 638–660.

Ganz, L., H.V.B. Hirsch, and S.B. Tieman 1972 The nature of perceptual deficits in visually deprived cats. Brain Res., *44* : 547–568.

Garber, B.B., and A.A. Moscona 1972a Reconstruction of brain tissue from cell suspensions. 1. Aggregation patterns of cells dissociated from different regions of the developing brain. Develop. Biol., *27* : 217–234.

Garber, B.B., and A.A. Moscona 1972b Reconstruction of brain tissue from cell suspensions. II. Specific enhancement of aggregation of embryonic cerebral cells by supernatant from homologous cell cultures. Develop. Biol., *27* : 235–243.

Garey, L., and C. Blakemore 1977 Monocular deprivation : morphological effects on different classes of neurons in the lateral geniculate nucleus. Science, *195* : 414–416.

Gaze, R.M. 1959 Regeneration of the optic nerve in *Xenopus laevis*. Quart. J. Exp. Physiol., *44* : 290–308.

Gaze, R.M. 1970 The Formation of Nerve Connections. Academic Press, New York.

Gaze, R.M. 1974 Neuronal specificity. Brit. Med. Bull., *30* : 116–121.

Gaze, R.M., and M. Jacobson 1963 A study of the retinotectal projection during regeneration of the optic nerve in the frog. Proc. Roy. Soc. Lond. B, *157* : 420–448.

Gaze, R.M., M. Jacobson, and G. Szekely 1963 The retinotectal projection in *Xenopus* with compound eyes. J. Physiol., Lond., *165* : 484–499.

Gaze, R.M., and M.J. Keating 1970 Further studies on the restoration of the contralateral retinotectal projection following regeneration of the optic nerve in the frog. Brain Res., *21* : 183–207.

Gaze, R.M., M.J. Keating, and S.H. Chung 1974 The evolution of the retinotectal map during development in *Xenopus*. Proc. Roy. Soc. Lond. B, *185* : 301–330.

Gaze, R.M., M.J. Keating, G. Szekely, and L. Beazley 1970 Binocular interaction in the formation of specific intertectal neuronal connexions. Proc. Roy. Soc. Lond. B, *175* : 107–147.

Gaze, R.M., and S.C. Sharma 1970 Axial differences in the reinnervation of the goldfish optic tectum by regenerating optic nerve fibers. Exp. Brain Res., *10* : 171–181.

Geffen, L.B., and B.G. Livett 1971 Synaptic vesicles in sympathetic neurons. Physiol. Rev., *51* : 98–157.

Gentschev, T., and C. Sotelo 1973 Degenerative patterns in the ventral cochlear nucleus of the rat after primary deafferentation. An ultrastructural study. Brain Res., *62* : 37–60.

Gilbert, D.S. 1975a Axoplasm architecture and physical properties as seen in the *Myxicola* giant axon. J. Physiol., Lond., *253* : 257–301.

Gilbert, D.S. 1975b Axoplasm chemical composition in *Myxicola* and solubility properties of its structural proteins. J. Physiol., Lond., *253* : 303–319.

Gilbert, D.S., B.J. Newby, and B.H. Anderton 1975 Neurofilament disguise, destruction, and discipline. Nature, London, *256* : 586–589.

Giroud, A., M. Martinet, and C. Roux 1962 Migration anormales des fibres optiques et considérations générales. Arch. Anat., Strasbourg, *45* : 177–190.

Globus, A., and A.B. Scheibel 1966 Loss of dendrite spines as an index of pre-synaptic terminal patterns. Nature, Lond., *212* : 463–465.

Globus, A., and A.B. Scheibel 1967a Pattern and field in cortical structure: the rabbit. J. Comp. Neurol., *131* : 155–172.

Globus, A., and A.B. Scheibel 1967b Synaptic loci in parietal cortical neurons : terminations of corpus callosum fibers. Science, *156* : 1127–1129.

Goldberg, S. 1974 Studies on the mechanics of development of the visual pathways in the chick embryo. Develop. Biol., *36* : 24–43.

Goldberg, S., and A.J. Coulombre 1972 Topographical development of the ganglion cell fiber layer in the chick retina. A whole mount study. J. Comp. Neurol., *146* : 507–518.

Goldberger, M.E., and M. Murray 1974 Restitution of function and collateral sprouting in cat spinal cord: deafferented animal. J. Comp. Neurol., *158* : 37–54.

Goldowitz, D., W.F. White, O. Steward, G. Lynch, and C. Cotman 1975 Anatomical evidence for a projection from the entorhinal cortex to the contralateral dentate gyrus of the rat. Exp. Neurol., *47* : 433–441.

Goldschneider, I., and A.A. Moscona 1972 Tissue-specific cell-surface antigens in embryonic cells. J. Cell Biol., *53* : 435–449.

Goodman, D.C., R.S. Bogdasarian, and J.A. Horel 1973 Axonal sprouting of ipsilateral optic tract following opposite eye removal. Brain, Behav. Evol., *8* : 27–50.

Goodman, D.C., and J.A. Horel 1966 Sprouting of optic tract projections in the brainstem of the rat. J. Comp. Neurol., *127* : 71–88.

Goodwin, B. and M.H. Cohen 1969 A phase-shift model for the spatial and temporal organization of living systems. J. Theor. Biol., *25* : 49–107.

Gottlieb, D.I., and W.M. Cowan 1972 Evidence for a temporal factor in the occupation of available synaptic sites during development of the dentate gyrus. Brain Res., *41* : 452–456.

Gottlieb, D.E., R. Merrell, and L. Glaser 1974 Temporal changes in embryonal cell surface recognition. Proc. Nat. Acad. Sci., Wash., *71* : 1800–1802.

Grafstein, B. 1975 Axonal transport : the intracellular traffic of the neuron. *In* The Handbook of the Nervous System. Vol. 1: Cellular Biology of Neurones, E.R. Kandel (Ed.), American Physiology Society, Washington, D.C.

Gray, E.G. 1975 Presynaptic microtubules and their association with synaptic vesicles. Proc. Roy. Soc. Lond. B, *190* : 369–372.

Greenough, W.T. 1976 Enduring brain effects of differential experience and training. *In* Neural mechanisms of learning and memory, M.R. Rosenzweig and E.L. Bennett (Eds.), M.I.T. Press, Cambridge, pp. 255–278.

Guillery, R.W. 1969 An abnormal retinogeniculate projection in Siamese cats. Brain Res., *14* : 739–741.

Guillery, R.W. 1972a Binocular competition in the control of geniculate cell growth. J. Comp. Neurol., *144* : 117–130.

Guillery, R.W. 1972b Experiments to determine whether retinogeniculate axons can form translaminar collateral sprouts in the dorsal lateral geniculate nucleus of the cat. J. Comp. Neurol., *146* : 407–419.

Guillery, R.W. 1974 Visual pathways in albinos. Sci. Amer., *230* pt. 5 : 44–54.

Guillery, R.W., and J. Kaas 1971 A study of normal and congenitally abnormal retinogeniculate projections in cats. J. Comp. Neurol., *143* : 73–100.

Guillery, R.W., and J.H. Kaas 1974 The effects of monocular lid suture upon the development of the lateral geniculate nucleus in squirrels. J. Comp. Neurol., *154* : 433–442.

Guillery, R.W., A.N. Okoro, and C.J. Witkop Jr. 1975 Abnormal visual pathways in the brain of a human albino. Brain Res., *96* : 373–377.

Guillery, R.W., and D.J. Stelzner 1970 The differential effects of unilateral lid closure upon the monocular and binocular segments of the dorsal lateral geniculate nucleus in the cat. J. Comp. Neurol., *139* : 413–422.

Guth, L. 1956 Regeneration in the mammalian peripheral nervous system. Physiol. Rev., *36* : 441–478.

Gyllensten, L., T. Malmfors, and M.L. Norrlin 1965 Effect of visual deprivation on the optic centers of growing and adult mice. J. Comp. Neurol., *124* : 149–160.

Hamberger, A., H.A. Hansson, and J. Sjostrand 1970 Surface structure of isolated neurons. Detachment of nerve terminals during axon regeneration. J. Cell Biol., *47* : 319–331.

Hamburger, V., and H. Hamilton 1951 A series of normal stages in the development of the chick embryo. J. Morph., *88* : 49–92.

Hamburger, V., and R. Levi-Montalcini 1949 Proliferation, differentiation and degeneration in the spinal ganglion of the chick embryo under normal and experimental conditions. J. Exp. Zool., *111* : 457–501.

Hanna, R.B., A. Hirano, and G.D. Pappas 1976 Membrane specializations of dendritic spines and glia in the weaver mouse cerebellum : a freeze-fracture study. J. Cell Biol., *68* : 403–410.

Harrison, R.G. 1910 The outgrowth of the nerve fiber as a mode of protoplasmic movement. J. Exp. Zool., *9* : 787.

Hart, J.R. 1969 Retrograde transneuronal degeneration in the ventral tegmental nucleus. Anat. Rec., *163* : 196.

Held, R. 1970 Two modes of processing spatially distributed visual stimulation. *In* The Neurosciences : Second Study Program. F. O. Schmitt (Ed.), Rockefeller Univ. Press, New York.

Hendrickson, A., and R. Boothe 1976 Morphology of the retina and dorsal lateral geniculate nucleus in dark-reared monkeys (*Macaca nemestrina*). Vis. Res., *16* : 517–521.

Hendrickson, A.E., and W.M. Cowan 1971 Changes in the rate of axoplasmic transport during postnatal development of the rabbits' optic nerve and tract. Exp. Neurol., *30* : 403–422.

Hendry, I.A. 1973 Trans-synaptic regulation of tyrosine hydroxylase activity in a developing mouse sympathetic ganglion : effects of nerve growth factor (NGF), antiserum and pempidine. Brain Res., *56* : 313–320.

Hendry, I.A. 1975 The retrograde trans-synaptic control of the development of cholinergic terminals in sympathetic ganglia. Brain Res., *86* : 483–487.

Hendry, I.A. 1976 Control in the development of the vertebrate sympathetic nervous system. *In* Reviews of Neuroscience, Vol. 2, S. Ehrenpreis and I.J. Kopin (Eds.) Raven Press, New York, pp. 149–194.

Hendry, I.A., and J. Campbell 1976 Morphometric analysis of rat superior cervical ganglion after axotomy and nerve growth factor treatment. J. Neurocytol., *5* : 351–360.

Hendry, I.A., and L.L. Iversen 1973 Changes in tissue and plasma concentrations of nerve growth factor following the removal of the submaxillary glands in adult mice and their effects on the sympathetic nervous system. Nature, Lond., *243* : 500–504.

Hendry, I.A., K. Stockel, H. Thoenen, and L.L. Iversen 1974 The retrograde axonal transport of nerve growth factor. Brain Res., *68* : 103–121.

Hendry, I.A., and H. Thoenen 1974 Changes of enzyme pattern in the sympathetic nervous system of adult mice after sub-maxillary gland removal : response to exogenous nerve growth factor. J. Neurochem., *22* : 999–1004.

Herman, C.J., and L.W. Lapham 1969 Neuronal polyploidy and nuclear volumes in the cat central nervous system. Brain Res., *15* : 35–48.

Herndon, R.M., G. Margolis, and L. Kilham 1971 The synaptic organization of the malformed cerebellum induced by perinatal infection with the feline panleukopenia virus (PLV). II. The Purkinje cell and its afferents. J. Neuropath. Exp. Neurol., *30* : 557–570.

Herndon, R.M., and M.L. Oster-Granite 1975 Effect of granule cell destruction on development and maintenance of the Purkinje cell dendrite. Adv. Neurol., *12*: 361–371.

Heuser, T.S., and J.E. Reese 1973 Evidence for recycling of synaptic vesicle membrane during transmitter release at the frog neuromuscular junction. J. Cell Biol., *57* : 315–344.

Hibbard, E. 1965 Orientation and directed growth of Mauthner's cell axons from duplicated vestibular nerve roots. Exp. Neurol., *13* : 289–301.

Hibbard, E. 1967 Visual recovery following regeneration of the optic nerve through the oculomotor root in *Xenopus*. Exp. Neurol., *19* : 350–356.

Hickey, T.L. 1975 Translaminar growth of axons in the kitten dorsal lateral geniculate nucleus following removal of one eye. J. Comp. Neurol., *161* : 359–382.

Hicks, S.P., and C.J. D'Amato 1970 Motor-sensory and visual behavior after hemispherectomy in newborn and mature rats. Exp. Neurol., *29* : 416–438.

Hicks. S.P., C.J. D'Amato, and M.J. Lowe 1959 The development of the mammalian nervous system. I. Malformation of the brain, especially the cerebral cortex, induced in rats by radiation. II. Some mechanisms of the malformations of the cortex. J. Comp. Neurol., *113* : 435–469.

Hinds, J.W., and P.L. Hinds 1972 Reconstruction of dendritic growth cones in neonatal mouse olfactory bulb. J. Neurocytol., *1* : 169–187.

Hinds, J.W., and P.L. Hinds 1974 Early ganglion cell differentiation in the mouse retina: an electron microscopic analysis utilizing serial sections. Develop. Biol., *37* : 381–416.

Hinds, J.W., and T.L. Ruffett 1971 Cell proliferation in the neural tube : an electron microscopic and Golgi analysis in the mouse cerebral vesicle. Z. Zellforsch., *115* : 226–264.

Hirano, A., and H.M. Dembitzer 1973 Cerebellar alterations in the weaver mouse. J. Cell Biol., *56* : 478–486.

Hirano, A., H.M. Dembitzer, and M. Jones 1972 An electron microscopic study of cycasin-induced cerebellar alterations. J. Neuropath. Exp. Neurol., *31* : 113–125.

Hirsch, H.V.B. 1972 Visual perception in cats after environmental surgery. Exp. Brain Res., *15* : 405–423.

Hirsch, H.V.B., and D.N. Spinelli 1970 Visual experience modifies distribution of horizontally and vertically oriented receptive fields in cats. Science, *168* : 869–871.

Hirsch, H.V.B., and D.N. Spinelli 1971 Modification of the distribution of receptive field orientation in cats by selective visual exposure during development. Exp. Brain Res., *13* : 509–527.

Hoffman, K.P., and S.M. Sherman 1975 Effects of early binocular deprivation on visual input to cat superior colliculus. J. Neurophysiol., *38* : 1049–1059.

Hofman, W.W., and S. Thesleff 1972 Studies on the trophic influence of nerve on skeletal muscle. Eur. J. Pharmacol., *20* : 256–261.

Hollyfield, J.G. 1972 Histogenesis of the retina in the Killifish, *Fundulus heteroclitus*. J. Comp. Neurol., *144* : 373–380.

Holtzman, E., S. Teichberg, S.J. Abrahams, E. Citkowitz, S.M. Crain, N. Kawai, and E.R. Peterson 1973 Notes on synaptic vesicles and related structures, endoplasmic reticulum, lysosomes and peroxisomes in nervous tissue and the adrenal medulla. J. Histochem. Cytochem., *21* : 349–385.

Hope, R.A., B.J. Hammond, and R.M. Gaze 1976 The arrow model : retinotectal specificity and map formation in the goldfish visual system. Proc. Roy. Soc. Lond. B, *194* : 447–466.

Horder, T.J. 1971 Retention, by fish optic nerve fibers regenerating to new terminal sites in the tectum, of "chemospecific" affinity for their original sites. J. Physiol., Lond., *216* : 53P–55P.

Horder, T.J. 1974a Electron microscopic evidence in goldfish that different optic nerve fibers regenerate selectively through specific routes into the tectum. J. Physiol., Lond., 241 : 84P–85P.

Horder, T.J. 1974b Changes of fibre pathways in the goldfish optic tract following regeneration. Brain Res., *72* : 41–52.

Hoshino, K., T. Matsuwaza, and O. Murakami 1973 Characteristics of the cell cycle of matrix cells in the mouse embryo during histogenesis of telencephalon. Exp. Cell Res., *77* : 89–94.

Howard, A., and S.R. Pelc 1953 Synthesis of deoxyribonucleic acid in normal and irradiated cells and its relation to chromosome breakage. Heredity, *6* : 261–273.

Hubel, D.H., and D.C. Freeman 1977 Projection into the visual field of ocular dominance columns in macaque monkey. Brain Res., *122* : 336–343.

Hubel, D.H., and T.N. Wiesel 1962 Receptive fields, binocular interaction and functional architecture in the cats' visual cortex. J. Physiol., Lond., *160* : 106–154.

Hubel, D.H., and T.N. Wiesel 1963 Receptive fields of cells in striate cortex of very young, visually inexperienced kittens. J. Neurophysiol., *26* : 994–1002.

Hubel, D.H., and T.N. Wiesel 1965 Binocular interaction in striate cortex of kittens reared with artificial squint. J. Neurophysiol., *28* : 1041–1059.

Hubel, D.H., and T.N. Wiesel 1970 The period of susceptibility to the physiological effects of unilateral eye closure in kittens. J. Physiol., Lond., *206* : 419–436.

Hubel, D.H., and T.N. Wiesel 1971 Aberrant visual projections in the Siamese cat. J. Physiol., Lond., *218* : 33–62.

Hubel, D.H., and T.N. Wiesel 1974 Uniformity of monkey striate cortex : a parallel relationship between field size, scatter, and magnification factor. J. Comp. Neurol., *158* : 267–294.

Hubel, D.H., T.N. Wiesel, and S. LeVay 1977 Plasticity of ocular dominance columns in monkey striate cortex. Phil. Trans. Roy. Soc. *B. 278* : 377–409.

Hughes, A. 1953 The growth of embryonic neurites. A study on cultures of chick neural tissue. J. Anat., Lond., *87* : 150–162.

Huneeus, F.C., and P.F. Davison 1970 Fibrillar proteins from squid axons. I. Neurofilament protein. J. Molec. Biol., *52* : 415–428.

Hunt, R.K. 1975a Developmental programming for retinotectal patterns. Ciba Found. Symp., *29* : 131–159. *In* Cell Patterning published by ASP (Elsevier Excerpta Medica, North-Holland), Amsterdam.

Hunt, R.K. 1975b Position dependent differentiation of neurons. *In* ICN-UCLA Symposia on molecular and cellular biology. Vol. 2, Developmental Biol., D. McMahon and C.F. Fox (Eds.), W.A. Benjamin, Menlo Park.

Hunt, R.K., and M. Jacobson 1972 Development and stability of positional information in *Xenopus* retinal ganglion cells. Proc. Nat. Acad. Sci., Wash., *69* : 780–783.

Hunt, R.K., and M. Jacobson 1973a Specification of positional information in retinal ganglion cells of *Xenopus* : assays for analysis of the unspecified state. Proc. Nat. Acad. Sci., Wash., *70* : 507–511.

Hunt, R.K., and M. Jacobson 1973b Neuronal locus specificity : altered pattern of spatial deployment in fused fragments of embryonic *Xenopus* eyes. Science, *180* : 509–511.

Hunt, R.K., and M. Jacobson 1974a Neuronal specificity revisited. Current Topics in Developmental Biology, *8* : 203–259.

Hunt, R.K., and M. Jacobson 1974b Specification of positional information in retinal ganglion cells of *Xenopus laevis:* intra-ocular control of the time of specification. Proc. Nat. Acad. Sci., Wash., *71* : 3616–3620.

Huttenlocher, P.R. 1974 Dendritic development in neocortex of children with mental defect and infantile spasms. Neurology (Minneap.), *24* : 203.

Hyden, H., and A. Pigon 1960 A cytophysiological study of the functional relationship between oligodendroglial cells and nerve cells of Deiters' nucleus. J. Neurochem, *6* : 57–72.

Isaacson, R.L., and K.H. Pribram (Eds.) 1975 The Hippocampus, Vols. 1 and 2. Plenum, New York.

Jacobson, M. 1968a Development of neuronal specificity in retinal ganglion cells of *Xenopus*. Develop. Biol., *17* : 202–218.

Jacobson, M. 1968b Cessation of DNA synthesis in retinal ganglion cells correlated with the time of specification of their central connections. Develop. Biol., *17* : 219–232.

Jacobson, M. 1970a Development, specification and diversification of neuronal connections. *In* The Neurosciences: Second Study Program, F.O. Schmitt (Ed.), Rockefeller Univ. Press, New York, pp. 116–119.

Jacobson, M. 1970b Developmental Neurobiology. Holt, Rinehart and Winston, New York.

Jacobson, M. 1974 A plenitude of neurons. *In* Studies of the Development of Behavior and Nervous System, Vol. 2, G. Gottlieb (Ed.). Academic Press, New York, pp.151–166.

Jacobson, M. 1976a Histogenesis of retina in the clawed frog with implications for the pattern of development of retinotectal connections. Brain Res., *103* : 541–545.

Jacobson, M. 1976b Premature specification of the retina in embryonic *Xenopus* eyes treated with Ionophore X537A. Science, *191* : 288–290.

Jacobson, M., and R.E. Baker 1969 Development of neuronal connections with skin grafts in frogs : behavioral and electrophysiological studies. J. Comp. Neurol., *137* : 121–142.

Jacobson, M., and R.K. Hunt 1973 The origins of nerve-cell specificity. Sci. Amer., *228* pt. 2 : 26–35.

Jacobson, M., and R.L. Levine 1975a Plasticity in the adult frog brain: filling the visual scotoma after excision or translocation of parts of the optic tectum. Brain Res., *88* : 339–345.

Jacobson, M., and R.L. Levine 1975b Stability of implanted duplicate tectal positional markers serving as targets for optic axons in adult frogs. Brain Res., *92* : 468–471.

Johnston, B.T., J.E. Schrameck, and R.F. Mark 1975 Re-innervation of axolotl limbs. II. Sensory nerves. Proc. Roy. Soc. Lond. B, *190* : 59–75.

Jones, D.G. 1975 Synapses and Synaptosomes, Morphological Aspects. Chapman and Hall, Ltd., London.

Kaas, J.H., and R.W. Guillery 1973 The transfer of abnormal visual field representations from the dorsal lateral geniculate nucleus to the visual cortex in Siamese cats. Brain Res., *59* : 61–95.

Kaas, J., R.W. Guillery, and J.M. Allman 1972 Some principles on organization in the dorsal lateral geniculate nucleus. Brain, Behav. Evol., *6* : 253–299.

Kadin, M.E., L.J. Rubinstein, and J.S. Nelson 1970 Neonatal cerebellar medulloblastoma originating from the fetal external granular layer. J. Neuropathol. Exp. Neurol., *29* : 583–600.

Kahn, A.J. 1973 Ganglion cell formation in the chick neural retina. Brain Res., *63* : 285–290.

Kaiserman-Abramof, I.R., A.M. Graybiel, and W.J.H. Nauta 1975 Neural connections of area 17 in an anophthalmic mouse strain. Neurosci. Abstr., *1* : 102.

Kalil, R.E. 1972 Formation of new retino-geniculate connections in kittens after removal of one eye. Anat. Rec., *172* : 339–340.

Kalil, R.E., and G.E. Schneider 1975 Abnormal synaptic connections of the optic tract in the thalamus after midbrain lesions in newborn hamsters. Brain Res., *100* : 690–698.

Karten, H.J. 1969 The organization of the avian telencephalon and some speculations on the phylogeny of the amniote telencephalon. Ann. N.Y. Acad. Sci., *167* : 164–179.

Kater, S.B., and C. Nicholson (Eds.) 1973 Intracellular Staining in Neurobiology, Springer-Verlag, New York.

Kato, H., P.O. Bishop, and G.A. Orban 1977 Binocular interaction on monocularly-discharged striate neurons in the cat. (unpublished manuscript)

Kauffman, S.L. 1968 Lengthening of the generation cycle during embryonic differentiation of the mouse neural tube. Exp. Cell Res., *49* : 420–424.

Kauffman, S.L. 1969 Cell proliferation in embryonic mouse neural tube following urethane exposure. Develop. Biol., *20* : 146–157.

Keating, M.J. 1974 The role of visual function in the patterning of binocular visual connexions. Brit. Med. Bull., *30* : 145–151.

Keating, M.J. 1976 The formation of visual neuronal connections: an appraisal of the present status of the theory of "neuronal specificity". *In* Studies on the development and behavior and the nervous system. Vol. 3 Neural and Behavioral Specificity. G. Gottleib (Ed.), Academic Press, New York, pp. 59–110.

Keating, M.J., and R.M. Gaze 1970 The depth distribution of visual units in the contralateral optic tectum following regeneration of the optic nerve in the frog. Brain Res., *21* : 197–206.

Kelly, J.P., and W.M. Cowan 1972 Studies on the development of the chick optic tectum. III. Effects of early eye removal. Brain Res., *42* : 263–288.

Kennedy, C., M.H. Des Rosiers, J.W. Jehle, M. Reivich, F. Sharpe, and L. Sokoloff 1975 Mapping of functional neural pathways by autoradiographic survey of local metabolic rate with [$^{14}$C] deoxyglucose. Science, *187* : 850–853.

Kennedy, C., M.H. Des Rosiers, O. Sakurada, M. Shinohara, M. Reivich, J.W. Jehle, and L. Sokoloff 1976 Metabolic mapping of the primary visual system of the monkey by means of the autoradiographic [$^{14}$C] deoxyglucose technique. Proc. Nat. Acad. Sci., Wash., *73* : 4230–4234.

Kerns, J.M., and E.J. Hinsman 1973 Neuroglial response to sciatic neurectomy II. Electron microscopy. J. Comp. Neurol., *151* : 255–280.

Kerr, F.W.L. 1972 The potential of cervical primary afferents to sprout in the spinal nucleus of V following long term trigeminal denervation. Brain Res., *43* : 547–560.

Kicliter, E., L.J. Misantone, and D.J. Stelzner 1974 Neuronal specificity and plasticity in frog visual system : anatomical correlates. Brain Res., *82* : 293–297.

Kirk, D.L. 1976 Projections of the Visual Field by the Axons of Cat Retinal Ganglion Cells. Section 2 : Decussation of Optic Axons in Siamese Cats, pp. 1–17. Ph.D. Thesis, Australian National University.

Kirk, D.L., W.R. Levick, B.G. Cleland, and H. Wässle 1976a Crossed and uncrossed representation of the visual field by brisk-sustained and brisk-transient cat retinal ganglion cells. Vis. Res., *16* : 225–231.

Kirk, D.L., W.R. Levick, and B.G. Cleland 1976b The crossed or uncrossed destination of axon of sluggish-concentric and non-concentric cat retinal ganglion cells, with an overall synthesis of the visual field representation. Vis. Res., *16* : 233–236.

Kjerulf, T.D., J.T. O'Neal, W.H. Calvin, J.D. Loeser, and L.E. Westrum 1973 Deafferentation effects in lateral cuneate nucleus of the cat : correlation of structural alterations with firing pattern changes. Exp. Neurol., *39* : 86–102.

Konyukhov., B.V., and M.W. Sazhina 1971 Genetic control over the duration of G1 phase. Experientia, *27* : 970–971.

Korr, I.M., P.N. Wilkinson, and F.W. Chornock 1967 Axonal delivery of neuroplasmic components to muscle cells. Science, *155* : 342–345.

Kratz, K.E., and P.D. Spear 1976 Effects of visual deprivation and alterations in binocular competition on responses of striate cortex neurons in the cat. J. Comp. Neurol., *170* : 141–152.

Kratz, K.E., P.D. Spear, and D.C. Smith 1976 Postcritical-period reversal of effects of monocular deprivation on striate cortex cells in the cat. J. Neurophysiol., *39* : 501–511.

Kruger, L., and J. Hamori 1970 An electron microscopic study of dendritic degeneration in the cerebral cortex resulting from laminar lesions. Exp. Brain Res., *10* : 1–16.

Kruger, L., and S. Saporta 1977 Axonal transport of [$^{3}$H] adenosine in visual and somatosensory pathways. Brain Res., *122* : 132–136.

Kuffler, S.W., and J.G. Nicholls 1966 The physiology of neuroglial cells. Ergebn. Physiol. Biol. Chem. Exp. Pharmakol., *57* : 1–90.

Kuffler, S.W., and J.G. Nicholls 1976 From Neuron to Brain: A Cellular Approach to the Function of the Nervous System. Sinauer Associates, Inc., Sunderland, Massachusetts.

Kuno, M., and R. Llinás 1970 Alterations of synaptic action in chromatolysed motoneurones of the cat. J. Physiol., Lond., *210* : 823–838.

Kurtzke, J.F. 1975 Epidemiology of spinal cord injury. Exp. Neurol., *48* : Suppl. pp. 163–236.

Kuwabara, T. 1975 Development of the optic nerve of the rat. Invest. Ophthal., *14* : 732–745.

Land, P.W., E.H. Polley, and M.M. Kernis 1976 Patterns of retinal projection to the lateral geniculate nucleus and superior colliculus of rats with induced unilateral congenital eye defects. Brain Res., *103* : 394–399.

Landis, D.M.D., and T.S. Reese 1977 Structure of the Purkinje cell membrane in staggerer and weaver mutant mice. J. Comp. Neurol., *171* : 247–260.

Landmesser, L., and G. Pilar 1970 Selective reinnervation of two cell populations in the adult pigeon ciliary ganglion. J. Physiol., Lond., *211* : 203–216.

Landmesser, L., and G. Pilar 1974a Synapse formation during embryogenesis on ganglion cells lacking a periphery. J. Physiol., Lond., *241* : 715–736.

Landmesser, L., and G. Pilar 1974b Synaptic transmission and cell death during normal ganglionic development. J. Physiol., Lond., *241* : 737–749.

Landmesser, L., and G. Pilar 1976 Fate of ganglionic synapses and ganglion cell axons during normal and induced cell death. J. Cell Biol., *68* : 357–374.

Lassek, A.M. 1954 The Pyramidal Tract: Its Status in Medicine. Charles C. Thomas, Springfield, Ill.

LaVail, J.H. 1975 Retrograde cell degeneration and retrograde transport techniques. *In* The Use of Axonal Transport for Studies of Neuronal Connectivity, W.M. Cowan and M. Cuénod (Eds.). Elsevier, Amsterdam, pp. 217–248.

LaVail, J.H., and W.M. Cowan 1971 The development of the chick optic tectum. I. Normal morphology and cytoarchitectonic development. Brain Res., *28* : 391–419.

LeDouarin N.M., D. Renaud, M.A. Teillet, and G.H. LeDouarin 1975 Cholinergic differentiation of presumptive adrenergic neuroblasts in interspecific chimeras after heterotopic transplantations. Proc. Nat. Acad. Sci., Wash., *72* : 728–732.

Lenneberg, E.H. 1967 Biological Foundations of Language. Wiley, New York.

Leong, S.K. 1976a A qualitative electron microscopic investigation of the anomalous corticofugal projections following neonatal lesions in the albino rat. Brain Res., *107* : 1–8.

Leong, S.K. 1976b An experimental study of the corticofugal system following cerebral lesions in the albino rat. Exp. Brain Res., *26* : 235–247.

Leong, S.K., and R.D. Lund 1973 Anomalous bilateral corticofugal pathways in albino rats after neonatal lesions. Brain Res., *62* : 218–221.

Letourneau, P.C. 1975a Possible roles for cell-to-substratum adhesion in neuronal morphogenesis. Develop. Biol., *44* : 77–91.

Letourneau, P.C. 1975b Cell-to-substratum adhesion and guidance of axonal elongation. Develop. Biol., *44* : 92–101.

Letourneau, P.C., and N.K. Wessels 1974 Migratory cell locomotion versus nerve axon elongation. J. Cell Biol., *61* : 56–69.

LeVay, S. 1973 Synaptic patterns in the visual cortex of the cat and monkey. Electron microscopy of Golgi preparations. J. Comp. Neurol., *150* : 53–86.

LeVay, S. 1977 Effects of visual deprivation on polyribosome aggregation in visual cortex of the cat. Brain Res., *119* : 73–86.

Leventhal, A.G., and H.V.B. Hirsch 1975 Cortical effect of early selective exposure to diagonal lines. Science, *190* : 902–904.

Levi-Montalcini, R., R.H. Angeletti, and P.U. Angeletti 1972 The nerve growth factor. *In* The Structure and Function of Nervous Tissue, Vol. V, Structure III and Physiology III, G.H. Bourne (Ed.). Academic Press, New York, pp. 1–38.

Levine, R.L., and M. Jacobson 1974 Deployment of optic nerve fibers is determined by positional markers in the frog's tectum. Exp. Neurol., *43* : 527–538.

Levine, R.L., and M. Jacobson 1975 Discontinuous mapping of retina onto tectum innervated by both eyes. Brain Res., *98* : 172–176.

Lieberman, A.R. 1971 The axon reaction : a review of the principal features of perikaryal responses to axon injury. Int. Rev. Neurobiol., *14* : 49–124.

Lilien, J.E. 1968 Specific enhancement of cell aggregation *in vitro*. Develop. Biol., *17* : 657–678.

Lilien, J.E. 1969 Toward a molecular explanation for specific cell adhesion. *In* Current Topics in Developmental Biology, Vol. 4, A.A. Moscona and A. Monroy (Eds.). Academic Press, New York, pp. 169–195.

Lim, K.H., and S.K. Leong 1975 Aberrant bilateral projections from the dentate and interposed nuclei in albino rats after neonatal lesions. Brain Res., *96* : 306–309.

Lindvall, O., A. Björklund, A. Nobin, and U. Stenevi 1974 The adrenergic innervation of the rat thalamus as revealed by the glyoxylic acid fluorescence method. J. Comp. Neurol., *154* : 317–348.

Liu, C.M., and W.W. Chambers 1958 Intraspinal sprouting of dorsal root axons. Arch. Neurol. Psychiat. *79* : 46–61.

Liu, C.N., and D. Scott 1958 Regeneration in the dorsal spinocerebellar tract of the cat. J. Comp. Neurol., *109* : 153–167.

Livett, B.G. 1976 Axonal transport and neuronal dynamics contributions to the study of neuronal connectivity. International Review of Physiology, Neurophysiology II, Vol. 10, R. Porter (Ed.). University Park Press, Baltimore, pp. 37–124.

Llinás, R. 1969 Neurobiology of Cerebellar Evolution and Development, R. Llinás (Ed.). Amer. Medical Association, Chicago.

Llinás, R., D.E. Hillman, and W. Precht 1973 Neuronal circuit reorganization in mammalian agranular cerebellar cortex. J. Neurobiol., *4* : 69–94.

Lloyd, C.W. 1975 Sialic acid and the social behaviour of cells. Biol. Rev., *50* : 325–350.

Lopresti, V., E.R. Macagno, and C. Levinthal 1973 Structure and development of neuronal connections in isogenic organisms. Cellular interactions in the development of the optic lamina of *Daphnia*. Proc. Nat. Acad. Sci., Wash., *70* : 433–437.

Loy, R., G. Lynch, and C.W. Cotman 1977 Development of afferent lamination in the fascia dentata of the rat. Brain Res., *121* : 229–244.

Lubinska, L. 1975 On axoplasmic flow. Int. Rev. Neurobiol., *17* : 241–296.

Ludueña, M.A. 1973 Cell nerve differentiation in vitro. Develop. Biol., *33* : 268–284.

Ludueña, M.A., and N.K. Wessells 1973 Cell locomotion, nerve elongation and microfilaments. Develop. Biol., *30* : 427–440.

Lund, J.S. 1973 Organization of neurons in the visual cortex, Area 17, of the monkey (*Macaca mulatta*). J. Comp. Neurol., *147* : 455–496.

Lund, J.S., and R.G. Boothe 1975 Interlaminar connections and pyramidal neuron organization in the visual cortex, area 17, of the Macaque monkey. J. Comp. Neurol., *159* : 305–334.

Lund, J.S., R.G. Boothe, and R.D. Lund 1977 Development of neurons in the visual cortex (area 17) of the monkey (*Macaca nemestrina*). A Golgi study from fetal day 127 to postnatal maturity. J. Comp. Neurol., *176* : 149–188.

Lund, J.S., R.D. Lund, A.E. Hendrickson, A.H. Bunt, and A.F. Fuchs 1975 The origin of efferent pathways from the primary visual cortex, Area 17, of the *Macaque* monkey as shown by retrograde transport of horseradish peroxidase. J. Comp. Neurol., *164* : 287–304.

Lund, J.S., F.L. Remington, and R.D. Lund 1976 Differential central distribution of optic nerve components in the rat. Brain Res., *116* : 83–100.

Lund, R.D. 1965 Uncrossed visual pathways of hooded and albino rats. *Science, 149* : 1506–1507.

Lund, R.D. 1972 Anatomic studies on the superior colliculus. Invest. Ophthal., *11* : 434–441.

Lund, R.D. 1975 Variations in the laterality of the central projections of retinal ganglion cells. Exp. Eye Res., *21* : 193–203.

Lund, R.D., and A.H. Bunt 1976 Prenatal development of central optic pathways in albino rats. J. Comp. Neurol., *165* : 247–264.

Lund, R.D., T.J. Cunningham, and J.S. Lund 1973 Modified optic projections after unilateral eye removal in young rats. Brain, Behav. Evol., *8* : 51–72.

Lund, R.D., and S.D. Hauschka 1976 Transplanted neural tissue develops connections with host rat brain. Science, *193* : 582–584.

Lund, R.D., and J.S. Lund 1971a Synaptic adjustment after deafferentation of the superior colliculus of the rat. Science, *171* : 804–807.

Lund, R.D., and J.S. Lund 1971b Modification of synaptic patterns in the superior colliculus of the rat during development and following deafferentation. Vision Res. *11* Suppl. 3 : 281–298.

Lund, R.D., and J.S. Lund 1972 Development of synaptic patterns in the superior colliculus of the rat. Brain Res., *42* : 1–20.

Lund, R.D., and J.S. Lund 1973 Reorganization of the retinotectal pathway in rats after neonatal retinal lesions. Exp. Neurol., *40* : 377–390.

Lund, R.D., and J.S. Lund 1976 Plasticity in the developing visual system : the effects of retinal lesions made in young rats. J. Comp. Neurol., *169* : 133–154.

Lund, R.D., J.S. Lund, and R.P. Wise 1974 The organization of the retinal projection to the dorsal lateral geniculate nucleus in pigmented and albino rats. J. Comp. Neurol., *158* : 383–404.

Lund, R.D., and B.F. Miller 1975 Secondary effects of fetal eye damage in rats on intact central optic projections. Brain Res., *92* : 279–289.

Lund, R.D., and M.J. Mustari 1977 Development of the geniculocortical pathway in rats. J. Comp. Neurol., *173* : 289–306.

Lynch, G., C. Gall, G. Rose, and C. Cotman 1976 Changes in the distribution of the dentate gyrus associational system following unilateral or bilateral entorhinal lesions in the adult rat. Brain Res., *110* : 57–71.

Lynch, G., D.A. Matthews, S. Mosko, T. Parks, and C. Cotman 1972 Induced acetylcholinesterase-rich layer in rat dentate gyrus following entorhinal lesions. Brain Res., *42* : 311–318.

Lynch, G.S., S. Mosko, T. Parks, and C.W. Cotman 1973a Relocation and hyper-development of the dentate gyrus commissural system after entorhinal lesions in immature rats. Brain Res., *50* : 174–178.

Lynch, G., B. Stanfield, and C.W. Cotman 1973b Developmental differences in post-lesion axonal growth in the hippocampus. Brain Res. *59* : 155–168.

Lynch, G., B. Stanfield, T. Parks, and C.W. Cotman 1974 Evidence for selective post-lesion axonal growth in the dentate gyrus of the rat. Brain Res., *69* : 1–11.

Macagno, E.R., V. Lopresti, and C. Levinthal 1973 Structure and development of neuronal connections in isogenic organisms : variations and similarities in the optic system of *Daphnia magna*. Proc. Nat. Acad. Sci., Wash., *70* : 57–61.

Maffei, L., and S. Bisti 1976 Binocular interaction in strabismic kittens deprived of vision. Science, *191* : 579–580.

Marin-Padilla, M. 1972 Structural abnormalities of the cerebral cortex in human chromosomal aberrations. Brain Res., *44* : 625–629.

Marin-Padilla, M. 1974 Structural organisation of the cerebral cortex (motor area) in human chromosomal aberrations. A Golgi study. I.D. (13–15) trisomy. Patau syndrome. Brain Res., *66* : 375–391.

Mark, R.F. 1974 Selective innervation of muscle. Brit. Med. Bull *30* : 122–125.

Mark, R. 1978 Silent synapses? Trends in Biochem. Sci. (in press)

Mark, R.F., L.R. Marotte, and P.E. Mart 1972 The mechanism of selective reinnervation of fish eye muscles. IV. Identification of repressed synapses. Brain Res., *46* : 149–157.

Marks, A.F. 1972 Regenerative reconstruction of a tract in a rat's brain. Exp. Neurol., *34* : 455–464.

Marotte, L.R., and R.F. Mark 1970 The mechanism of selective reinnervation of fish eye muscle. II. Evidence from electronmicroscopy of nerve endings. Brain Res., *19* : 53–62.

Matinian, L.A., and W.F. Windle 1975 Functional and structural restitution after transection of the spinal cord. Anat. Rec., *181* : 423.

Matthews, D.A., C. Cotman, and G. Lynch 1976 An electron microscopic study of lesion-induced synaptogenesis in the dentate gyrus of the adult rat. II. Reappearance of morphologically normal synaptic contacts. Brain Res., *115* : 23–41.

Matthews, M.A. 1974 Microglia and reactive "M" cells of degenerating central nervous system: Does similar morphology and function imply a common origin? Cell Tiss. Res., *148* : 477–491.

Matthews, M.R., W.M. Cowan, and T.P.S. Powell 1960 Transneuronal cell degeneration in the lateral geniculate nucleus of the macaque monkey. J. Anat., *94* : 145–169.

Matthews, M.R., and V.H. Nelson 1975 Detachment of structurally intact nerve endings from chromatolytic neurones of rat superior cervical ganglion during the depression of synaptic transmission induced by post-ganglionic axotomy. J. Physiol., Lond.,

*245* : 91–135.

Matthey, R. 1925 Recuperation de la vue apres resection des nerfs optiques chez le *Triton*. C.R. Séanc. So. Biol., *93* : 904–906.

Maturana, H.R., J.T. Lettvin, W.S. McCulloch, and W.H. Pitts 1959 Physiological evidence that cut optic nerve fibers in the frog regenerate to their proper place in the tectum. Science, *130* : 1709–1710.

Maturana, H.R., J.T. Lettvin, W.S. McCulloch, and W.H. Pitts 1960 Anatomy and physiology of vision in the frog. J. Gen. Physiol., *43* Suppl. : 129–175.

Matus, A.I., B.B. Walters, and S. Mughal 1975 Immunohistochemical demonstration of tubulin associated with microtubules and synaptic junctions in mammalian brain. J. Neurocytol., *4* : 733–744.

McClay, D.R., and A.A. Moscona 1974 Purification of the specific cell-aggregating factor from embryonic neural retinal cells. Exp. Cell Res., *87* : 438–443.

McCouch, G.P., G.M. Austin, C.N. Liu, and C.Y. Liu 1958 Sprouting as a cause of spasticity. J. Neurophysiol., *21* : 205–216.

McIlwain, H. 1973 Adenosine in neurohumoral and regulatory roles in the brain. *In* Central Nervous System—Studies on Metabolic Regulation and Function, E. Genazzani and H. Herken (Eds.), Springer, Berlin, pp. 3–11.

McMahon, D. 1973 A cell contact model for cellular position determination in development. Proc. Nat. Acad. Sci., Wash., *70* : 2396–2400.

McQuarrie, I.G., and B. Grafstein 1973 Axon outgrowth enhanced by a previous nerve injury. Arch. Neurol., Chicago, *29* : 53–55.

Meisami, E. 1976 Effects of olfactory deprivation on postnatal growth of the rat olfactory bulb utilizing a new method for production of neonatal unilateral anosmia. Brain Res., *107* : 437–444.

Mendell, L.M., J.B. Munson, and J.G. Scott 1974 Connectivity changes of Ia afferents on axotomized motoneurons. Brain Res., *73* : 338–342.

Meyer, R.L. 1975 Tests for regulation in the goldfish retinotectal system. Anat. Rec., *181* : 426.

Meyer, R.L., and R.W. Sperry 1976 Retinotectal specificity : Chemoaffinity theory. *In* Studies on the Development of Behavior and the Nervous System, Vol. 3, Neural and behavioral specificity, G. Gottlieb (Ed.). Academic Press, New York, pp. 111–149.

Millar, J., A.I. Basbaum, and P.D. Wall 1976 Restructuring of the somatotopic map and appearance of abnormal neuronal activity in the gracile nucleus after partial deafferentation. Exp. Neurol., *50* : 658–672.

Miller, B.F., and R.D. Lund 1975 The pattern of retinotectal connections in albino rats can be modified by fetal surgery. Brain Res., *91* : 119–125.

Miner, N. 1956 Integumental specification of sensory fibers in the development of cutaneous local sign. J. Comp. Neurol., *105* : 161–170.

Mire, J.J., W.J. Hendelman, and R.P. Bunge 1970 Observations on a transient phase of focal swelling in degenerating unmyelinated nerve fibers. J. Cell Biol., *45* :9–22.

Mitchell, D.E., M. Cynader, and J.A. Movshon 1977 Recovery from the effects of monocular deprivation in kittens. J. Comp. Neurol. (submitted).

Mitchell, D.E., F. Giffin, F. Wilkinson, P. Anderson, and M.L. Smith 1976 Visual resolution in young kittens. Vis. Res., *16* : 363–366.

Model, P.G., M.B. Bornstein, S.M. Crain, and G.D. Pappas 1971 An electron microscopic study of the development of snyapses in cultured fetal mouse cerebrum continuously exposed to xylocaine J. Cell Biol., *49* : 362–371.

Moore, R.Y., A. Björklund, and U. Stenevi 1971 Plastic changes in the adrenergic innervation of the rat septal area in response to denervation. Brain Res., *33* : 13–35.

Moore, R.Y., A. Björklund, and U. Stenevi 1974 Growth and plasticity of adrenergic neurons. *In* The Neurosciences Third Study Program, F.O. Schmitt and F.G. Worden (Eds.). MIT Press, Cambridge, Mass, pp. 961–977.

Morest, D.K. 1968 The growth of synaptic endings in the mammalian brain : a study of the calyces of the trapeziod body. Z. Anat. EntwGesch., *127* : 201–220.

Morest, D.K. 1969a The differentiation of cerebral dendrites: a study of the postmigratory neuroblast in the medial nucleus of the trapezoid body. Z. Anat. EntwGesch., *128* : 271–289.

Morest, D.K. 1969b The growth of dendrites in the mammalian brain. Z.Anat. EntwGesch. *128* : 290–317.

Morest, D.K. 1970 A study of neurogenesis in the forebrain of opossum pouch young. Z. Anat. EntwGesch., *130* : 265–305.

Moscona, A.A. 1974 Surface specification of embryonic cells: lectin receptors, cell recognition, and specific cell ligands. *In* The Cell Surface in Development, A.A. Moscona (Ed.). Wiley, New York, pp. 67–90.

Movshon, J.A. 1976a Reversal of the physiological effects of monocular deprivation in the kitten's visual cortex. J. Physiol., Lond., *261* : 125–174.

Movshon, J.A. 1976b Reversal of the behavioural effects of monocular deprivation in the kitten. J. Physiol., Lond., *261* : 175–187.

Mugnaini, E. and P.F. Forstrønen 1967 Ultrastructural studies on the cerebellar histogenesis. I. Differentiation of granule cells and development of glomeruli in the chick embryo. Z. Zellforsch., *77* : 115–143.

Muir, D.W. and D.E. Mitchell 1973 Visual resolution and experience : acuity deficits in cats following early selective visual deprivation. Science, *180* : 420–422.

Murray, M. 1976 Regeneration of retinal axons into the goldfish optic tectum. J. Comp. Neurol., *168* : 175–196.

Murray, M., and M.E. Goldberger 1974 Restitution of function and collateral sprouting in cat spinal cord - partially hemisected animal. J. Comp. Neurol., *158* : 19–36.

Musick, J., and J.I. Hubbard 1972 Release of protein from mouse motor nerve terminals. Nature, Lond., *237* : 279–281.

Mustari, M.J. 1976 Cell genesis and patterns of cell migration in the albino rat superior colliculus: a $^3$H-thymidine study. Neurosci. Abstr., *2* : 221.

Mustari, M.J., and R.D. Lund 1976 An aberrant crossed visual corticotectal pathway in albino rats. Brain Res., *112* : 37–42.

Nadler, J.V., C.W. Cotman, and G.S. Lynch 1973 Altered distribution of choline acetyltransferase and acetylcholinesterase activities in the developing rat dentate gyrus following entorhinal lesion. Brain Res., *63* : 215–230.

Nadler, J.V., C.W. Cotman, and G.S. Lynch 1974 Biochemical plasticity of short axon interneurons : increased glutamate decarboxylase activity in the denervated area of rat dentate gyrus following entorhinal lesion. Exp. Neurol., *45* : 403–413.

Nakai, J. 1960 Studies on the mechanism determining the course of nerve fibers in tissue culture. II. The mechanism of fasciculation. Z. Zellforsch. *52* : 427–449.

Nakai, J., and Y. Kawasaki 1959 Studies on the mechanism determining the course of nerve fibres in tissue culture. I. The reaction of the growth cone to various obstructions. Z. Zellforsch. *51* : 108–122.

Nakajima, S. 1965 Selectivity in fasciculation of nerve fibres in vitro. J. Comp. Neurol., *125* : 193–205.

Nakamura, Y., N. Mizuno, A. Konishi, and M. Sato 1974 Synaptic reorganization of the red nucleus after chronic deafferentation from cerebellorubral fibers : an electron microscopic study in the cat. Brain Res., *82* : 298–301.

Nauta, W.J.H., and S.O.E. Ebbesson (Eds.) 1970 Contemporary Research Methods in Neuroanatomy. Springer-Verlag, New York.

Nauta, W.J.H., and P.A. Gygax 1951 Silver impregnation of degenerating axon terminals in the central nervous system: (1) Technic, (2) Chemical notes. Stain Technol., *26* : 5–11.

Nauta, W.J.H., and P.A. Gygax 1954 Silver impregnation of degenerating axons in the central nervous system : a modified technique. Stain Technol., *29* : 91–93.

Nauta, W.H., and H.J. Karten 1970 A general profile of the vertebrate brain, with sidelights on the ancestry of cerebral cortex. *In* The Neurosciences Second Study Program, F.O. Schmitt (Ed.). Rockefeller Univ. Press, New York, pp. 7–26.

Nicolson, G.L. 1974 The interactions of lectins with animal cell surfaces. Int. Rev. Cytol., *39* : 89–190.

Nieuwkoop, P.D., and J. Faber 1956 Normal table of *Xenopus laevis* (Daudin). Amsterdam : North-Holland.

Nikara, T., P.O. Bishop, and J.D. Pettigrew 1968 Analysis of retinal correspondence by studying receptive fields of binocular single units in cat striate cortex. Exp. Brain Res., *6* : 353–372.

Novikoff, P.M., A.B. Novikoff, N. Quintana, and J.J. Hauw 1971 Golgi apparatus, GERL, and lysosomes of neurons in rat dorsal root ganglia, studied by thick section and thin section cytochemistry. J. Cell Biol., *50* : 859–886.

Nowakowski, R.S., and P. Rakic 1974 Clearance rate of exogenous $^3$H - thymidine from the plasma of pregnant Rhesus monkeys. Cell Tissue Kin., *7* : 189–194.

O'Lague, P.H., P.R. Macleish, C.A. Nurse, P. Claude, E.J. Furshpan, and D.D. Potter 1976 Physiological and morphological studies on developing sympathetic neurons in dissociated cell culture. Cold Spr. Harb. Symp. Quant. Biol., *40* : 399–408.

Olson, C.R., and J.V. Pettigrew 1974 Single units in visual cortex of kittens reared in stroboscopic illumination. Brain Res., *70* : 189–204.

Oppenheim, R.W., and M.B. Heaton 1975 The retrograde transport of horseradish peroxidase from the developing limb of the chick embryo. Brain Res., *98* : 291–302.

Oppenheimer, J.M. 1942 The decussation of Mauthner's fibers in *Fundulus* embryos. J. Comp. Neurol., *77* : 577–587.

Oster-Granite, M.L., and R.M. Herndon 1976 The pathogenesis of parvovirus-induced cerebellar hypoplasia in the Syrian hamster, *Mesocricetus auratus*. Fluorescent antibody, foliation, cytoarchitectonic, Golgi and electron microscopic studies. J. Comp. Neurol. *169* : 481–522.

Palay, S.L., and V. Chan-Palay 1974 Cerebellar Cortex Cytology and Organisation, Springer-Verlag, Berlin, Heidelberg.

Patterson, P.H., L.F. Reichardt, and L.L.Y. Chun 1976 Biochemical studies on the development of primary sympathetic neurons in cell culture. Cold Spr. Harb. Symp. Quant. Biol., *40* : 389–398.

Perisic, M., and M. Cuénod 1972 Synaptic transmission depressed by colchicine blockages of axoplasmic flow. Science, *175* : 1140–1142.

Pettigrew, J.D. 1974 The effect of visual experience on the development of stimulus specificity by kitten cortical neurons. J. Physiol., Lond., *237* : 49–74.

Pettigrew, J.D., and R.D. Freeman 1973 Visual experience without lines : effect on developing cortical neurons. Science, *182* : 599–601.

Pettigrew, J.D., and L.J. Garey 1974 Selective modification of single neuron properties in the visual cortex of kittens. Brain Res., *66* : 160–164.

Pettigrew, J.D. and M. Konishi 1976 Effect of monocular deprivation on binocular neurones in the owl's visual Wulst. Nature *264* : 753–754.

Pettigrew, J.D., C. Olson, and H.V.B. Hirsch 1973 Cortical effect of selective visual experience : degeneration or reorganization? Brain Res., *51* : 345–351.

Pfenninger, K.H., and R.B. Bunge 1974 Freeze-fracturing of nerve growth cones and young fibers. A study of developing plasma membrane. J. Cell Biol., *63* : 180–196.

Pfenninger, K.H., and M.F. Maylié-Pfenninger 1976 Differential lectin receptor content on the surface of nerve growth cones of different origin. Neurosci. Abstr., *2* : 224.

Phelps, C.H. 1969 Comparative ultrastructure of spinal cord regeneration in teleosts and mammals. Anat. Rec., *163* : 244.

Piatt, J. 1943 The course and decussation of ectopic Mauthner's fibers in *Amblystoma punctatum*. J. Comp. Neurol., *79* : 165–183.

Piatt, J. 1944 Experiments on the decussation and course of Mauthner's fibers in *Amblystoma punctatum*. J. Comp. Neurol., *80* : 335–353.

Piatt, J. 1947 A study of the factors controlling the differentiation of Mauthner's cell in *Amblystoma*. J. Comp. Neurol., *86* : 199–235.

Pickel, V.M., M. Segal, and F.E. Bloom 1974 Axonal proliferation following lesions of cerebellar peduncles. A combined fluorescence microscopic and radioautographic study. J. Comp. Neurol., *155* : 43–60.

Pilar, G., and L. Landmesser 1972 Axotomy mimicked by localized colchicine application. Science, *177* : 1116–1118.

Pilar, G., and L. Landmesser 1976 Ultrastructural differences during embryonic cell death in normal and peripherally deprived ciliary ganglia. J. Cell Biol., *68* : 339–356.

Plum, F., A. Gjedde, and F.E. Samson 1976 Neuroanatomical functional mapping by the radioactive 2-deoxy-D-glucose method. Neurosci. Res. Prog. Bull., *14* : 457–518.

Prendergast, J., and D.J. Stelzner 1976 Increases in collateral axonal growth rostral to a thoracic hemisection in neonatal and weanling rat. J. Comp. Neurol., *166* : 163–172.

Prestige, M.C. 1970 Differentiation, degeneration, and the role of the periphery : quantitative considerations. *In* The Neurosciences, Second Study Program, F.O. Schmitt (Ed.). Rockefeller Univ. Press, New York, pp. 73–82.

Privat, A. 1975 Dendritic growth *in vitro*. *In* Advances in Neurology, Vol. 12, Physiology and Pathology of Dendrites, G.W. Kreutzberg (Ed.). Raven Press, New York, pp. 201–216.

Purpura, D.P. 1974 Dendritic spines "dysgenesis" and mental retardation. Science, *186* : 1126–1128.

Purves, D., and A. Nja 1976 Effect of nerve growth factor on synaptic depression after axotomy. Nature, Lond., *260* : 535-536.

Pysh, J.J., and R.G. Wiley 1974 Synaptic vesicle depletion and recovery in cat sympathetic ganglia electrically stimulated *in vivo*. Evidence for transmitter secretion by exocytosis. J. Cell Biol., *60* : 365-374.

Raffin, J.P., and J. Repérant 1975 Étude expérimentale de la spécificité des projections visuelles d'embryons et de poussins de *Gallus domesticus* L. microphtalmes et monophtalmes. Arch. Anat. Micr. Morph. Exp., *64* : 93-111.

Raisman, G. 1969 Neuronal plasticity in the septal nuclei of the rat. Brain Res., *14* : 25-48.

Raisman, G., and P.M. Field 1971 Sexual dimorphism in the preoptic area of the rat. Science, *173* : 731-733.

Raisman, G., and P.M. Field 1973 A quantitative investigation of the development of collateral reinnervation after partial deafferentiation of the septal nuclei. Brain Res., *50* : 241-264.

Rakic, P. 1971 Neuron-glia relationship during granule cell migration in developing cerebellar cortex. A Golgi and electronmicrographic study in *Macacus rhesus*. J. Comp. Neurol., *141* : 283-312.

Rakic, P. 1972a Mode of cell migration to the superficial layers of fetal monkey neocortex. J. Comp. Neurol., *145* : 61-84.

Rakic, P. 1972b Extrinsic cytological determinants of basket and stellate cell dendritic pattern in the cerebellar molecular layer. J. Comp. Neurol., *146* : 335-354.

Rakic, P. 1974 Neurons in rhesus monkey visual cortex : systematic relation between time of origin and eventual disposition. Science, *183* : 425-427.

Rakic, P. 1975a Cell migration and neuronal ectopias in the brain. Birth Defects : Original Article Series, *11* : 95-129.

Rakic, P. 1975b Role of cell interaction in development of dendritic patterns. Adv. Neurol., *12* : 117-134.

Rakic, P. 1975c Timing of major ontogenetic events in the visual cortex of the rhesus monkey. *In* Brain Mechanisms in Mental Retardation, N.A. Buchwald and M.A.B. Brazier (Eds.). Academic Press, New York, pp. 3-40.

Rakic, P. 1976 Prenatal genesis of connections subserving ocular dominance in the rhesus monkey. Nature, Lond., *261* : 467-471.

Rakic, P. 1977 Prenatal development of the visual system in the rhesus monkey. Phil. Trans. Roy. Soc. Lond., B *278* : 245-260.

Rakic, P., and R.L. Sidman 1970 Histogenesis of cortical layers in human cerebellum, particularly the lamina dissecans. J. Comp. Neurol., *139* : 473-500.

Rakic, P., and R.L. Sidman 1973a Sequence of developmental abnormalities leading to granule cell deficit in cerebellar cortex of weaver mutant mice. J. Comp. Neurol., *152* : 103-132.

Rakic, P., and R.L. Sidman 1973b Organization of cerebellar cortex secondary to deficit of granule cells in weaver mutant mice. J. Comp. Neurol., *152* : 133-162.

Rakic, P., and R.L. Sidman 1973c Weaver mutant mouse cerebellum: defective neuronal migration secondary to abnormality of Bergmann glia. Proc. Nat. Acad. Sci., Wash., *70* : 240-244.

Rakic P., L.J. Stensaas, E.P. Sayre, and R.L. Sidman 1974 Computer-aided three-dimensional reconstruction and quantitative analysis of cells from serial electron microscopic montages of fetal monkey brain. Nature, Lond., *250* : 31-34.

Ralston, H.J., and K.L. Chow 1973 Synaptic reorganisation in the degenerating lateral geniculate nucleus of the rabbit. J. Comp. Neurol., *147* : 321-350.

Ramon y Cajal, S. 1928 Degeneration and Regeneration of the Nervous System. R.M. May (trans.). Oxford University Press, London.

Ramon y Cajal, S. 1960 Studies on vertebrate neurogenesis. L. Guth (trans.). Charles C. Thomas, Springfield, Ill.

Rasch, E., H. Swift, A.H. Riesen, and K.L. Chow 1961 Altered structure and composition of retinal cells of dark-reared mammals. Exp. Cell Res., *25* : 348-363.

Rasmussen, G.L. 1957 Selective silver impregnation of synaptic endings *In* New Research Techniques of Neuroanatomy. W.F. Windle (Ed.) Charles C. Thomas, Springfield, Ill. pp. 27-39.

Redfern, P.A. 1970 Neuromuscular transmission in new-born rats. J. Physiol., Lond., *209* : 701-709.

Rees, R.P., M.B. Bunge, and R.P. Bunge 1976 Morphological changes in the neuritic growth cone and target neuron during synaptic junction development in culture. J. Cell Biol., *68* : 240-263.

Regal, D.M., R. Boothe, D.Y. Teller, and G.P. Sackett 1976 Visual acuity and visual responsiveness in dark-reared monkeys (*Macaca nemestrina*). Vis. Res., *16* : 523-530.

Reichardt, L.F., P.H. Patterson, and L.L.Y. Chun 1976 Norepinephrine and acetylchoine synthesis by individual sympathetic neurons under various culture conditions. Neurosci. Abstr., *2* : 225.

Reier, P.J., and H. de F. Webster 1974 Regeneration and remyelination of *Xenopus* tadpole optic nerve fibers following transection or crush. J. Neurocytol., *3* : 591-618.

Rezai, Z., and C.H. Yoon 1972 Abnormal rate of granule cell migration in the cerebellum of "weaver" mutant mice. Develop. Biol., *29* : 17-26.

Rhoten, W.B. 1973 Emiocytosis of B granules from Saurian pancreatic islets perfused *in vitro*. Amer. J. Anat., *138* : 481-498.

Richards, W., and R. Kalil 1974 Dissociation of retinal fibers by degeneration rate. Brain Res., *72* : 288-293.

Riesen, A.H. 1966 Sensory deprivation. Prog. Physiol. Psychol., *1* : 117-147.

Rizzolatti, B., and V. Tradari 1971 Pattern discrimination in monocularly reared cats. Exp. Neurol., *33* : 181-194.

Rose, J.E., L.I. Malis, L. Kruger, and C.P. Baker 1960 Effects of heavy, ionizing monoenergic particles on the cerebral cortex. II. Histological appearance of laminar lesions and growth of nerve fibers after laminar destructions. J. Comp. Neurol., *115* : 243-296.

Rose, S.M. 1962 Tissue arc control of regeneration in the amphibian limb. *In* Regeneration, D. Rudnick (Ed.). Ronald Press, New York, pp. 153-176.

Roseman, S. 1974 Complex carbohydrates and intercellular adhesion. *In* The Cell Surface in Development, A. Moscona (Ed.). Wiley, New York, pp. 255-271.

Rubinstein, L.J. 1972 Cytogenesis and differentiation of primitive central neuroepithelial tumors. J. Neuropath. Exp. Neurol., *31* : 7-26.

Ruiz-Marcos, A., and F. Valverde 1969 The temporal evolution of the distribution of dendritic spines in the visual cortex of normal and dark raised mice. Exp. Brain Res., 8 : 284–294.

Rustioni, A., and C. Sotelo 1974 Some effects of chronic deafferentation in the ultrastructure of the nucleus gracillis of the cat. Brain Res., 73 : 527–533.

Rutishauser, U., J.P. Thiery, R. Brackenbury, B.A. Sela, and G.M. Edelman 1976 Mechanisms of adhesion among cells from neural tissues of the chick embryo. Proc. Nat. Acad. Sci., Wash., 73 : 577–581.

Ryugo, D.K., R. Ryugo, and H.P. Killackey 1975b Changes in pyramidal cell density consequent to vibrissae removal in the newborn rat. Brain Res., 96 : 82–87.

Ryugo, R., D.K. Ryugo, and H.P. Killackey 1975a Differential effect of enucleation on two populations of layer V pyramidal cells. Brain Res., 88 : 554–559.

Samaha, F.J., L. Guth, and R.W. Albers 1970 The neural regulation of gene expression in the muscle cell. Exp. Neurol., 27 : 276–282.

Sanderson, K.J. 1975 Retinogeniculate projections in the rabbits of the albino allelomorphic series. J. Comp. Neurol., 159 : 15–28.

Sanderson, K.J., R.W. Guillery, and R.M. Shackelford 1974 Congenitally abnormal visual pathways in mink (Mustela vision) with reduced retinal pigment. J. Comp. Neurol., 154 : 225–248.

Sauer, F.C. 1936 The interkinetic migration of embryonic epithelial nuclei. J. Morph., 60 : 1–11.

Sauer, M.E., and A.C. Chittenden 1959 Deoxyribonucleic acid and content of cell nuclei in the neural tube of the chick embryo : evidence for intermitotic migration of nuclei. Exp. Cell Res., 16 : 1–16.

Sauer, M.E., and B.E. Walker 1959 Radioautographic study of interkinetic nuclear migration in the neural tube. Proc. Soc. Exp. Biol. Med., 101 : 557–560.

Saunders, N.R. 1972 Lack of effect of nerve growth factor on peripheral sensory nerve regeneration. In Nerve Growth Factor and its Antiserum, E. Zaimis and J. Knight (Eds.). The Athlone Press of the Univ. of London.

Schapero, M. 1971 Amblyopia. Chilton Book Co., Philadelphia.

Schimke, R.T. 1959 Effects of prolonged light deprivation for the development of retinal enzymes in the rabbit. J. Biol. Chem., 234 : 700–703.

Schlaepfer, W.W. 1974 Calcium-induced degeneration of axoplasm in isolated segments of rat peripheral nerve. Brain Res., 69 : 203–215.

Schmechel, D.E., and P. Rakic 1973 Evolution of radial glial cells in developing monkey telencephalon : a Golgi study. Anat. Rec., 175 : 436.

Schmitt, F.O. 1968 The molecular biology of neuronal fibrous proteins. Neurosci. Res. Prog. Bull., 6 : 119–144.

Schneider, G.E. 1970 Mechanisms of functional recovery following lesions of the visual cortex or superior colliculus in neonate and adult hamsters. Brain, Behav., Evol., 50 : 295–323.

Schneider, G.E. 1973 Early lesions of superior colliculus: factors affecting the formation of abnormal retinal projections. Brain, Behav. Evol., 8 : 73–109.

Schneider, G.E., and S.R. Jhaveri 1974 Neuroanatomical correlates of spared or altered function after early brain lesions in the newborn hamster. In Plasticity and Recovery of Function in the Central Nervous System, D.G. Stein, J.J. Rosen, and N. Butters

(Eds.). Academic Press, New York, pp. 65–109.

Schneider, G.E., D.A. Singer, B.L. Finlay, and K.G. Wilson 1975 Abnormal retinotectal projections in hamster with unilateral neonatal tactum lesions : topography and correlated behavior. Anat. Rec., *181* : 472.

Schubert, P.E. 1976 Characteristics of dendritic and axonal transport. *In* Molecular and Functional Neurobiology, W.H. Gispen (Ed.). Elsevier, Amsterdam, pp. 87–109.

Schubert, P., and G.W. Kreutzberg 1974 Axonal transport of adenosine and uridine and transfer to postsynaptic neurons. Brain Res., *76* : 525–530.

Schubert, P., K. Lee, M. West, S. Deadwyler, and G. Lynch 1976 Stimulation-dependent release of ³H-adenosine derivatives from central axon terminals to target neurons. Nature, Lond., *260* : 541–542.

Scott, Jr., D., and C.D. Clemente 1955 Regeneration of spinal cord fibers in the cat. J. Comp. Neurol., *102* : 633–669.

Scott Jr., D., and C.N. Liu 1964 Factors promoting regeneration of spinal neurons: positive influence of nerve growth factor. Prog. Brain Res., *13* : 127–150.

Scott, M.Y. 1975 Functional capacity of compressed retinotectal projection in goldfish. Anat. Rec., *181* : 474.

Scott, S.A. 1975 Persistence of foreign innervation on reinnervated goldfish extraocular muscles. Science, *189* : 644–646.

Scott, S.A. 1976 On competitive innervation of goldfish eye muscles. Science, *192* : 1145–1146.

Seymour, R.M., and M. Berry 1975 Scanning and transmission electron microscope studies of interkinetic nuclear migration in the cerebral vesicles of the rat. J. Comp. Neurol., *160* : 105–126.

Sharma, S.C. 1972a Redistribution of visual projections in altered optic tecta of adult goldfish. Proc. Nat. Acad. Sci., Wash., *69* : 2637–2639.

Sharma, S.C. 1972b Reformation of retinotectal projections after various tectal ablations in adult goldfish. Exp. Neurol., *34* : 171–182.

Sharma, S.C. 1973 Anomalous retinal projection after removal of contralateral optic tectum in adult goldfish. Exp. Neurol., *41* : 661–669.

Sharma, S.C. 1975 Visual projection in surgically created 'compound' tectum in adult goldfish. Brain Res., *93* : 497–501.

Sharma, S.C., and R.M. Gaze 1971 The retinotopic organization of visual responses from tectal reimplants in adult goldfish. Arch. Ital. Biol., *109* : 357–366.

Shatz, C. 1977 Abnormal interhemispheric connections in the visual system of Boston Siamese cats : a physiological study. J. Comp. Neurol., *171* : 229–246.

Shatz, C., S. Lindstrom, and T. Wiesel 1975 Ocular dominance columns in the cat's visual cortex. Neurosci. Abstr., *1* : 56.

Shepherd, G.M. 1974 The Synaptic Organization of the Brain. An Introduction. Oxford University Press, New York.

Sherk, H., and M.P. Stryker 1976 Quantitative study of cortical orientation selectivity in visually inexperienced kitten. J. Neurophysiol., *39* : 63–70.

Sherman, S.M. 1973 Visual field defects in monocular and binocular deprived cats. Brain Res., *49* : 25–42.

Sherman, S.M., R.W. Guillery, J.H. Kaas, and K.J. Sanderson 1974 Behavioral, electrophysiological and morphological studies of binocular competition in the development of the geniculo-cortical pathway of cats. J. Comp. Neurol., *158* : 1–18.

Sherman, S.M., K.P. Hoffman, and J. Stone 1972 Loss of a specific cell type from the dorsal lateral geniculate nucleus in visually deprived cats. J. Neurophysiol., *35* : 532–541.

Sherman, S.M., and J. Stone 1973 Physiological normality of the retina in visually deprived cats. Brain Res., *60* : 224–230.

Sherman, S.M., and J.R. Wilson 1975 Behavioral and morphological evidence for binocular competition in the postnatal development of the dog's visual system. J. Comp. Neurol., *161* : 183–196.

Shimada, M., and J. Langman 1970 Repair of the external granular layer after postnatal treatment with 5-fluorodeoxyuridine. Amer. J. Anat., *129* : 247–260.

Sidman, R.L. 1968 Development of interneuronal connections in brains of mutant mice. *In* Physiological and Biochemical Aspects of Nervous Integration, F.D. Carlson (Ed.). Prentice-Hall, Englewood Cliffs, N.J., pp. 163–193.

Sidman, R.L. 1970 Autoradiographic methods and principles for study of the nervous system with thymidine-H³. *In* Contemporary Research Methods in Neuroanatomy, W.J.H. Nauta and S.O.E. Ebbesson (Eds.). Springer-Verlag, New York, pp. 252–274.

Sidman, R.L. 1972 Cell interactions in developing mammalian central nervous system. *In* Cell Interaction, Proceedings of the Third Lepetit Colloquium, L.G. Silvestri (Ed.). North-Holland, Amsterdam, 1972c, pp. 1–13.

Sidman, R.L. 1974 Contact interaction among developing mammalian brain cells. *In* The Cell Surface in Development, A.A. Moscona (Ed.). Wiley, New York, pp. 221–253.

Sidman, R.L., M.C. Green, and S.H. Appel 1965 Catalog of the neurological mutants of the mouse. Harvard Univ. Press, Cambridge, Mass.

Sidman, R.L., and P. Rakic 1973 Neuronal migration, with special reference to developing human brain : a review. Brain Res., *62* : 1–35.

Sillito, A.M. 1975 The contribution of inhibitory mechanisms to the receptive field properties of neurones in the striate cortex of the cat. J. Physiol., Lond., *250* : 305–329.

Singer, S.J., and G.L. Nicolson 1972 The fluid mosaic model of the structure of cell membranes. Science, *175* : 720–731.

Singer, W., and F. Tretter 1976 Receptive-field properties and neuronal connectivity in striate and parastriate cortex of contour-deprived cats. J. Neurophysiol., *39* : 613–630.

Skoff, R.P., and V. Hamburger 1974 Fine structure of dendritic and axonal growth cones in embryonic chick spinal cord. J. Comp. Neurol., *153* : 107–148.

So, K.F., and G.E. Schneider 1976 Abnormal recrossing retinotectal projections after early lesions in Syrian hamsters : a critical-age effect. Anat. Rec., *184* : 535–536.

Sobkowicz, H.M., R.W. Guillery, and M.B. Bornstein 1968 Neuronal organization in long term cultures of the spinal cord of the fetal mouse. J. Comp. Neurol., *132* : 365–396.

Sosula, L., and P.H. Glow 1971 Increase in number of synapses in the inner plexiform layer of light deprived rat retinae: quantitative electron microscopy. J. Comp. Neurol., *141* : 427–452.

Sotelo, C. 1975 Anatomical, physiological and biochemical studies of the cerebellum from mutant mice. II. Morphological study of cerebellar cortical neurons and circuits in the weaver mouse. Brain Res., *94* : 19–44.

Sotelo, C. and J.P. Changeux 1974a Transsynaptic degeneration 'en cascade' in the cerebellar cortex of staggerer mutant mice. Brain Res., *67* : 519–526.

Sotelo, C., and J.P. Changeux 1974b Bergmann fibers and granular cell migration in the cerebellum of homozygous weaver mutant mice. Brain Res., *77* : 484–491.

Sotelo, C., and S.L. Palay 1971 Altered axons and axon terminals in the lateral vesibular nucleus of the rat. Lab. Invest., *25* : 653–671.

Špaček, J., J. Pařízek, and A.R. Lieberman 1973 Golgi cells, granule cells and synaptic glomeruli in the molecular layer of the rabbit cerebellar cortex. J. Neurocytol., *2* : 407–428.

Spear, P.D., and L. Ganz 1975 Effects of visual cortex lesions following recovery from monocular deprivation in the cat. Exp. Brain Res., *23* : 181–201.

Speidel, C.C. 1964 *In vivo* studies of myelinated nerve fibers. Int. Rev. Cytol. *16* : 173–231.

Sperry, R.W. 1943a Effect of 180-degree rotation of the retinal field on visuomotor coordination. J. Exp. Zool., *92* : 263–279.

Sperry, R.W. 1943b Visuomotor coordination in the newt (*Triturus viridescens*) after regeneration of the optic nerve. J. Comp. Neurol., *79* : 33–55.

Sperry, R.W. 1944 Optic nerve regeneration with return of vision in anurans. J. Neurophysiol., *7* : 57–70.

Sperry, R.W. 1945 Restoration of vision after crossing of optic nerves and after contralateral transplantation of eye. J. Neurophysiol., *8* : 15–28.

Sperry, R.W. 1951 Mechanisms of neural maturation. *In* Handbook of Experimental Psychology, S.S. Stevens (Ed.). Wiley, New York, pp. 236–280.

Sperry, R.W. quoted by W.F. Windle 1956 Regeneration of axons in the vertebrate central nervous system. Physiol. Rev., *36* : 427–440.

Sperry, R.W. 1963 Chemoaffinity in the orderly growth of nerve fiber patterns of connections. Proc. Nat. Acad. Sci., Wash., *50* : 703–710.

Sperry, R.W. 1965 Embryogenesis of behavioral nerve nets. *In* Organogenesis, R.L. De Haan and H. Ursprung (Eds.). Holt, Reinhardt and Winston, New York, pp. 161–186.

Spooner, B.S., J.F. Ash, J.T. Wrenn, R.B. Frater, and N.K. Wessells 1973 Heavy meromyosin binding to microfilaments involved in cell and morphogenic movements. Tissue Cell, *5* : 37–46.

Spooner, B.S., M.A. Ludueña, and N.K. Wessells 1974 Membrane fusion in the growth cone-microspike region of the embryonic nerve cells undergoing axon elongation in cell culture. Tissue Cell, *6* : 399–409.

Sréter, F.A., A.R. Luff, and J. Gergely 1975 Effect of cross reinnervation on physiological parameters and on properties of myosin and sarcoplasmic reticulum of fast and slow muscles of the rabbit. J. Gen. Physiol., *66* : 811–821.

Stanfield, B., and W.M. Cowan 1976 Evidence for a change in the retino-hypothalamic projection in the rat following early removal of one eye. Brain Res., *104* : 129–136.

Stell, W.K., and P. Witkovsky 1973 Retinal structure in the smooth dogfish, *Mustelus canis* : general description and light microscopy of giant ganglion cells. J. Comp. Neurol., *148* : 1–32.

Stenevi, U., A. Björklund, and R.Y. Moore 1972 Growth of intact central adrenergic neurons in the denervated lateral geniculate body. Exp. Neurol., *35* : 290–299.

Stensaas, L.J. 1967 The development of hippocampal and dorsolateral pallial regions of the cerebral hemisphere in fetal rabbit. III. Twenty-nine millimeter stage, marginal lamina. J. Comp. Neurol., *130* : 149–162.

Stephens, R.E., and K.T. Edds 1976 Microtubules : structure, chemistry, and function. Physiol. Rev., *56* : 709–777.

Steward, O., C.W. Cotman, and G.S. Lynch 1973 Re-establishment of electrophysiologically functional entorhinal cortical input to the dentate gyrus deafferented by ipsilateral entorhinal lesions : innervation by the contralateral entorhinal cortex. Exp. Brain Res., *18* : 396–414.

Steward, O., C. Cotman, and G. Lynch 1976 A quantitative autoradiographic and electrophysiological study of the reinnervation of the dentate gyrus by the contralateral entorhinal cortex following ipsilateral entorhinal lesions. Brain Res., *114* : 181–200.

Stewart, R.M., D.P. Richman, and V.S. Caviness, Jr. 1975 Lissencephaly and pachygyria:an architectonic and topographical analysis. Acta neuropath., *31* : 1–12.

Stone, J., and Y. Fukuda 1974 The naso-temporal division of the cats' retina re-examined in terms of Y-, X- and W- cells. J. Comp. Neurol., *155* : 377–394.

Stone, L.S. 1948 Functional polarization in developing and regenerating retinae of transplanted eyes. Ann. N.Y. Acad. Sci., *49* : 856–865.

Stone, L.S. 1953 Normal and reversed vision in transplanted eyes. Arch. Ophthal., Chicago., *49* : 28–35.

Stone, L.S., and I.S. Zaur 1940 Reimplantation and transplantation of adult eyes in the salamander (*Triturus viridescens*) with return of vision. J. Exp. Zool., *85* : 243–269.

Strassman, R.J., and N.K. Wessells 1973 Orientational preference shown by microspikes of growing nerve cells *in vitro*. Tissue Cell, *5* : 401–412.

Straznicky, K. 1973 The formation of the optic fibre projection after partial tectal removal in *Xenopus*. J. Embryol. exp. Morph., *29* : 397–409.

Straznicky, K., and R.M. Gaze 1971 The growth of the retina in *Xenopus laevis* : an autoradiographic study. J. Embryol. Exp. Morph., *26* : 67–79.

Straznicky, K., and R.M. Gaze 1972 The development of the tectum in *Xenopus laevis* : an autoradiographic study. J. Embryol. Exp. Morph., *28* : 87–115.

Straznicky, K., R.M. Gaze, and M.J. Keating 1971 The retinotectal projections after uncrossing the optic chiasma in *Xenopus* with one compound eye. J. Embryol. Exp. Morph., *26* : 523–542.

Stryker, M.P., and C.J. Shatz 1976 Ocular dominance in layer IV of the normal and deprived cat's visual cortex. Neurosci. Abstr., *2* : 1137.

Stryker, M.P., and H. Sherk 1975 Modification of cortical orientation selectivity in the cat by restricted visual experience : a re-examination. Science, *190* : 904–905.

Sumner, B.E.H., and W.E. Watson 1971 Retraction and expansion of the dendritic tree of motor neurones of adult rats induced *in vivo*. Nature, Lond., *233* : 273–275.

Szekely, G. 1954 Zur ausbildung der lokulen funktionellen spezifität der retina. Acta morph. Acad. Sci. Hung., *5* : 157–167.

Taber-Pierce, E. 1973 Time of origin of neurons in the brain stem of the mouse. Prog. Brain Res., *40* : 53–65.

Taylor, A.C., and J.J. Kollros 1946 Stages in the normal development of *Rana pipiens* larvae. Anat. Rec., *94* : 7–23.

Tees, R.C. 1967a Effects of early auditory restriction in the rat on adult pattern discrimination. J. Comp. Physiol. Psychol., *63* : 389–393.

Tees, R.C. 1967b The effects of early auditory restriction in the rat on adult duration discrimination. J. Aud. Res., *7* : 195–207.

Tees, R.C. 1976 Perceptual development in mammals. *In* Studies on the Development of Behavior and the Nervous System, Vol. 3, Neural and Behavioral Specificity, G. Gottlieb (Ed.). Academic Press, New York, pp. 282–326.

Tennyson, V.M. 1970 The fine structure of the axon and growth cone of the dorsal root neuroblast of the rabbit embryo. J. Cell Biol., *44* : 62–79.

Thoenen, H. 1974 Trans-synaptic enzyme induction. Life Sci., *14* : 223–235.

Thorn, F., M. Gollender, and P. Erikson 1976 The development of the kittens' visual optics. Vis. Res., *16* : 1145–1149.

Thuline, D.N., and R.P. Bunge 1972 Preliminary observations on the transplantation of spinal cord tissue in rats. Anat. Rec., *172* : 418.

Tilney, F. 1933 Behavior in its relation to the development of the brain. Part II. Correlation between the development of the brain and behavior in the albino rat from embryonic states to maturity. Bull. Neurol. Inst. N.Y., *3* : 252–358.

Tretter, F., M. Cynader, and W. Singer 1975 Modification of direction selectivity of neurons in the visual cortex of kittens. Brain Res., *84* : 143–149.

Tsukahara, N., H. Hultborn, and F. Murakami 1974 Sprouting of cortico-rubral synapses in red nucleus neurones after destruction of the nucleus interpositus of the cerebellum. Experientia, *30* : 57–58.

Tsukahara, N., H. Hultborn, F. Murakami, and Y. Fujito 1975 Electrophysiological study of formation of new synapses and collateral sprouting in red nucleus neurons after partial denervation. J. Neurophysiol., *38* : 1359–1372.

Tuchmann-Duplessis, H., and L. Mercier-Parot 1960 The teratogenic action of the antibiotic actinomycin D. Ciba Foundation. Symposium on Congenital malformation. London, J. & A. Churchill, Ltd., London, pp. 115–128.

Turner, J.E., and M. Singer 1974 The ultrastructure of regeneration in the severed newt optic nerve. J. Exp. Zool., *190* : 249–268.

Uchizono, K. 1965 Characteristics of excitatory and inhibitory synapses in the central nervous system of the cat. Nature, Lond., *207* : 642–643.

Valdivia, O. 1971 Methods of fixation and the morphology of synaptic vesicles. J. Comp. Neurol., *142* : 257–274.

Valverde, F. 1967 Apical dendritic spines of the visual cortex and light deprivation in the mouse. Exp. Brain Res., *3* : 337–352.

Valverde, F. 1968 Structural changes in the area striata of the mouse after enucleation. Exp. Brain Res., *5* : 274–292.

Van Buren, J.M. 1963 Trans-synaptic retrograde degeneration in the visual system of primates. J. Neurol. Neurosurg. Psychiat., *26* : 402–409.

Van der Loos, H. 1965 The "improperly" oriented pyramidal cell in the cerebral cortex and its possible bearing on problems of neuronal growth and cell orientation. Bull. Johns Hopkins Hosp., *117* : 228–250.

Van Hof-Van Duin, J. 1976 Development of visuomotor behavior in normal and dark-reared cats. Brain Res., *104* : 233–241.

Varon, S. 1975 Nerve growth factor and its mode of action. Exp. Neurol., *48* : No. 3, part 2, 75–92.

Varon, S., C. Raiborn, and P.A. Burnham 1974 Implication of a nerve growth factor-like antigen in the support derived by ganglionic neurons from their homologous glia in dissociated cultures. Neurobiol., *4* : 317–327.

Vaughan, D.W. 1977 Age-related deterioration of pyramidal cell basal dendrites in rat auditory cortex. J. Comp. Neurol., *171* : 501–516.

Vaughn, J.E., C.K. Henrickson, and J.A. Grieshaber 1974 A quantitative study of synapses on motor neuron dendritic growth cones in developing mouse spinal cord. J. Cell Biol., *60* : 664–672.

Vaughn, J.E., and R.P. Skoff 1972 Neuroglia in experimentally altered central nervous system. *In* Structure and function of nervous system, Vol. 5., G.H. Bourne (Ed.). Academic Press, New York, pp. 39–72.

Vital-Durand, F., and M. Jeannerod 1974 Maturation of the optokinetic response: genetic and environmental factors. Brain Res., *71* : 249–257.

Von Noorden, G.K. 1967 Classification of amblyopia. Amer. J. Ophthal., *63* : 238–244.

Von Noorden, G.K. 1973 Experimental amblyopia in monkeys. Further behavioral observations and clinical correlations. Invest. Ophthal., *12* : 721–738.

Von Noorden, G.K., M.L.J. Crawford, and P.R. Middleditch 1977 Effect of lid suture on retinal ganglion cells in *Macaca mulatta*. Brain Res., *122*, 437–444.

Von Senden, M. 1960 Space and sight. Trans. P. Heath, Methuen, London.

Vrensen, G., and D. de Groot 1974 The effect of dark rearing and its recovery on synaptic terminals in the visual cortex of rabbits. A quantitative electron microscopic study. Brain Res., *78* : 263–278.

Vrensen, G., and D. de Groot 1975 The effect of monocular deprivation on synaptic terminals in the visual cortex of rabbit. A quantitative electron microscopic study. Brain Res., *93* : 15–24.

Wahlsten, D. 1974 Heritable aspects of anomalous myelinated fibre tracts in the forebrain of the laboratory mouse. Brain Res., *68* : 1–18.

Wall, P.D., and D. Eggers 1971 Neural connexions—functional reorganization after partial deafferentation in the rat. Nature, Lond., *232* : 542–544.

Watterson, R.L., P. Veneziano, and A. Bartha 1956 Absence of a true germinal zone in neural tubes of young chick embryos as demonstrated by the colchicine technique. Anat. Rec., *124* : 379.

Weiler, I.J. 1966 Restoration of visual acuity after optic nerve section and regeneration in *Astronotus acellatus*. Exp. Neurol., *15* : 377–386.

Weiss, P. 1934 In vitro experiments on the factors determining the course of outgrowing nerve fiber. J. Exp. Zool., *68* : 393–448.

Weiss, P. 1936 Selectivity controlling the central peripheral relations in the nervous system. Biol. Rev., *11* : 494–531.

Weiss, P. 1941 Nerve patterns : mechanics of nerve growth. Growth, *5* : 163–203.

Weiss, P. 1947 The problems of specificity in growth and development. Yale J. Biol. Med., *19* : 235–278.

Weiss, P., and H.B. Hiscoe 1948 Experiments on the mechanism of nerve growth. J. Exp. Zool., *107* : 315–396.

Weiss, P., and A.C. Taylor 1944 Further experimental evidence against neurotropism in nerve regeneration. J. Exp. Zool., *95* : 233–257.

Wessells, N.K., B.S. Spooner, J.F. Ash, M.O. Bradley, M.A. Ludueña, E.L. Taylor, J.T. Wrenn, and K.M. Yamada 1971 Microfilaments in cellular and developmental processes. Science, *171* : 135–143.

Wessells, N.K., B.S. Spooner, and M.A. Ludueña 1973 Surface movements, microfilaments and cell locomotion. *In* Locomotion of Tissue Cells, Ciba Foundation Symposium 14, W. Porter and D.W. Fitzsimmons (Eds.). Elsevier, Amsterdam, pp. 53–82.

West, R.W. 1976 Light and electron microscopy of the ground squirrel retina : functional considerations. J. Comp. Neurol., *168* : 355–378.

West, R.W., and J.E. Dowling 1972 Synapses onto different morphological types of retinal ganglion cells. Science, *178* : 510–512.

Westrum, L.E. 1975 Axonal patterns in olfactory cortex after olfactory bulb removal in newborn rats. Exp. Neurol., *47* : 442–447.

Westrum, L.E., and R.G. Black 1971 Fine structural aspects of the synaptic organization of the spinal trigeminal nucleus (pars interpolaris) of the cat. Brain Res., *25* : 265–287.

Westrum, L.E., R.C. Canfield, and R.G. Black 1976 Transganglionic degeneration in the spinal trigeminal nucleus following removal of tooth pulps in adult cats. Brain Res., *101* : 137–140.

Wickelgren, B.G., and P. Sterling 1969 Effect on the superior colliculus of cortical removal in visually deprived cats. Nature, Lond., *224* : 1032–1033.

Wiesel, T.N., and D.H. Hubel 1963a Effects of visual deprivation on morphology and physiology of cells in the cats' lateral geniculate body. J. Neurophysiol., *26* : 978–993.

Wiesel, T.N., and D.H. Hubel 1963b Single-cell responses in striate cortex of kittens deprived of vision in one eye. J. Neurophysiol., *26* : 1003–1017.

Wiesel, T.N., and D.H. Hubel 1965a Comparison of the effects of unilateral and bilateral eye closure on cortical unit responses in kittens. J. Neurophysiol., *28* : 1029–1040.

Wiesel, T.N., and D.H. Hubel 1965b Extent of recovery from the effects of visual deprivation in kittens. J. Neurophysiol., *28* : 1060–1072.

Wiesel, T.N., and D.H. Hubel 1974 Ordered arrangement of orientation columns in monkeys lacking visual experience. J. Comp. Neurol., *158* : 307–318.

Wiitanen, J.T. 1969 Selective silver impregnation of degenerating axons and axon terminals in the central nervous system of the monkey (*Macaca mulatta*). Brain Res., *14* : 546–548.

Willard, M., W.M. Cowan, and P.R. Vagelos 1974 The polypeptide composition of intra-axonally transported proteins : evidence for four transport velocities. Proc. Nat. Acad. Sci., Wash., *71* : 2183–2187.

Williams, R.S., R.J. Ferrante, and V.S. Caviness Jr. 1975 Neocortical organization in human cerebral malformation : a Golgi study. Neurosci. Abstr., *1* : 776.

Wilson, D.B. 1973 Chronological changes in the cell cycle of chick neuroepithalial cells. J. Embryol. Exp. Morph., *29* : 745–751.

Wilson, D.B. 1974 Proliferation in the neural tube of the splotch (Sp) mutant mouse. J. Comp. Neurol., *154* : 249–256.

Wilson, D.B., and E.M. Center 1974 The neural cell cycle in the looptail (Lp) mutant mouse. J. Embryol. Exp. Morph., *32* : 697–705.

Wilson, J.R., S.V. Webb, and S.M. Sherman 1976 The influence of differential illumination and temporal light modulation on binocular competition in the developing kitten's visual system. Neurosci. Abstr., *2* : 1140.

Windle, W.F. 1956 Regeneration of axons in the vertebrate central nervous system. Physiol. Rev., *36* : 426–440.

Wise, R.P., and R.D. Lund 1976 The retina and central projections of heterochromic rats. Exp. Neurol., *51* : 68–77.

Wise, S.P. 1975 The laminar organization of certain afferent and efferent fiber systems in the rat somatosensory cortex. Brain Res., *90* : 139–142.

Wise, S.P. 1976 The development and specificity of connections in the somatic sensory cortex of the rat. Anat. Rec., *184* : 565–566.

Wise, S.P., and E.G. Jones 1976 The organization and postnatal development of the commissural projection of the rat somatic sensory cortex. J. Comp. Neurol., *168* : 313–344.

Wolpert, L. 1969 Positional information and the spatial pattern of cellular differentiation. J. Theoret. Biol., *25* : 1–47.

Wolpert, L. 1971 Positional information and pattern formation. Curr. Top. Develop. Biol., *6* : 183–224.

Wuerker, R.B., and J.B. Kirkpatrick 1972 Neuronal microtubules, neurofilaments, and microfilaments. Int. Rev. Cytol., *33* : 45–75.

Yamada, K.M., B.S. Spooner, and N.K. Wessells 1970 Axon growth: roles of microfilaments and microtubules. Proc. Nat. Acad. Sci., Wash., *66* : 1206–1212.

Yamada, K.M., B.S. Spooner, and N.K. Wessells 1971 Ultrastructure and function of growth cones and axons of cultured nerve cells. J. Cell Biol., *49* : 614–635.

Yinon, U., and E. Auerbach 1974 The electroretinogram of children deprived of pattern vision. Invest. Ophthal., *13* : 538–543.

Yip, J.W., and M.J. Dennis 1976a Suppression of transmission at foreign synapses in adult newt muscle involves reduction in quantal content. Nature, Lond., *260* : 350–352.

Yip, J.W. and M.J. Dennis 1976b Formation and subsequent suppression of incorrect synapses during salamander limb regeneration. Neurosci. Abstr., *2* : 841.

Yoon, M. 1971 Reorganization of retinotectal projection following surgical operations on the optic tectum in goldfish. Exp. Neurol., *33* : 395–411.

Yoon, M.G. 1972a Reversibility of the reorganization of retinotectal projections in goldfish. Exp. Neurol., *35* : 565–577.

Yoon, M.G. 1972b Transposition of the visual projection from the nasal hemiretina onto the foreign rostral zone of the optic tectum in goldfish. Exp. Neurol., *37* : 451–462.

Yoon, M.G. 1973 Retention of the original topographic polarity by the 180° rotated tectal reimplant in young adult goldfish. J. Physiol., Lond., *233* : 575–588.

Yoon, M.G. 1976a Topographic polarity of the optic tectum studied by reimplantation of the tectal tissue in adult goldfish. Cold Spr. Harb. Symp. Quant. Biol., *40* : 503–519.

Yoon, M.G. 1976b Progress of topographic regulation of the visual projection in the halved optic tectum of adult goldfish. J. Physiol., Lond., *257* : 621–643.

Yoon, M.G. 1977 Induction of compression in the re-established visual projections onto a rotated tectal reimplant that retains its original topographic polarity within the halved optic tectum of adult goldfish. J. Physiol., Lond., *264* : 379–410.

Zimmer, J. 1973a Extended commissural and ipsilateral projections in postnatally deentorhinated hippocampus and fascia dentata demonstrated in rats by silver impregnation. Brain Res., *64* : 293–311.

Zimmer, J. 1973b Changes in the Timm sulfide silver staining pattern of the rat hippocampus and fascia dentata following early postnatal deafferentation. Brain Res., *64* : 313–326.

Zimmer, J. 1974a Proximity as a factor in the regulation of aberrant axonal growth in postnatally deafferented fascia dentata. Brain Res., *72* : 137–142.

Zimmer, J. 1974b. Long term synaptic reorganization in rat fascia dentata deafferented at adolescent and adult stages : observations with the Timm method. Brain Res., *76* : 336–342.

# INDEX